AF566656

Theory and Evaluation of Single-Molecule Signals

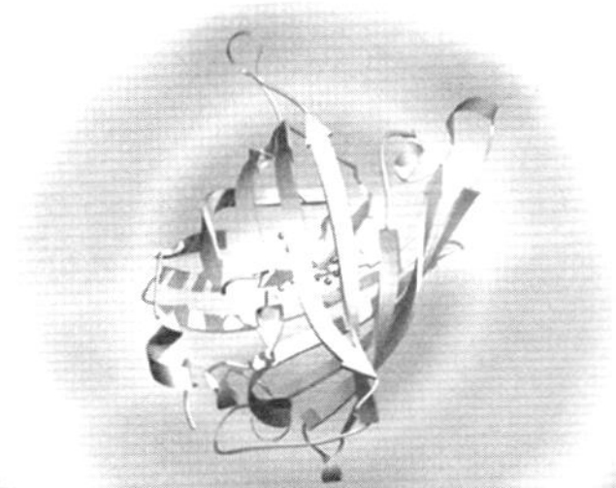

Theory and Evaluation of

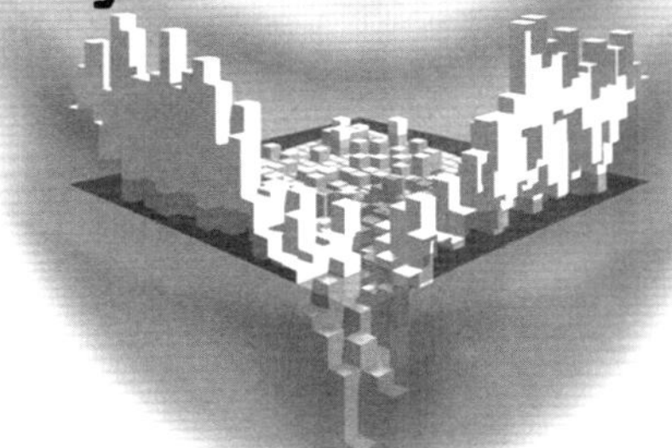

Single-Molecule Signals

Editors

Eli Barkai
Bar Ilan University, Israel

Frank L H Brown
University of California at Santa Barbara, USA

Michel Orrit
Leiden University, Netherlands

Haw Yang
University of California at Berkeley, USA

World Scientific

NEW JERSEY • LONDON • SINGAPORE • BEIJING • SHANGHAI • HONG KONG • TAIPEI • CHENNAI

Published by

World Scientific Publishing Co. Pte. Ltd.
5 Toh Tuck Link, Singapore 596224
USA office: 27 Warren Street, Suite 401-402, Hackensack, NJ 07601
UK office: 57 Shelton Street, Covent Garden, London WC2H 9HE

British Library Cataloguing-in-Publication Data
A catalogue record for this book is available from the British Library.

THEORY AND EVALUATION OF SINGLE-MOLECULE SIGNALS

ISBN-13 978-981-279-348-5
ISBN-10 981-279-348-8

Typeset by Stallion Press
Email: enquiries@stallionpress.com

Printed in Singapore by World Scientific Printers

Acknowledgment

The cover picture shows the beta-barrel structure of a green fluorescent protein. When a single fluorescent protein is excited at the focus of a confocal microscope, intermittent fluorescence signals can be observed. This so-called blinking is characteristic for the emission of single nano-objects. The pattern of blinking events can be characterized by correlating the lengths of the bright emission periods (the "on"-times) for variable lag. The simulated two-dimensional histogram shown in the lower part of the figure indicates an excess of events close to the axes, corresponding to sudden changes between fast and slow blinking dynamics (see Lippitz *et al.*, *Chem. Phys. Chem.* **6** (2005) 770–789). Such patterns can be used as signatures of memory effects in the protein's dynamics.

The cover picture was created by Dr Markus Lippitz. He used the Protein Data Bank record 1EMB to generate a 3D representation of the green fluorescent protein with Molscript (P. J. Kraulis, *J. Appl. Crystallogr.* **24** (1991) 946–950; www.avatar.se). He converted the VRML output of Molscript with vrml2pov (www.chemicalgraphics.com) and incorporated it in a script for the ray-tracing program POVray (www.povray.org) in order to render the combined scene.

Preface

Steady progress in experimental techniques and imaging capabilities has made the observation of individual atoms and molecules a routine practice in many laboratories worldwide. Despite this fact, single-molecule measurements remain difficult to perform and interpret. The exceedingly weak signals arising from individual molecules are more readily degraded and distorted by macroscopic-scale noise sources than are most ensemble averaged signals. We accept these difficulties since single-molecule experiments offer qualitatively new information about the behavior of physical systems, information unobtainable by traditional means. Observations of individual molecules, nanoparticles and other nano-objects reveal the full extent of their heterogeneity, be it for enzymes at physiological conditions, or for fluorescent probes in supercooled molecular liquids. Even more importantly, single-molecule signals open a window to directly observe dynamical fluctuations at the nanoscale.

A variety of modern techniques providing single-molecule sensitivity have been reviewed in the literature. The first methods to observe single molecules were scanning probe microscopies: STM (scanning tunnelling microscopy)[1] and AFM (atomic force microscopy).[2] In biophysics, electrical conductivity measurements can monitor the function of single ion-channels.[3] More recently, optical spectroscopy and fluorescence microscopy[4,5] have reached single-molecule sensitivity. In spite of the differences in the experimental approaches, the obtained single-molecule data share the same characteristic features; the data is noisy and contains large and irreproducible fluctuations. Depending on the experiment, these fluctuations may be of primary interest or may frustratingly serve to obscure a more interesting dynamics. Stated differently, single-molecule data streams are inherently stochastic — a given single molecule data trajectory is apparently random and is irreproducible. However, as we may extract statistically meaningful information from these data streams, their fundamental irreproducibility is an asset as well as a curse. While complex data treatments, analysis methods, statistical indicators, etc., are required to ensure the robustness of our conclusions, the information gained is unique and is inaccessible by ensemble-averaged measurements.

Statistical methods of data analysis applied to single molecules should, to the extent possible, be unbiased and model-free. They should eliminate the effects of noise, while retaining the significant features essential to understanding molecular dynamics. A number of procedures are currently in use to aid in this analysis, as summarized in Refs. 6 and 7. The simplest way to evaluate the time-trace of a single-molecule signal is to bin the signals by a low-frequency pass filter, and compare the average signal value to thresholds. This works well for slow variations, but the choice of bins and thresholds is somewhat arbitrary. More systematic analysis tools, either time-dependent (such as delay distributions, lifetimes, correlation functions) or signal-strength dependent (such as photon-counting histograms) do not involve any arbitrary parameter, but do average the data in ways which can sometimes erase the useful information. Finally, simple statistical indicators can sometimes answer a question about the data. Examples are the Mandel parameter indicating deviations from Poisson statistics in photon-counting traces, or indicators of the renewal character of series of events. (Is the data consistent with all events having been drawn from the same probability distribution?) Of course, an ideal analysis scheme would let the data "speak for itself", i.e. determine a model for the molecular dynamics based solely on the collected data without prior hypotheses influencing the final interpretation. This is seldom possible.

Since it is typically not possible to directly invert single-molecule data into a complete model for molecular dynamics, it is important to be able to model single-molecule experiments theoretically and computationally. Theoretical predictions for a given model may be compared to observables obtained experimentally, which allows for the refinement of hypothetical models via the scientific process. While most of the models currently used in the description of single-molecule experiments are based on traditional physical pictures (kinetics, diffusion, quantum mechanics, etc.), the observables available in single-molecule measurements are inherently different from those found in ensemble-averaged experiments. The tools used to predict ensemble-averaged experimental observables for a given model's dynamics are typically not well suited to determining the statistical information obtained via single-molecule experiments.

The aim of this book is to discuss some of the theoretical approaches to single-molecule data evaluation and interpretation developed over the last 10 years. We focus on fluorescence and mechanical measurements. We do not explicitly consider the case of ion channels, which have been considered in some detail previously.[8,9] The following chapters consider specific theoretical and statistical problems unique to single-molecule systems and single-molecule experiments. These problems range from the conditioning of single-molecule data to enable effective analysis, to the analysis of single-molecule data streams to the modelling of single-molecule systems by simulation and/or analytical theory.

Faced with a raw data stream, the first problem encountered in analysis is to identify which features are statistically significant. For example, how many distinct levels or states are resolvable in fluorescence traces or mechanical displacement measurements. Yang (Chapter 1) explains how to apply maximum likelihood criteria to objectively assess experimental time series.

A similar problem is discussed by Plakhotnik in Chapter 2. In particular, he outlines the procedure of Bayesian probability analysis to estimate the probability that a given hypothesis or model is correct, given experimental data and additional background information. This contribution stresses the importance of a proper statistical analysis in single-molecule measurements, where this step is much more crucial than with ensembles.

In Chapter 3, Brown presents the generating function formalism, which allows for the calculation of single-molecule experimental observables using tools that are only slightly modified from traditional ensemble-averaged theories. The generating function method applies to kinetic, stochastic and quantum-mechanical dynamics schemes.

A similar approach is proposed in Chapter 4 by Sanda and Mukamel to access multiphoton event probabilities and multipoint correlation functions, which are connected to the susceptibilities found in nonlinear optics. Here too, the formalism may be applied to classical kinetic schemes or to quantum-mechanical problems described by a Liouville equation. Sanda and Mukamel also discuss how to extend the formalism to non-Markovian and continuous-time randow walk processes.

The contribution of Hummer and Szabo (Chapter 5) discusses thermodynamic and kinetic properties of single molecules in mechanical force measurements. They show how to extract information on the free energy profile through non-equilibrium measurements, and how potential energy landscapes translate into time-dependent rates for breaking or unfolding reactions.

A common result of single-molecule measurements is long time traces, in which transitions between different levels can be distinguished with suitable methods. The next step in the evaluation is to interpret these changes as random transitions between states, for example between conformational substates of a protein. Describing these transitions as Markovian jumps between levels leads to kinetic schemes, which can be evaluated theoretically to derive statistical properties for comparison to experimental data.

The next chapters focus on relating experimental results to classical kinetic schemes. Gopich and Szabo (Chapter 6) show how the kinetic-scheme formulation can be generalized to include the many processes present in a single-molecule measurement. Among these processes are: the complex photophysics of fluorescence, conformational evolution of single molecules, and translational diffusion of the molecules through the focus of the exciting beam. They show how experimental observables — the photon counting intensity distribution, waiting time distribution and related correlation functions — can be generated in a consistent way accounting for all these effects.

Cao (Chapter 7) examines the general solution of kinetic schemes represented by rate matrices and focuses on specific features in these solutions, such as echoes, and other indicators, which can directly give *a priori* information on the system, without the need for a full solution.

Kinetic schemes obeying detailed balance give rise to exponential relaxation, as do overdamped stochastic models. Several experiments have found surprising oscillating behavior in the dynamics of single protein molecules, which suggest that these simple models may be inappropriate for the description of proteins. For example Lu *et al.*[10] found accumulations of diagonal correlations in the consecutive on-times of cholesterol oxidase enzyme, Edman and Rigler found oscillations in the fluorescence of labeled horse radish peroxidase,[11] and Baldini *et al.* in the dynamics of fluorescent proteins excited with two photons.[12] Vlad and Ross, in Chapter 8, discuss two possible theoretical models that may explain these intriguing experimental results.

Chapter 9, by Kolomeisky, considers the motion of molecular motors. The stepping of the motors is represented as a sequence of transitions in an open kinetic scheme, which leads to prediction of their dynamical and kinetic properties.

In the preceding chapters, the complex dynamics of biomolecules is described by kinetic schemes, i.e. a network of states connected with rate constants. In most cases, however, even the topology of this network is unknown. The determination of the relevant scheme from experimental data alone is difficult, even impossible because the same experimental data can correspond to several kinetic schemes with different topologies and connectivities. This follows from the information loss taking place when the molecule's evolution in a multidimensional space is projected onto the few dimensions accessible to experiment (e.g. intensity in the case of a fluorescence trace). In Chapter 10, Flomenbom and Silbey discuss how to classify kinetic schemes and reduce them to canonical forms, so as to select the simplest class of schemes compatible with the data.

Single molecules provide unique insights into heterogeneities and fluctuation phenomena which cannot be seen directly in ensemble measurements. As Barkai discusses in Chapter 11, ergodicity breaking is a perfect example of a property which directly appears from the behavior of blinking quantum dots and molecules in disordered systems. Ergodicity is broken when the time-average of a single molecule's property, e.g. its emission rate, does not coincide with the ensemble average of this property over many equivalent molecules. These differences, obviously stand out in single-molecule data. Ergodicity breaking can be strong, when each individual molecule explores only a very small, disconnected region of the phase space, or weak, when a large part of the phase space is explored by each molecule, but sojourn times in microstates of the system obey fractal power law statistics.

Statistical analysis, modeling and understanding of single-molecule fluorescence and mechanical experiments have expanded spectacularly in scope and power during the last ten years. We hope that the theoretical chapters collected in this book will be of use to experimentalists as well as to theoreticians, and will facilitate the necessary dialogue between them. Only a back-and-forth interaction between experiments and theory can lead to a deeper understanding of fluorescence intermittency, conformational fluctuations, and many other fascinating signatures of individual molecules and nano-objects.

E. Barkai, F. Brown, M. Orrit and H. Yang

References

1. C. Julian Chen, *Introduction to Scanning Tunneling Microscopy*, 2nd edn. (Oxford University Press, 2007).
2. J. Howard, *Mechanics of Motor Proteins and the Cytoskeleton* (Sinauer Press, Sunderland, Massachusetts, 2001).
3. *Single-Channel Recording*, B. Sackmann and E. Neher, eds. (Plenum, New York, 1995).
4. *Single-Molecule Optical Detection, Imaging and Spectroscopy*, Th. Basché, W. E. Moerner, M. Orrit and U. P. Wild, eds. (Wiley-VCH, 1997).
5. *Single-Molecule Detection in Solution*, Ch. Zander, J. Enderlein and R. A. Keller, eds. (Wiley-VCH, 2002).
6. M. Lippitz, F. Kulzer and M. Orrit, Fluctuating fluorescence of single nano-objects, *Chem. Phys. Chem.* **6** (2005) 770.
7. E. Barkai, Y. Jung and R. Silbey, Theory of single molecule spectroscopy: Beyond the ensemble average, *Ann. Rev. Phys. Chem.* **55** (2004) 457.
8. D. Colquhoun and F. J. Sigworth, Fitting and statistical analysis of single channel records, in *Single-Channel Recording*, Chap. 19, B. Sackmann and E. Neher, eds. (Plenum, New York, 1995), pp. 483–587.
9. L. Venkataramanan and F. J. Sigworth, Applying hidden Markov models to the analysis of single ion channel activity, *Biophys. J.* **82** (2002) 1930–1942.
10. H. P. Lu, Y. Xun and X. S. Xie, Single-molecule enzymatic dynamics, *Science* **282** (1998) 1877.
11. L. Edman and R. Rigler, Memory landscapes of single-enzyme molecules, *Proc. Nat. Acad. Sci. USA* **97** (2000) 8266.
12. G. Baldini, F. Cannone and G. Chirico, Pre-unfolding resonant oscillations of single green fluorescent protein molecules, *Science* **309** (2005) 1096.

Contents

Acknowledgment **v**

Preface **vii**

1. Model-Free Statistical Reduction of Single-Molecule Time Series **1**

Haw Yang

1 Introduction 1
2 General theoretical background: Statistical likelihood 3
3 The transition between molecular states occurs much faster than the experimental time resolution — Intermittency 12
4 Experimental time resolution is sufficient to follow the transition between molecular states 15
5 Distributions 17
6 Bursts 21
7 Conclusion 24
Acknowledgments 24
References 25

2. Testing Hypothesis with Single Molecules: Bayesian Approach **31**

Taras Plakhotnik

1 Introduction to the Bayesian approach 31
2 Hidden parameters 39
3 Finding quasi-stationary states 41
4 Finding distances between molecules 49
References 59

3. Generating Functions for Single-Molecule Statistics **61**

Frank L. H. Brown

1 Introduction 61
2 The Poisson process and introduction of generating functions 64
3 Generating functions for photon emission: More complex kinetic models 73
4 Generating functions for photon emission: Quantum treatment 79
5 Quantum dynamics examples 81
6 Conclusion 90
Acknowledgments 91
References 91

4. Multipoint Correlation Functions for Photon Statistics in Single-Molecule Spectroscopy: Stochastic Dynamics in Liouville Space **93**

František Šanda and Shaul Mukamel

1 Photon statistics: Factorial moments vs. correlation functions 93
2 Photon statistics in weakly driven systems: Analogy with four wave mixing 99
3 Multipoint correlation functions for slow fluctuations 108
4 Summary 134
Acknowledgments 134
References 135

5. Thermodynamics and Kinetics from Single-Molecule Force Spectroscopy **139**

Gerhard Hummer and Attila Szabo

1 Introduction 139
2 Thermodynamics from single-molecule pulling experiments 141
3 Kinetics from single-molecule pulling experiments 155
4 Concluding remarks 175

Acknowledgments 176
References 176

6. Theory of Photon Counting in Single-Molecule Spectroscopy 181

Irina V. Gopich and Attila Szabo

1 Introduction 181
2 General formalism 186
3 Fluorescence quenching and conformational dynamics 203
4 Influence of translational diffusion 225
5 Concluding remarks 238
Acknowledgments 239
References 239

7. Memory Effects in Single-Molecule Time Series 245

Jianshu Cao

1 Introduction 245
2 Modulated reaction model: General formalism . . 247
3 On–off blinking time series 257
4 Photon emission time series 269
5 Data analysis 278
6 Concluding remarks 280
Acknowledgments 281
References 281

8. Analysis of Experimental Observables and Oscillations in Single-Molecule Kinetics 287

Marcel O. Vlad and John Ross

1 Introduction 287
2 Correlation functions and oscillations 290
3 On–off time distributions and oscillations 304
4 Reaction event statistics and oscillations 307
5 Conclusions 310
Acknowledgments 311
References 311

9. Discrete Stochastic Models of Single-Molecule Motor Proteins Dynamics **313**

Anatoly B. Kolomeisky

1 Introduction 313
2 Single-molecule experiments 315
3 Theoretical models 316
4 Conclusions 329
References 330

10. Unique Mechanisms From Finite Two-State Trajectories **337**

Ophir Flomenbom and Robert J. Silbey

1 Introduction 337
2 Mathematical formulations 340
3 RD forms 344
4 Constructing the RD form from the data 348
5 Concluding remarks 352
Appendix A 353
Appendix B 354
References 360

11. Weak Ergodicity Breaking in Single-Particle Dynamics **365**

E. Barkai

1 Introduction 365
2 Blinking nanocrystals 369
3 Continuous time random walk 380
4 The quenched trap model 383
5 Discussion 387
Acknowledgment 389
References 389

About the Editors **393**

Index **395**

CHAPTER 1

Model-Free Statistical Reduction of Single-Molecule Time Series

Haw Yang

Department of Chemistry, University of California at Berkeley, and Physical Biosciences Division, Lawrence Berkeley National Laboratory, Berkeley, CA 94720, USA

1. Introduction

Studying individual molecules allows an experimentalist to follow the time-dependent evolution of molecular states in real time. Yet, single-molecule experiments can be difficult and time-consuming; it is important to identify the potential benefits and limitations of particular measurements before designing new experiments. The new information that can be obtained includes the distribution of molecular properties, the mechanism and kinetics of complicated chemical reactions, and, most importantly, the local dynamics of a microscopic system. The nature of single-molecule data, however, is also markedly different from that of bulk experiments.

As illustrated in Fig. 1, suppose one is interested in understanding the physical principles that govern the fluctuations of a molecular dipole embedded in a condensed phase host medium. Because bulk experiments measure the mean of an experimental observable over many molecules, the uncertainties will follow Gaussian statistics by virtue of the large-number principle (Central-Limit Theorem). The "true" value for the mean of a physical parameter (in this case the

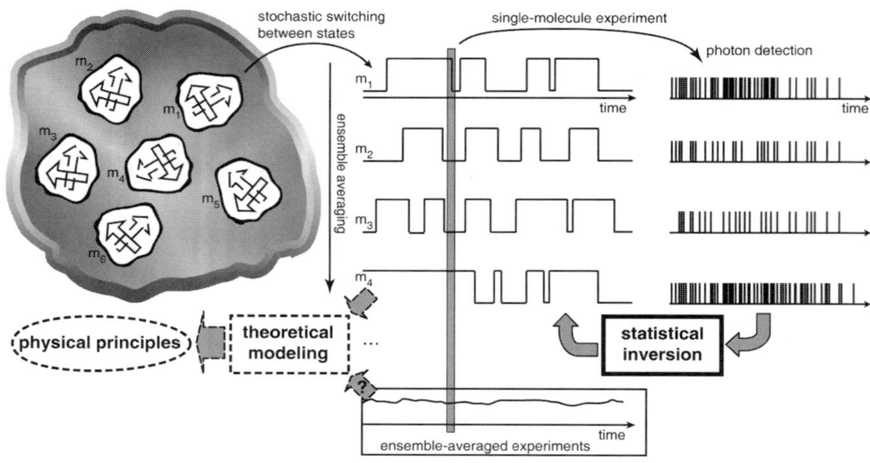

Fig. 1. Thermal fluctuations drive individual molecules $\{m_1, m_2, \ldots\}$ to change stochastically. The gray vertical bar represents a measurement, the width of which illustrates the time it takes to conduct a measurement. For bulk experiments, both the switching dynamics and detection noise are averaged out to give a time-independent mean value. At the single-molecule level, the statistics of photon detection adds an additional layer of complexity in our attempt to understand the microscopic system. This chapter discusses practical statistical inversion methods that remove photon-counting noise without any assumed models.

average dipole) can almost always be converged to by taking more data points; therefore, in a sense, ensemble-averaged experimental data are deterministic.

In contrast, single-molecule data are stochastic; time-averaging instead of number-averaging is used to improve the statistics. The physical picture, in particular the dynamics that can be corroborated by a single-molecule data set, may be greatly influenced by how the data is averaged. At the single-molecule level, the detailed microscopic interactions between the molecule and its environment become noticeable. If an experimentalist were to follow the dipole of a single molecule as a function of time, for instance, he will find that the molecular dipole does not maintain a fixed value, but randomly fluctuates in a time-dependent manner (cf. idealized dipole-time traces in the middle column of Fig. 1). From the perspective of interpreting single-molecule data, these fluctuations encode a layer of "randomness" not often present in bulk experiments. There is more. The noiseless traces just described are not the ones that an experimentalist measures; instead, the molecular states are read out spectroscopically by detecting the molecule's emitted photons. Since the detection events are also probabilistic, they comprise an additional layer of "randomness" that masks the true dynamics. The two layers of "randomness encoding" — stochastic molecular scale dynamics and probabilistic photon detection — ensure that no two single-molecule time traces are identical. New abstraction schemes that can cope with this kind of data are therefore necessary.

In this chapter, the reduction of a single-molecule time series is treated as a statistical inversion problem, to which statistically robust and unbiased solutions are outlined. Aiming to provide a link between experimental data and physical models, the methods described herein attempt to remove the randomness introduced during the photon detection step, while preserving the physical randomness due to molecular scale dynamics. The principal criterion driving these developments is that reduction be accomplished without any presumed kinetic models or distributions — after all, it is the (presumably unknown) physics that one wishes to characterize. Several issues that are pertinent to the quantitative abstraction of physical information from single-molecule experiments will be discussed: They are (1) intermittency, when the transition between molecular states occurs on a timescale much faster than the experimental resolution; (2) time-dependent FRET (Förster-type resonance energy transfer) efficiency modulation, when the experimental time resolution is sufficient to follow the changes in molecular state; (3) distributions; and (4) photon burst detection. The discussion starts with a brief overview of the ideas and terminologies used in the statistical approaches that address these issues (also see the appendices in Chap. 2).

2. General theoretical background: Statistical likelihood

Consider a typical fluorescence experiment in which one acquires spectroscopic signals as a function of time from individual molecules by detecting the photons emanating from them. The emission can be detected by imaging using cameras with enhanced charge-coupled devices (CCD).[1] This detection scheme, while extremely convenient and easy to set up, is limited in time resolution to the camera's frame rate, typically 30 frames per second. On the other hand, single-photon counting devices such as avalanche photodiodes (APD) or photon multiplier tubes (PMT) do not impose such limitations, potentially providing more information. The following discussions therefore assume that photon-counting detectors are used in experiments. To a very good approximation, the photon detection events under most single-molecule experimental conditions follow Poisson statistics. This can be verified for individual experiments (Ref. 2; also see Chap. 3, Sec. 2). The task at hand is thus to extract physical parameters from the stream of time-dependent photon detection events, a single-molecule time series. To this end, the experimental measurement is treated as an applied statistics problem. The connection between physical measurements and applied statistics can be seen in commonly used expressions for ensemble-averaged experiments. For example, the mean and the standard deviation of a certain physical quantity x are, respectively, $\bar{x} \approx \frac{1}{N}\sum_{i=1}^{N} x_i$ and $\sigma_x \approx \sqrt{\frac{1}{N}\sum_{i=1}^{N}(x_i - \bar{x})^2}$, where x_i are the observables collected in each independent measurement. As one can see from this simple example, measurement of a physical parameter is almost always a statistical estimation of the physical parameter's true value as inferred from multiple observations.

2.1. *Probability density, likelihood, and information*

A probability density function that properly describes the behavior of experimental observables will be needed to connect these observables to the physical parameters inferred from the experimental data. For bulk experiments, the Gaussian probability density function is usually used for this purpose. For single-molecule photon-counting experiments, on the other hand, the time intervals between consecutive photon detection events, Δ, at a given detected emission intensity, I, are characterized by an exponential function,

$$f(\Delta|I) = I \exp[-\Delta I]. \tag{1}$$

Here, the intensity is defined as the photon detection rate and has a unit of number of photons per unit time. The number of photons detected, n, within a time interval, T, follows Poisson statistics,

$$g(n|T, I) = \frac{(IT)^n \exp[-IT]}{n!}. \tag{2}$$

The time series of inter-photon durations, $\{\Delta_i(t_i)\}$, and the number of photons detected during the time interval T_j, $\{n_j(T_j)\}$, are the primary single-molecule experimental observables (cf. Fig. 2). Whether these observables follow the prescribed statistical distribution can be easily verified by control experiments (and they usually do).

These functions, however, only describe the probability of obtaining *one* data point. To describe the probability of obtaining many data points (photons), one needs another function, the likelihood function. If the acquisition of these experimental data points is independent, which can also be verified experimentally, one can describe the likelihood of acquiring a series of inter-photon durations $\{\Delta_1, \ldots, \Delta_N\}$, for example, as

$$L_N(\{\Delta_1, \ldots, \Delta_N\}|I) = f(\Delta_1|I) \times \cdots \times f(\Delta_N|I) = \prod_{i=1}^{N} f(\Delta_i|I). \tag{3}$$

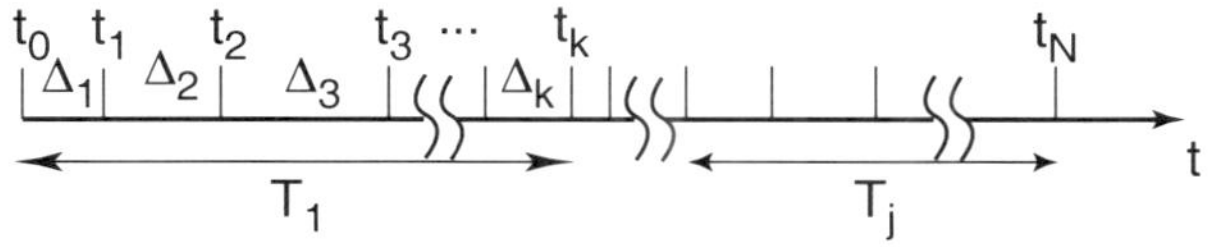

Fig. 2. Timing of photon detection. t_i are the chronological times at which the i-th photon is detected, Δ_i are the durations between two sequentially detected photons, and T_j are the time intervals over which the emission intensity is to be estimated.

Oftentimes the intensity I is the physical parameter that one wishes to measure. Thus, as a statistical inversion problem, one may ask: what is the most likely value of I that gives rise to the *observed* inter-photon duration sequence $\{\Delta_1, \ldots, \Delta_N\}$? The solution is straightforward: one simply finds the I value that maximizes the likelihood function in Eq. (3). Since all the probability functions are positive, this operation is the same as carrying out a maximization procedure for $\ln L$ — finding the I value that leads to $\frac{\partial}{\partial I} \ln L = 0$. The likelihood function Eq. (3) is maximized when $I = \hat{I} = N/\sum_{i=1}^{N} \Delta_i = N/T_{\text{total}}$, where T_{total} is the total time duration for acquiring N photons — a very sensible result. This procedure, called maximum likelihood estimation (MLE),[3] in fact gives the optimal unbiased estimate, that is, in the limit of infinite number of observables, the MLE follows Gaussian statistics and converges to the "true" value.[4]

Intuitively, one would expect that the more photons there are, the more accurately the physical parameter, say ξ (which can, for example, be the emission intensity, the distance between the donor and the acceptor dye in a FRET experiment, or the emission wavelength), can be estimated. To formulate this physical intuition quantitatively, ideas from information theory are used (also see Chap. 7).[5] The amount of information regarding ξ that is contained in a data set can be evaluated by Fisher information,[6] defined as

$$J(\xi) = \left\langle \left(\frac{\partial}{\partial \xi} \ln L_N(\{\Delta_i\}|\xi) \right)^2 \right\rangle_{\Delta}, \tag{4}$$

where $\langle \cdots \rangle_{\Delta}$ denotes averaging overall possible realizations of the inter-photon durations, $\{\Delta_i\}$, and $\{\Delta_i\} \equiv \{\Delta_1, \Delta_2, \ldots, \Delta_N\}$ is introduced as a shorthand notation. The variance of a given measured ξ is related to the Fisher information by[7,8]

$$\text{var}(\xi) \geq J(\xi)^{-1}, \tag{5}$$

where the equality is true when the Fisher information is evaluated at the ξ value calculated using MLE, that is, $\text{var}(\hat{\xi}) = J(\hat{\xi})^{-1}$.

Equation (5) thus provides a consistent likelihood-based framework for statistical estimation and the associated uncertainties. As will be discussed later, Eq. (5) also forms the basis for the information bounds on the time resolution achievable by single-molecule FRET measurements.[9]

2.2. *Statistical tests*

In many cases, it is necessary to determine whether there is a change in the experimental observable. For example, one may wish to assign an "on" or an "off" emission state to a single molecule that exhibits a blinking behavior. This seemingly simple task turns out to be non-trivial for noisy single-molecule data, if a quantitative and unbiased assignment is to be made. The determination of sudden intensity changes can also be achieved by recasting the task as a statistical test problem. Likelihood functions can be used to evaluate whether there is sufficient evidence to support the assertion that there is a change in the physical parameter. To this end, one compares a likelihood function that assumes a change in the time series with the one that assumes there is no change by taking the ratio of the two: likelihood ratio $= L_N(\text{there is a change})/L_N(\text{there is no change})$. The likelihood ratio is also probabilistic, because it is computed from two probability functions. Therefore, statistically, if the data support the notion that there is a change in the time series, the ratio will give a greater numerical value. On the other hand, if the data does not support the assertion that there is a change in the physical parameter, the ratio will be small. To make this comparison quantitatively, a critical value is used for making the decision whether or not there is a change in the emission intensity. Due to the probabilistic nature of the likelihood ratio function, there is a finite probability that the decision may not be correct. To describe the probability of making a wrong decision, a confidence interval, denoted by α, may be associated with a given critical value. For example, if the likelihood ratio (evaluated on the log scale using N observables) is greater than the critical value $\lambda_C(N, \alpha)$ (depending on both α and N), one may say that there is a change in ξ within the time series:

$$\lambda(N) = \ln\left[\frac{L_N(\text{there is a change in }\xi)}{L_N(\text{there is no change in }\xi)}\right] > \lambda_C(\alpha, N). \quad (6)$$

In general, $\lambda_C(\alpha, N)$ is a complicated function such that rigorous analytical expressions are often very difficult to obtain. Even if the exact solutions have been obtained, it is often necessary to evaluate them numerically. For example, the critical values for determining abrupt intensity changes have been calculated and tabulated in Ref. 10. Indeed, the critical values are often evaluated by Monte Carlo computer simulations; only in special cases where asymptotic approximations are applicable such as in the detection of changes in diffusive behavior in single-particle tracking experiments,[11] can analytical expressions be derived. With a test statistic like Eq. (6), one is able to make a statement as, "with 95% confidence, there is a change in emission intensity for this molecule." Similar ideas can also be applied to determining whether there is a single-molecule "burst" in diffusion-type experiments,[12] or whether the mode of a moving particle has changed as in single-particle tracking experiments.[11]

At this point, a few words about some notions related to statistical tests are in order. Because the decision as to whether there is a physically important change in the single-molecule signal is made based on the statistical assessment, there is always a finite probability that the decision is *wrong*. Using the detection of photon bursts as an example, if a segment of a time series is determined to have a burst due to photon counting noise but that in fact does not have any (e.g., in a control experiment), then this kind of mistake is termed Type-I error (false-positive) and occurs probabilistically, the magnitude of which depends on the size of the critical value, $\lambda_C(\alpha, N)$.[13] On the other hand, if the statistical test does not detect a photon burst that is known to occur, then this kind of mistake is termed Type-II error (false-negative), which also happens probabilistically and is related to the "power" of the statistical test. These two characteristics of a statistical test are interrelated. The more powerful a statistical test is (does not miss many bursts), the more likely that it will give false-positive results, and vice versa. Understanding the frequency of both error types is important, especially if one wishes to place error bars on derived quantities based on the intensity changes or number of bursts, say, kinetic parameters of a functioning enzyme.

2.3. *Bayesian information criterion and maximum entropy method*

As mentioned earlier, one of the uniquely powerful aspects of the single-molecule approach is that it provides the distribution of molecular properties — new information that cannot be readily provided using ensemble-based approaches. Obtaining such a distribution quantitatively from single-molecule data, however, is, once again, not an easy task. Fortunately, suitable tools have already been developed in the applied statistics and statistical physics communities. Two representative cases are discussed below.

In the first case, one would like to determine the number and the relative weight of discretely distributed intensity states from a single-molecule (or single-particle) emission time series. This knowledge is important for elucidating the reaction mechanisms of complicated chemical reactions. The discussion is perhaps best given in terms of a specific example. Figure 3(A) displays a computer simulation of a single molecule randomly switching between five distinct intensity states. The noisy trace in Fig. 3(B) illustrates how photon-counting detection corrupts the signal, making it very difficult to identify the time instances of state switching (intensity change points) and to measure the distribution. The likelihood ratio test (cf. Eq. (6)) allows one to locate the intensity change points, from which one can reconstruct the trajectory as shown in Fig. 3(C). In this example, $N_{\text{cp}} = 68$ intensity jumps are found. Connecting these intensity change points yields $(N_{\text{cp}} + 1)$ estimated intensity states. In a way, such a change-point analysis can be understood as likelihood-based fitting the data to a discrete-jump model without presumed kinetics or number of states.

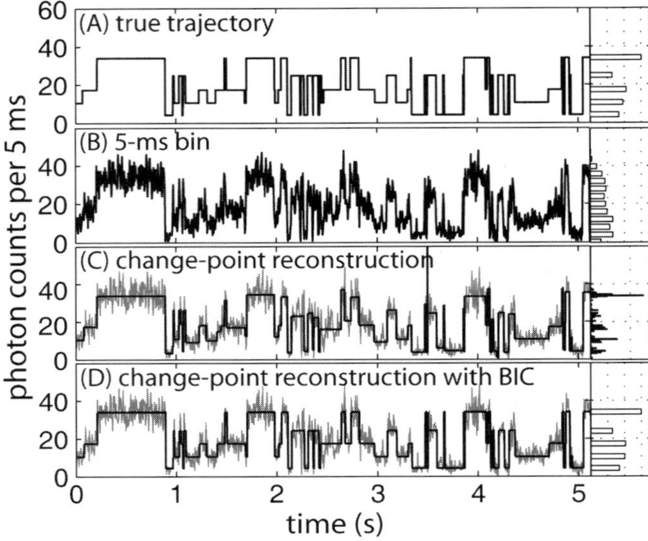

Fig. 3. Extraction of state distribution combining likelihood ratio change point analysis and Bayesian information criterion. Histograms for the intensity-state distribution are displayed at the right of each trajectory. See text for details.

Due to the photon-counting noise present in the data series, all the $(N_{\mathrm{cp}} + 1)$ intensities are numerically distinct (cf. intensity histogram of Fig. 3(C)). To avoid over interpretation of the data, one would like to determine the minimum number of states needed to describe the intensity trace. To this end, one may use a model selection measure called the Bayesian Information Criterion (BIC),[14,15] which quantifies the amount of evidence in a data set supporting a particular model. Implementing BIC to the present example amounts to maximizing the following function (see Ref. 10 for details of implementation):

$$\ln h(\{\Delta_1, \ldots, \Delta_N\}, n_G) \approx 2 \ln L_G - [2(n_G - 1) \ln N_{\mathrm{cp}} + N_{\mathrm{cp}} \ln N], \tag{7}$$

where n_G is the number of intensity states, N is the number of photons in the time series, and L_G is the likelihood function evaluated with a particular n_G. Thus, the number of intensity states can be determined objectively by the maximum in a plot of Eq. (7) as a function of n_G.

The result is displayed in Fig. 3(D), which is in excellent agreement with the true time series.

In the second case, one would like to determine the entire distribution of a continuously distributed physical parameter. For example, if the physical parameter is the intra-molecular distance (x) measured using FRET, then the distribution in x can be related to the apparent potential of mean force (Helmholtz free energy). However, the raw distribution in x as constructed from single-molecule measurements ($\hat{x}$) is greatly broadened, because the measured $\hat{x}$ contain uncertainties resulting from photon-counting detection. If the distances are estimated under the likelihood framework, then the Fisher information in Eq. (5) can be used to quantitatively account for the uncertainty due to counting noise. The remaining task is thus to deconvolve out the noise component from the raw distribution. As in the previously discussed discrete-state case, straight deconvolution will give an overfitted result that produces many modes. Following the same principle to describe the experimental observation with minimum parameters, one may use the Maximum Entropy Method[16–18] to do the deconvolution by minimizing the merit function,[19]

$$M[p(x), \Lambda] = \chi^2 + \Lambda \int_{-\infty}^{\infty} p(x) \ln p(x) dx, \tag{8}$$

where χ^2 is the standard chi-squared measure, Λ is the Lagrange undetermined multiplier to be optimized during deconvolution, and $p(x)$ is the deconvolved probability density function that one seeks. To determine the uncertainties associated with the distribution,[20] one can use the nonparametric (model-free) Bootstrap method[21,22] to resample the single-molecule trajectory. The ensemble of n resampled distributions is subjected to the same maximum-entropy deconvolution to form a collection of probability densities, $\{p_i(x)\}$. These bootstrapped probability densities are subsequently used to estimate the uncertainties in $p(x)$ at a given x by $\mathrm{var}[p(x)] = \frac{1}{n}\sum_{i=1}^{n}(p_i(x) - \langle p_i(x)\rangle)^2$.

So far, it has been shown how tools and ideas from the statistics and information theory communities can be applied to solving problems that may prevent the single-molecule approach from reaching its full potential. In most cases, all the necessary derivations and numerical treatments have been developed and can be found in the literature as indicated in the preceding discussion. However, there are also cases where additional derivations are needed in order to rigorously address issues unique to single-molecule experiments.[10] The remainder of this chapter will focus on the applications of the basic ideas introduced in this section.

3. The transition between molecular states occurs much faster than the experimental time resolution — Intermittency

Intensity intermittency is one of the characteristics of single-molecule emission. It signifies an abrupt change in the molecular state (e.g., the electronic state of a chromophore,[24–26] or the conformation of an oligonucleotide[27,28]) that occurs on a timescale that is much shorter than the experimental time resolution. It is one of the many phenomena that can only be revealed by single-molecule experiments. One remarkable example is the CdSe/ZnS core/shell quantum dot (QD), the bulk sample of which emits a vividly bright color following photoexcitation. At the single-particle level, quite surprisingly, intermittent luminescence was observed.[29] It turns out that there are very interesting statistical properties associated with the time-dependent intensity changes,[30] whose underlying physics is currently under intense experimental and theoretical studies (see, for example, Chap. 11). The single QD example will be used to illustrate the problems involved in analyzing this type of data, and a solution to resolve them.

Figure 4(A) shows the intensity–time trajectory of a typical QD averaged (binned) over different time resolutions. It is clear that there is a trade-off between time resolution and apparent noise level. More importantly, it is evident that different binning times will change the

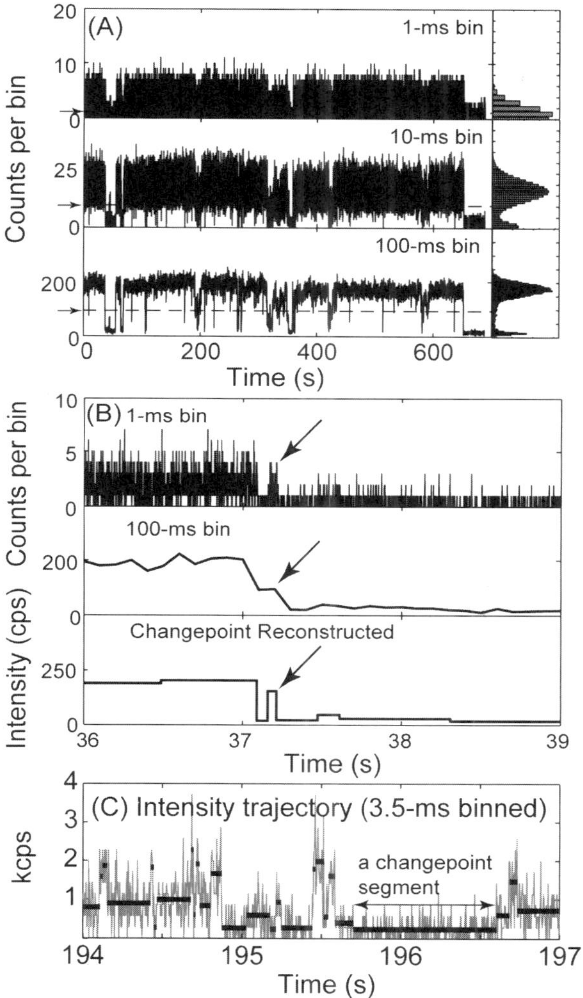

Fig. 4. Intensity–time trajectory from a single quantum dot (QD) binned at different time resolutions. The intensity distribution for each binning time is illustrated on the right. For each panel, a threshold for assigning on–off states that correspond to 1 count/ms is indicated by a dashed line as well as by an arrow. It is apparent that the on–off state assignment depends both on the binning time and on the threshold value. (B) Change-point reconstruction is able to detect transiently rare events (indicated by an arrow) that would not be picked up otherwise. (C) Intermediate intensity states from a blinking quantum dot are unambiguously determined. Reprinted in part with permission from Ref. 23. Copyright (2006) American Chemical Society.

intensity distribution dramatically. How, then, does one proceed to choose the binning time, and to determine whether the QD is in its "on" or "off" state? A commonly used approach is to decide on an intensity threshold above which the particle is considered to be "on," and that with intensities below threshold designated as "off." As illustrated in the figure, however, the state assignment strongly depends on the choice of binning time as well as the threshold value. Note also that this initial analysis to determine the intensity states will critically impact subsequent studies,[31] for example, the on- or off-time distribution and the kinetics. The dynamics of QD blinking also presents an interesting example in which the underlying kinetic scheme remains unknown,[30,32–38] rendering inadequate many well-developed analysis schemes that require a presumed kinetic model such as the Hidden Markov method,[39–41] which has also received significant development for the analysis of single ion-channel recording.[42]

Realizing that the problem of resolving intensity intermittencies in a single-molecule time trace is isomorphic to the change-point problem in applied statistics,[43–49] one may use the change-point ideas developed in that community to provide a general solution to the problem.[10] In particular, the solution has the following salient advantages: (1) it does not require such physical models as a kinetic scheme; (2) it is objective in that no bias is introduced through user-adjustable parameters as in the aforementioned binning–thresholding scheme; (3) it is able to deal with short trajectories commonly observed in most single-molecule data (also see Chap. 10 for extracting kinetics scheme from finite-length data); and (4) finally, it is quantitative, so that a confidence interval can be associated with derivative parameters such as the lifetime of a particular molecular state. The change-point method uses a generalized likelihood ratio test (cf. Eq. (6)) to determine if there is a change in the emission intensity at a particular time point along the single-molecule time series. The numerical values for the critical value (λ_α) at different confidence levels (α) are used to assess the uncertainties associated with the method. The procedure is applied to the time series photon by photon so that excellent time resolution can be achieved. For example, if the intensity changes by a factor of 5 or greater, the time instance at which the intensity changes can be located to within one to two photons with 90% confidence.

Applying this method to the QD experimental data, one sees that the change-point method is able to resolve transient and rare excursions to a variety of intensity levels that the QD exhibits (cf. Fig. 4(B)). This experiment also provides strong evidence to the fact that the commonly adopted two-state model describing QD intermittency is insufficient to account for the observed power-law like intensity correlation[50–52] and power spectral density[53] for QDs. A close interaction between theoretical modeling and experimental investigation can therefore further our understanding of this intriguing phenomenon.

4. Experimental time resolution is sufficient to follow the transition between molecular states

Consider a biological macromolecule or a polymer that undergoes conformational changes on the millisecond to second timescale. One may use single-pair FRET from an energy donor to acceptor to probe this type of motion.[54] By simultaneously monitoring the intensities of the donor and the acceptor fluorescence, and assuming that the energy transfer between them occurs via a dipole–dipole interaction, one may deduce the manner in which the donor–acceptor distance changes over time. The challenges involved in quantitatively analyzing this type of experiment are similar to those discussed in Sec. 3: How does one determine the binning time in an unbiased way?[55] Since single-molecule experiments rely on time averaging, one makes more precise measurements by averaging more (longer binning time), but at the expense of time resolution. Therefore, the solution to this problem should also allow one to strike a balance between time resolution and measurement uncertainties.

The desired solution can be obtained by a combined use of maximum likelihood estimation and Fisher information (cf. Eqs. (4) and

(5)). The idea is that one would analyze the single-molecule time trajectory in a way that each distance measurement from a "bin" of photons will give the same uncertainty.[9] In other words, the time series is "binned" adaptively according to the amount of information contained in each time bin. Each bin, T_j, satisfies the following equation:

$$J(\hat{x}_j) = \frac{36\hat{x}_j^{10}}{(1+\hat{x}_j^6)^3}\left[I_d^\beta T_j \frac{(1-\beta_d^{-1})^2}{(\hat{x}_j^6+\beta_d^{-1})} + I_a^\beta T_j \frac{(1-\beta_a^{-1})^2}{(\hat{x}_j^6+\beta_a^{-1})}\right] = \frac{1}{\mathrm{var}(\hat{x}_j)}, \tag{9}$$

where $\hat{x} \equiv \hat{R}/R_0$ is the normalized distance estimated using MLE with $\hat{R}$ being the donor–acceptor distance and R_0 the Förster radius at which the energy transfer efficiency is 50%, and $I_d^\beta (I_a^\beta)$ is the donor intensity in the absence of the acceptor (donor) with $\beta_d (\beta_a)$ being the signal-to-background ratio for the donor (acceptor) channel. Depending on the nature of the question that the experiment is designed to address, an experimentalist may decide on the precision ($\mathrm{var}(\hat{x})$) with which the measurement is to be accomplished. Equation (9) also provides a quantitative relationship between measurement uncertainties, $\mathrm{var}(\hat{x}_j)$, and the time resolution for realistic experimental conditions, T_j, for the j-th time bin. From an information theory viewpoint, this method yields the optimal achievable time resolution.

Figure 5 illustrates the application of this approach.[19] The time and distance uncertainties associated with each adaptive time bin help to assess any statements made about the movements of this polypeptide. In the present case, polyprolines are expected to appear static[56–60]; therefore, one does not expect to observe any dynamical changes in the end-to-end distance. The distance trajectory shown in Fig. 5(C), for example, does indicate that all the apparent fluctuations in the distance are well within the measurement uncertainties. However, the time-weighted distance distribution appears very broad, as opposed to a δ-function-like sharp distribution that one would have expected. This is because the distributions constructed from single-molecule

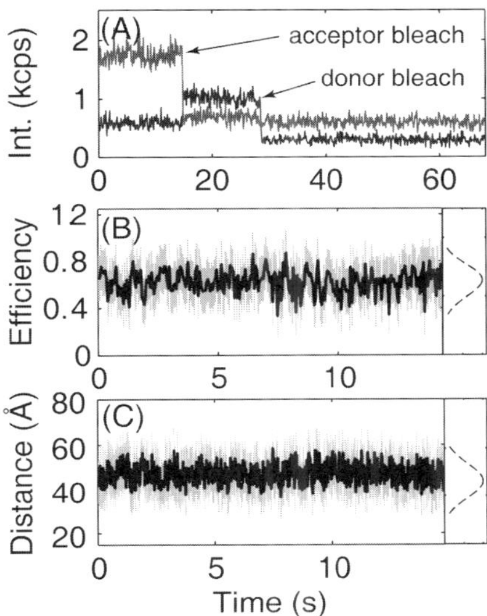

Fig. 5. Intensity–time traces of donor (black) and acceptor (gray) emissions for a single poly-L-proline molecule with 15 proline repeats. (B) and (C) Efficiency–time and distance–time traces (thick black lines) obtained using the information-theoretical analysis. The width and height of each gray box along the traces represent uncertainties in time and distance (or efficiency), respectively. The time-weighted distributions for the efficiency and distance are displayed to the right. Reprinted in part with permission from Ref. 19. Copyright (2006) American Chemical Society.

time series tend to be severely broadened by photon-counting noise. It turns out that to be able to uncover quantitatively the distribution of a certain molecular property from single-molecule measurements is quite challenging. Section 5 describes how this problem could be resolved in an unbiased way.

5. Distributions

The estimation of the probability density of an unknown distribution in its most general form remains an unsolved problem in statistics.[61] For example, if one were to estimate the probability density using histograms by binning the measured values, the choice of the number of

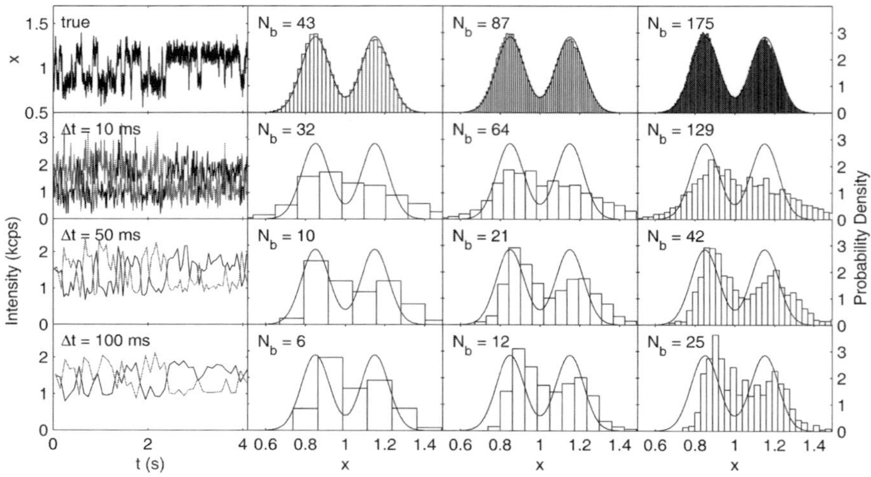

Fig. 6. Issues involved in recovering distance distribution from single-molecule FRET measurements. The computer-generated trajectory simulates a Langevin dynamics between two wells. The leftmost column shows how the apparent FRET intensity traces, also simulated using realistic experimental conditions, are changed by different binning times. Columns 2–4 show how different binning times along the FRET trajectory, and the number of bins used in constructing the distribution, can dramatically alter the resulting distribution. The solid line represents the true distribution used to generate the Langevin dynamics. Reprinted with permission from Ref. 19. Copyright (2006) American Chemical Society.

bins and the bin width can dramatically alter the shape of the resultant distribution, especially in the presence of experimental uncertainties. Figure 6 illustrates this issue in the context of single-molecule measurements. Suppose x denotes the distance between two points in a protein that can be measured by FRET. Its distribution, $p(x)$, can be related to the potential of mean force by $V(x) = -k_B T \ln p(x)$, where k_B is the Boltzmann constant and T is the temperature in Kelvin. Therefore, the single-molecule approach, in principle, allows one to directly explore the potential energy surface that governs a molecule's conformational dynamics. Because the exact shape of the underlying potential energy surface is unknown, any method used to recover such a distribution from single-molecule data should be general; an assumption of the number and shape of modes in the distribution should not be required. Furthermore, it should take into account the contributions from photon-counting noise and, if possible, remove them. While the majority of the well-established density estimation methods have been developed under the assumption that there are no uncertainties associated with the location of the abscissa,[20] this is certainly not the case for single-molecule data. New strategies will have to be developed in order to deal with single-molecule time series.

One begins by recognizing that the distribution of maximum likelihood estimated parameters are asymptotically Normal (Gaussian) distributed, the width of which is given by the Fisher information (cf. Eq. (5)). Thus, a distribution can be constructed using Gaussian kernel density estimation that is expected to converge to the "true" distribution in the limit of infinite number of data points.[62] This way, the measurement uncertainties (in the form of Fisher variance) can be incorporated into the construction of the probability density in a very natural fashion. To recover the probability density $p(x)$, the maximum entropy method (cf. Eq. (8); Ref. 63) can be used to quantitatively remove the noise component. This procedure, in principle, allows the recovery of the *entire* distribution without assuming any modality in $p(x)$. In practice, however, one can only recover the distribution for the distance range where FRET-based measurements are applicable (roughly 2.5–8 nm), limited by the validity of dipole approximation for the fluorescent dyes in the short-distance limit, and by the capability of detecting the acceptor photons in the long-distance limit. As an example, this method is applied to the polyproline experiment discussed earlier. Short polyprolines are expected to exhibit sharp, δ-function-like end-to-end distributions on the single-molecule experimental timescale. The results are shown in Fig. 7.

At this point, while practical solutions have been provided for analyzing certain types of single-molecule time series, there remain several cases that await solutions. For example, a general solution that goes beyond the discrete-state model[64,65] for the intermediate case in which the timescale of the dynamics coincides with that of the experiment is not yet available.[66,67] In addition, the primary assumption in the preceding development is that all the detected photons are

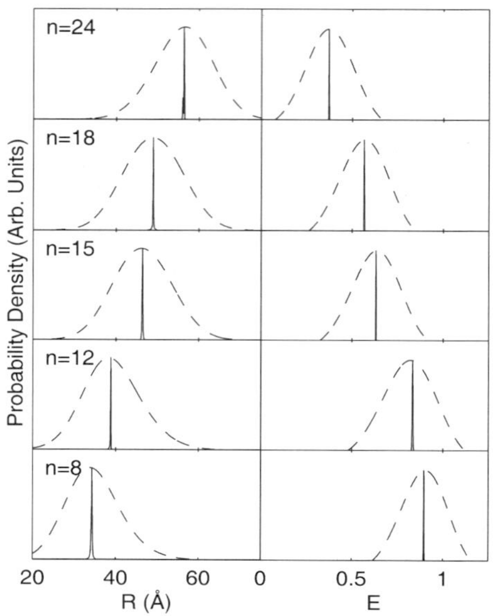

Fig. 7. Distribution of end-to-end distances from individual poly-proline molecules, where *n* denotes the number of proline repeat units in the molecule. The dashed line is the raw distribution before deconvolution whereas the solid line is the delta-function like distribution after deconvolution. *E* is the FRET efficiency. Reprinted with permission from Ref. 19. Copyright (2006) American Chemical Society.

uncorrelated. This assumption breaks down when the photon detections are correlated as in the case of detecting a diffusing molecule. Therefore, it would be greatly helpful if a solution can be found to treat such correlated experimental results in a consistent and objective way without explicitly modeling the dynamics of the molecule.

The statistical analysis tools discussed so far have covered experiments that study immobilized molecules. There is another class of experimental approach that studies the spectroscopic signatures of individual fluorescently labeled molecules as they diffuse through the focal volume of a microscope. While the nature of the problems involved in extracting quantitative information from such experiments is similar to those with immobilized molecules, it is considerably more challenging, and several key aspects of the problem remain unsolved. Section 6 provides one example showing how the ideas of unbiased analysis can be applied to this kind of problem.

6. Bursts

In a diffusion-type experiment, a fluorescent molecule will give a photon burst as it passes through the detection volume of a confocal microscope, as illustrated by the simulated trace in Fig. 8. These photon bursts are then collected from individual molecules for further analysis. The first step in interpreting these experiments is to determine whether there is a photon burst or not. For this purpose, one may decide on a specific binning time and an intensity threshold value, above which the molecule is considered to be on its "on" state. As illustrated in Fig. 8, however, the binning time and the threshold value — both are parameters subjectively chosen in this example — can affect the number of bursts collected in a time series. It should be pointed out that some of the bursts identified this way might be due to

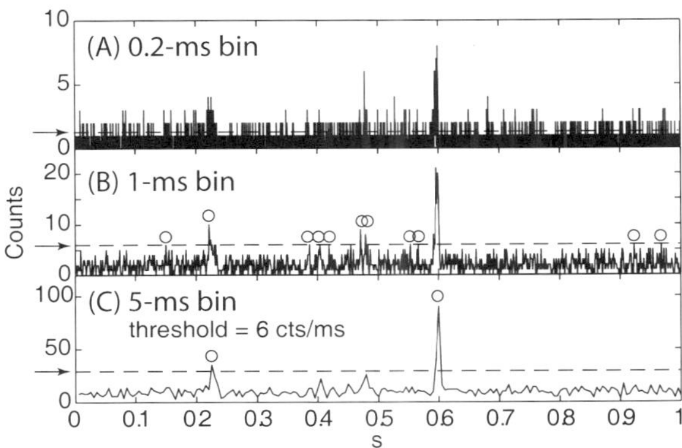

Fig. 8. Binning time changes the appearance of a simulated single-molecule burst time trace. A threshold equal to 6 counts per millisecond (dashed line; indicated by an arrow on each panel) is used to illustrate how a photon burst can be detected and collected. Each open circle indicates a binned intensity level that rises above the threshold (circles are omitted in panel A due to the large number of bursts). Reprinted in part with permission from Ref. 12. Copyright (2006) American Chemical Society.

photon-counting noise (false-positive identification) so that the collected photons are from the background and not from the molecule of interest. Therefore, the binning–thresholding process impacts the efficiency with which one may extract information from photon bursts. Several research groups have contributed to improving this initial step.[68–74] A likelihood-based approach is outlined in the following paragraph.

The problem is isomorphic to the sequential test problem in applied statistics, which has been used to solve such practical problems as quality control in manufacturing factories and clinical tests. For the current problem, the idea is to use the likelihood ratio test to examine a pool of photon detection intervals, $\{\Delta_1, \ldots, \Delta_N\}$, for whether the detected intensity level can be categorized as background, photon burst, or indeterminate. If the test, R_N, falls below the critical value for the background, $\lambda^{\text{background}}$, the pool of photons can be categorized as background. On the other hand, if the likelihood ratio rises above the λ^{signal} value, there is sufficient evidence to justify assigning the photon pool to signal. If, however, the likelihood ratio falls in between these two critical values, there is insufficient evidence to warrant any assignment with confidence. In this case, additional data points (photons) will be needed to improve the statistical confidence and the test is repeated. These ideas can be summarized as follows:[12]

$$\begin{aligned}
&\lambda(N) \leq \lambda_C^{\text{background}}\text{—assign to background},\\
&\lambda(N) \geq \lambda_C^{\text{signal}}\text{—assign to signal,}\\
&\lambda_C^{\text{background}} < \lambda(N) < \lambda_C^{\text{signal}}\text{—take an additional photon,}
\end{aligned} \tag{10}$$

where $\lambda(N)$ is the log-likelihood ratio (cf. (6)) calculated using N photons. The critical values, $\lambda_C^{\text{background}}$ and $\lambda_C^{\text{signal}}$, depend on the number of photons being considered and also on the previously discussed false-positive and false-negative errors; they can be constructed based on the assumption of photon-counting Poisson statistics and the known intensity levels for background count

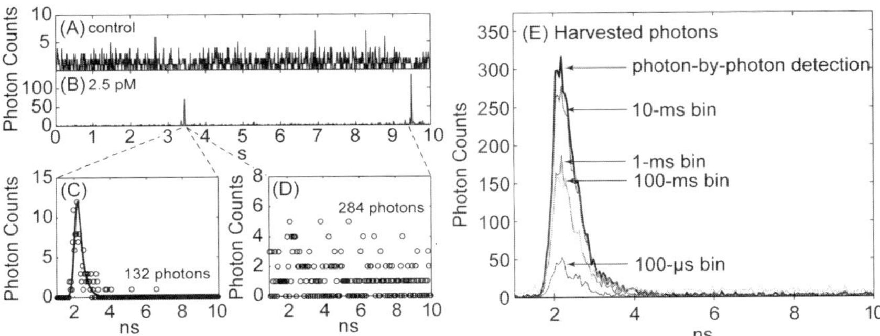

Fig. 9. Detection of two-photon excited luminescence from individual 60-nm gold nanoparticles diffusing through the detection volume in water. (A) A typical trace from control samples that contain only water. (B) 2.5 pM gold nanoparticles. (C) TCSPC lifetime trace harvested from the indicated photon burst that contains 132 photons. The solid line represents instrument response function of the spectrometer. (D) TCSPC lifetime trace for a segment of background photons showing no discernable features. (E) Comparison of different photon-harvesting methods. (Reprinted in part with permission from Ref. 12. Copyright (2006) American Chemical Society.)

and single-molecule emission — both can be characterized independently by control experiments.

Figure 9 shows the results of applying this method to the detection of the two-photon excited luminescence from diffusing single 60-nm gold nanoparticles. The experiment was carried out using a time-correlated single-photon counting (TCSPC) configuration so that both the emission intensity and luminescence lifetime of a chromophore can be simultaneously recorded. One may also use this lifetime information to characterize the efficiency of the photon-by-photon burst harvesting method. As shown in Figs. 9(C) and 9(D), the signal and the background can be clearly distinguished by their TCSPC data. To examine the effectiveness of this approach, one may compare the number of signal photons extracted using this method with those extracted using other binning–thresholding methods by comparing their TCSPC data. As shown in Fig. 9(E), the likelihood-based method does appear to be most efficient.

Luminescence lifetime is not the only example of the spectroscopic characteristics that can be extracted from photon bursts from individual molecules. Other information includes the time-dependent distance changes between a FRET donor–acceptor pair. However, since the excitation intensity and detection efficiency are not constant as the molecule diffuses stochastically through the confocal detection volume, it remains very challenging to recover the kind of information discussed in the previous sections (see also Chap. 6). Yet another unsolved problem along these lines is the quantitative extraction of the time lapse between consecutive bursts without explicitly modeling the path by which the chromophore traverses through the focal volume.

7. Conclusion

The importance of an objective assessment of noisy experimental time series cannot be overemphasized. Any scientific conclusion must be supported by rigorous treatment of the experimental results. This chapter has discussed several aspects that may prevent optical single-molecule spectroscopy from reaching its full potential, and has provided solutions that are unbiased, requiring no presumed models either in the kinetic scheme or in the distribution modality. These methods effectively remove the uncertainties associated with photon-counting detection, and afford the direct application of powerful theoretical methods such as those included in this book to unravel the dynamics of microscopic systems. It is hopeful that as new experimental approaches continue to be developed, concurrent development of new theoretical approaches will keep on pushing our scientific knowledge to new frontiers.

Acknowledgments

The research discussed herein was supported by the National Science Foundation and the US Department of Energy. Lucas P. Watkins and Kai Zhang are acknowledged for their contributions, Jeffrey A. Hanson for helpful discussions, and Frank L. H. Brown, JAH, and Chia-Ying Wang for critical reading of and helpful comments on the manuscript.

References

1. W. E. Moerner and D. P. Fromm, Methods of single-molecule fluorescence spectroscopy and microscopy, *Review of Scientific Instruments* **74**(8) (2003) 3597–3619.
2. L. Mandel and E. Wolf, *Optical Coherence and Quantum Optics* (Cambridge University Press, New York, 1995).
3. R. A. Fisher, On the mathematical foundations of theoretical statistics, *Philosophical Transactions of the Royal Society of London Series A* **22** (1922) 309–368.
4. M. J. Schervish, *Theory of Statistics*, Springer Series in Statistics (Springer, New York, 1995).
5. T. M. Cover and J. A. Thomas, *Elements of Information Theory* (John Wiley & Sons, Inc., New York, 1991).
6. R. A. Fisher, Theory of statistical estimation, *Proceedings of the Cambridge Philosophical Society* **22** (1925) 700–725.
7. H. Cramér, *Mathematical Methods of Statistics* (Princeton University Press, Princeton, NJ, 1946).
8. C. R. Rao, Sufficient statistics and minimum variance estimates, *Proceedings of the Cambridge Philosophical Society* **45**(2) (1949) 213–218.
9. L. P. Watkins and H. Yang, Information bounds and optimal analysis of dynamic single molecule measurements, *Biophysical Journal* **86**(6) (2004) 4015–4029.
10. L. P. Watkins and H. Yang, Detection of intensity change points in time-resolved single-molecule measurements, *Journal of Physical Chemistry B* **109**(1) (2005) 617–628.
11. D. Montiel, H. Cang and H. Yang, Quantitative characterization of changes in dynamical behavior for single-particle tracking studies, *Journal of Physical Chemistry B* **110**(40) (2006) 19763–19770.
12. K. Zhang and H. Yang, Photon-by-photon determination of emission bursts from diffusing single chromophores, *Jounal of Physical Chemistry B* **109**(46) (2005) 21930–21937.
13. P. G. Hoel, S. S. Port and C. C. Stone, *Introduction to Statistical Theory* (Houghton Mifflin Company, Boston, MA, 1971).
14. G. Schwarz, Estimating the dimension of a model **6**(2) (1978) 461–464.

15. A. D. Lanterman, Schwarz, Wallace and Rissanen: Intertwining themes in theories of model selection, *International Statistical Review* **69**(2) (2001) 185–212.
16. E. T. Jaynes, Information theory and statistical mechanics, *Physical Review* **108**(2) (1957) 171–190.
17. E. T. Jaynes, Information theory and statistical mechanics, *Physical Review* **106**(4) (1957) 620–630.
18. E. T. Jaynes, On the rationale of maximum-entropy methods, *Proceedings of the Institute of Electrical and Electronics Engineers* **70**(9) (1982) 939–952.
19. L. P. Watkins, H. Chang and H. Yang, Quantitative single-molecule conformational distributions: A case study with poly-L-proline, *Journals of Physical Chemistry A* **110**(15) (2006) 5191–5203.
20. B. W. Silverman, *Density Estimation for Statistics and Data Analysis* (Chapman & Hall, New York, 1986).
21. B. Efron and G. Gong, A leisurely look at the bootstrap, the jackknife, and cross-validation, *Annals of Statistics* **37**(1) (1983) 36–48.
22. B. Efron, 1977 Rietz lecture — Bootstrap methods — Another look at the jackknife, *Annals of Statistics* **7**(1) (1979) 1–26.
23. K. Zhang *et al.*, Continuous distribution of emission intensity and its nonlinear correlation to luminescence decay rates from single CdSe/ZnS quantum dots, *Nano Letters* **6** (2006) 843–847.
24. R. M. Dickson *et al.*, On/off blinking and switching behaviour of single molecules of green fluorescent protein, *Nature* **388**(6640) (1997) 355–358.
25. W. T. Yip *et al.*, Classifying the photophysical dynamics of single- and multiple-chromophoric molecules by single molecule spectroscopy, *Journals of Physical Chemistry A* **102**(39) (1998) 7564–7575.
26. P. Tinnefeld *et al.*, Direct observation of collective blinking and energy transfer in a bichromophoric system, *Journals Physical Chemistry A* **107**(3) (2003) 323–327.
27. X. W. Zhuang *et al.*, A single-molecule study of RNA catalysis and folding, *Science* **288**(5473) (2000) 2048–2051.
28. S. A. McKinney *et al.*, Structural dynamics of individual Holliday junctions, *Nature Structural Biology* **10**(2) (2003) 93–97.
29. M. Nirmal *et al.*, Fluorescence intermittency in single cadmium selenide nanocrystals, *Nature* **383**(6603) (1996) 802–804.
30. M. Kuno *et al.*, Nonexponential "blinking" kinetics of single CdSe quantum dots: A universal power law behavior, *Journal of Chemical Physics* **112**(7) (2000) 3117–3120.
31. M. Lippitz, F. Kulzer and M. Orrit, Statistical evaluation of single nano-object fluorescence, *ChemPhysChem* **6** (2005) 770–789.
32. M. Kuno *et al.*, Modeling distributed kinetics in isolated semiconductor quantum dots, *Physical Review B* **67**(12) (2003) 125304.

33. M. Kuno *et al.*, "On"/"off" fluorescence intermittency of single semiconductor quantum dots, *Journal of Chemical Physics* **115**(2) (2001) 1028–1040.
34. S. A. Empedocles and M. G. Bawendi, Quantum-confined stark effect in single CdSe nanocrystallite quantum dots, *Science* **278**(5346) (1997) 2114–2117.
35. R. G. Neuhauser *et al.*, Correlation between fluorescence intermittency and spectral diffusion in single semiconductor quantum dots, *Physical Review Letters* **85**(15) (2000) 3301–3304.
36. K. T. Shimizu *et al.*, Blinking statistics in single semiconductor nanocrystal quantum dots, *Physical Review B* **6320**(20) (2001) 205316.
37. I. H. Chung and M. G. Bawendi, Relationship between single quantum-dot intermittency and fluorescence intensity decays from collections of dots, *Physical Review B* **70**(16) (2004) 165304.
38. B. R. Fisher *et al.*, Emission intensity dependence and single-exponential behavior in single colloidal quantum dot fluorescence lifetimes, *Journal of Physical Chemistry B* **108**(1) (2004) 143–148.
39. L. E. Baum *et al.*, A maximization technique occurring in statistical analysis of probabilistic functions of Markov chains, **41**(1) (1970) 164–171.
40. L. E. Baum and T. Petrie, Statistical inference for probabilistic functions of finite state markov chains, **37**(6) (1966) 1554–1563.
41. M. Andrec, R. M. Levy and D. S. Talaga, Direct determination of kinetic rates from single-molecule photon arrival trajectories using hidden Markov models, *Journal of Physical Chemistry A* **107**(38) (2003) 7454–7464.
42. B. Sakmann, *Single-Channel Recording* 2nd edn. (Plenum Press, New York, 1995).
43. D. V. Hinkley, Inference in 2-phase regression, *Journal of American Statistical Association* **66**(336) (1971) 736–743.
44. D. V. Hinkley, Inference about change-point from cumulative sum tests, *Biometrika* **58**(3) (1971) 509–523.
45. D. V. Hinkley, Inference about change-point in a sequence of random variables, *Biometrika* **57**(1) (1970) 1–17.
46. L. Horváth, The maximum-likelihood method for testing changes in the parameters of normal observations, *Annals of Statistics* **21**(2) (1993) 671–680.
47. E. Gombay and L. Horváth, Rates of convergence for U-statistic processes and their bootstrapped versions, *Journal of Statical Planning and Inference* **102**(2) (2002) 247–272.
48. E. Gombay and L. Horváth, On the rate of approximations for maximum likelihood tests in change-point models, *Jounal of Multivariate Analysis* **56**(1) (1996) 120–152.

49. V. K. Jandhyala and S. B. Fotopoulos, Capturing the distributional behaviour of the maximum likelihood estimator of a changepoint, *Biometrika* **86**(1) (1999) 129–140.
50. R. Verberk *et al.*, Environment-dependent blinking of single semiconductor nanocrystals and statistical aging of ensembles, *Physica E* **26**(1–4) (2005) 19–23.
51. R. Verberk and M. Orrit, Photon statistics in the fluorescence of single molecules and nanocrystals: Correlation functions versus distributions of on- and off-times, *Journal of Chemical Physics* **119**(4) (2003) 2214–2222.
52. R. Verberk, A. M. van Oijen and M. Orrit, Simple model for the power-law blinking of single semiconductor nanocrystals, *Physical Review B* **66**(23) (2002) 233202.
53. M. Pelton, D. G. Grier and P. Guyot-Sionnest, Characterizing quantum-dot blinking using noise power spectra, *Applied Physics Letters* **85**(5) (2004) 819–821.
54. T. Ha *et al.*, Probing the interaction between two single molecules: Fluorescence resonance energy transfer between a single donor and a single acceptor, *Proceedings of the Natural Academy of Science, USA* **93**(13) (1996) 6264–6268.
55. X. Michalet, S. Weiss and M. Jager, Single-molecule fluorescence studies of protein folding and conformational dynamics, *Chemical Review* **106**(5) (2006) 1785–1813.
56. L. Stryer and R. P. Haugland, Energy transfer — A spectroscopic ruler, *Proceedings of the Natural Academy of Science, USA* **58**(2) (1967) 719–726.
57. P. R. Schimmel and P. J. Flory, Conformational energy and configurational statistics of poly-L-proline, *Proceedings of the Natural Academy of Sciences USA* **58**(1) (1967) 52–59.
58. W. L. Mattice and L. Mandelkern, Conformational properties of poly-L-proline form II in dilute solution, *Journal of American Chemical Society* **93**(7) (1971) 1769–1777.
59. S. Tanaka and H. A. Scheraga, Calculation of characteristic ratio of randomly coiled poly(L-proline), *Macromolecules* **8**(5) (1975) 623–631.
60. B. Schuler *et al.*, Poly-proline and the "spectroscopic ruler" revisited with single-molecule fluorescence, *Proceedings of the Natural Academy of Sciences* **102**(8) (2005) 2754–2759.
61. A. J. Izenman, Recent developments in nonparametric density-estimation, *Journal of the American Statistical Association* **86**(413) (1991) 205–224.
62. B. W. Silverman, Weak and strong uniform consistency of kernel estimate of a density and its derivatives, *Annals of Statistics* **6**(1) (1978) 177–184.
63. J. Skilling and R. K. Bryan, Maximum-entropy image-reconstruction — general algorithm, *Monthly Notices of the Royal Astronomical Society* **211**(1) (1984) 111–124.

64. E. Geva and J. L. Skinner, Two-state dynamics of single biomolecules in solution, *Chemical Physics Letters* **288**(2–4) (1998) 225–229.
65. I. V. Gopich and A. Szabo, Single-macromolecule fluorescence resonance energy transfer and free-energy profiles, *Journal of Physical Chemistry B* **107**(21) (2003) 5058–5063.
66. P. W. Anderson, A mathematical model for the narrowing of spectral lines by exchange or motion, *Jounal of the Physical Society of Japan* **9**(3) (1954) 316–339.
67. R. Kubo, Note on the stochastic theory of resonance absorption, *Journal of the Physical Society of Japan* **9**(6) (1954) 935–944.
68. K. Peck *et al.*, Single-molecule fluorescence detection — Auto-correlation criterion and experimental realization with phycoerythrin, *Proceedings of the Natural Academy of Sciences USA* **86**(11) (1989) 4087–4091.
69. J. Enderlein *et al.*, Statistics of single-molecule detection, *Journal of Physical Chemistry B* **101**(18) (1997) 3626–3632.
70. J. Enderlein *et al.*, A maximum likelihood estimator to distinguish single molecules by their fluorescence decays, *Chemical Physics Letters* **270**(5–6) (1997) 464–470.
71. J. R. Fries *et al.*, Quantitative identification of different single molecules by selective time-resolved confocal fluorescence spectroscopy, *Journal of Physical Chemistry A* **102**(33) (1998) 6601–6613.
72. C. Eggeling *et al.*, Data registration and selective single-molecule analysis using multi-parameter fluorescence detection, *Journal of Biotechnology* **86** (2001) 163–180.
73. R. W. Clarke, A. Orte and D. Klenerman, Optimized threshold selection for single-molecule two-color fluorescence coincidence spectroscopy, *Analytical Chemistry* **79**(7) (2007) 2771–2777.
74. E. Nir *et al.*, Shot-noise limited single-molecule FRET histograms: Comparison between theory and experiments, *Journal of Physical Chemistry B* **110**(44) (2006) 22103–22124.

CHAPTER 2

Testing Hypothesis with Single Molecules: Bayesian Approach

Taras Plakhotnik

School of Physical Sciences, The University of Queensland, Brisbane, St Lucia, QLD4072, Australia

— What is the probability to meet a dinosaur in the streets of New York today?
— This equals 50%. Either you meet him or you don't.
Author unknown

1. Introduction to the Bayesian approach

Do you believe that you know the correct answer to the question in the quotation above? Bayesian statistics is based on the paradigm that no such answer exists and that every opinion about the likelihood of meeting a dinosaur in the streets of New York is actually based on personal beliefs, background, and experience and cannot be justified from the first principles. Although most people are likely to suggest a small probability for the meeting, the numerical value cannot be derived theoretically or verified experimentally. And yet, dealing with rare events in a quantitative manner is, for example, an essential part of risk assessment in insurance companies. Are you prepared to insure someone against meeting a dinosaur for the benefit of 100 billion dollars? What should be the premium? Do not rush with any unreasonable commitment. Your choice may depend on an infinite number of data and factors such as, for example, your awareness of birds (technically dinosaurs) or the knowledge that the disastrous event was not fatal for crocodiles which shared this planet with dinosaurs for about 150 million years and then happily lived after. It is also important to emphasize that although data and experiments may have a profound influence on the probabilities of certain events, the data analysis (where statistics is a tool) is not an experimental/observational science but belongs to the realm of logic.

It is perhaps unexpectedly parallel with the example above, but logical analysis of experiments with single-molecules also requires a due attention to rare events. This will be considered in the following sections, but first we need to establish other motivations for writing/reading this chapter.

Irrespective of your choice for the probability, once a prior opinion is expressed and experimental results and observations are gathered, there will be certain rules how the prior probabilities should change to avoid logical inconsistencies. These rules also provide the best possible use of the available information. Since the data supplied by experiments with single molecules are usually restricted due to the limits set by the survival time of the molecules and other experimental difficulties, it is very important to process the obtained information most efficiently. Therefore, the rules which turn out to be the rules of Bayesian statistics should be the learned.

Although Bayesian methods are getting more and more popular in general, surprisingly they have had a limited attention in relation to single-molecule data, although several recent publications (see, for example, Refs. 1–5 and Chap. 1) are starting to fill the gap.

The subject we discuss here necessarily involves some math which is moved partially to the Appendices. The Appendices also minimize the need for external references, and provide some information beyond the narrow scope of this chapter. However, the available space limits the coverage, and those who are interested in other related topics could refer to an excellent introduction to Bayesian statistics by Sivia[6] or a comprehensive account on the subject written by Jaynes,[7] which, in my opinion, stands somewhere between science and fiction taking the best from the two. A good text with a large number of examples is offered by Gelman *et al.*[8]

Hypothesis making, model testing, and evaluation of their posterior likelihoods given experimental data make the essence of modern science and can be conveniently formalized using the following expression called the Bayes theorem

$$P(M_1|D, B) = \frac{P(D|M_1, B)P(M_1|B)}{P(D|B)}. \tag{1}$$

This equation relates $P(M_1|D, B)$, the posterior likelihood of model M_1 (probability that the model is correct given data D and background information B) to $P(M_1|B)$, the likelihood of M_1 prior to the data availability and $P(D|M_1, B)$, the data likelihood function conditional on the selected model (that is the probability of obtaining data set D if M_1 is correct). According to the marginalization rule (see Appendix 1), the denominator $P(D|B)$ equals $\sum_{k=1}^{\infty} P(D|M_k, B)P(M_k|B)$, where summation runs over a complete and mutually exclusive set of models $\{M_k\}$. Because all the probabilities in this paper are conditional on the background information, we will omit any explicit reference to this (symbol B after the vertical separation line) in all the following equations. A special case of the hypotheses testing which is usually called *parameter estimation* assesses competing hypotheses which differ from each other only by the values of some deducible (or fitted) parameters. In other words, we assume that the fundamentals of a model are generally correct and only some parameters are to be determined. Therefore, the success of the parameter estimation logically tells very little about the general validity of the model.

An immediate observation which can be made after a short inspection of Eq. (1) exploits Shannon's entropy which is defined as $H = -\sum_{k=1}^{\infty} P(M_k)\ln[P(M_k)]$. The change of this quantity resulting from evaluation of the new values of $P(M_k)$ after new data have become available can be used as a measure of the informational content of the data. If the change of H is small, the data can be valued as noninformative. If the set of the competing hypothesis contains only M_1, then $P(M_1|D) = P(M_1) = 1$ no matter what the data tell and how large/small $P(D|M_1)$ is (provided that $P(D|M_1) \neq 0$). Any theory/model is unshakable if it does not have an alternative which one considers conceivable, and any experiment "testing" such a hypothesis is useless because informational content of any related data is zero *a priory*. Religious beliefs form one class of such unshakable theories, and science is also not completely immune from the lack of alternatives and the presence of unjustified overconfidence.

It appears that the problem of lacking alternatives can be resolved by considering the possibility that M_1 is false as an option. Unfortunately, in many important cases this leads to an uncertainty or to $P(D|\text{NOT } M_1) = 0$ when one attempts to calculate the likelihood of the data. For example, let us consider a simple model M_1 stating that the value of physical variable V equals E_1 (that is, E_1 is a prediction of the model for V). Calculation of $P(D|M_1)$ for the result of a single measurement is simple if one believes that normally distributed noise is the only reason for a disagreement between the predictions of M_1 and the experimental data

$$\rho(D|M_1) = \frac{1}{(2\pi)^{1/2}\sigma} \exp\left(-\frac{(D - E_1)^2}{2\sigma^2}\right), \tag{2}$$

where σ stands for the standard deviation of the measurements, and the use of $\rho(D|M_1)$, the probability density function (PDF), instead of $P(D|M_1)$ emphasizes that datum D refers to a continuous variable. The probability density and the probability are related as $P(D|M_1) = \rho(D|M_1)dD$, where dD is a small interval around D. In the future, we will not distinguish between the probability density and the probability and will use P to label any of them. Actually, one may think of a continuous variable as being discreet and countable and therefore all the integrations (which still will be used in equations for briefness) can be thought of as summations over an index labeling the discreet values of the corresponding variable. The difference between the probability density and the probability is thus just a normalization factor. This factor is of no importance if only a ratio of two probabilities depending on the same normalization factor affects the analysis. We will frequently use the proportionality sign instead of equality in the following equations and will omit all irrelevant normalization constants.

What can we say about $P(D|\text{NOT } M_1)$? Logically, although NOT M_1 follows from $V \neq E_1$, no conclusion can be drawn from NOT M_1. We have to be more specific about the NOT M_1 hypothesis. For example, if NOT M_1 means that V may be with a uniform probability any number from a large in comparison to σ but finite interval of length L (the interval must include the point $V = D$, otherwise $P(D|\text{NOT } M_1) = 0$), then $P(D|\text{NOT } M_1) \propto 1/L$. $P(D|\text{NOT } M_1) = 0$ in the limit of $L \to \infty$. Because NOT M_1 does not place any restrictions on V in this limit, the ratio of the posterior likelihoods of M_1 and of NOT M_1 favours M_1 unless $P(M_1) = 0$ or $P(D|M_1) = 0$ (as can be seen from the Bayes theorem). If the model M_1 states a specific functional dependence of V on time and NOT M_1 is interpreted as a different but unspecified functional dependence, the space of *equally probable* functions becomes so large that we again should prefer more specific M_1.

When a scientific publication discusses the data exploiting only one model (this happens more frequently in physical sciences, where a sufficient number of established theories exists), the probability $P(D|M)$ is frequently considered as a measure of the theory "goodness" without any explicit reference to the plausible alternatives. Although this looks odd in the view of Eq. (1) and the above discussion, some rationale can be given to such an approach.

Let us consider model U, which provides the best possible value for $P(D|U)$ and call it a great ultimate model. Although such a model may never be known, the belief that the ultimate theory U exists is the basic paradigm of modern science and is the major force driving scientific progress. The comparison of the working model M to the great U could be made if the estimate for $P(D|M)/P(D|U)$ were known.

The value of $P(D|M)$ in Eq. (2) critically depends on the standard deviation σ and, in practice, accurate estimation of this parameter represents a nontrivial exercise, but let us ignore this complication for the moment. In other words, we assume that the accurate value of σ is given to us by an unquestionable authority. In principle, we could repeat the measurements infinite number of times to verify that the distribution of D follows the distribution described by Eq. (2). But, how can one get the likelihood of the data given U is correct? The best that the ultimate model can do is to predict the observable to be equal to its measured value. Therefore, we declare that $P(D|U) = (2\pi)^{-1/2}\sigma^{-1}$. Consequently, the estimator for the goodness of M reads

$$\frac{P(D|M)}{P(D|U)} = \exp\left(-\frac{(D-E)^2}{2\sigma^2}\right). \tag{3}$$

If $P(D|M)/P(D|U) \approx 1$, any alternative model will not gain much support from the data and therefore it is not worth the effort to think of any such alternative until more data becomes available. Contrarily, a small value of $P(D|M)/P(D|U)$ puts the theory in a dangerous position and reduces its reliability because at least in principle it leaves room for alternatives which can perform much better. So far, it seems that we can do very well without specifying the alternative models. But let us now consider a situation when the prediction of a theory is not a single value but a continuous function. In any real experiment, this continuous function will be represented by a set of N data points $\{D_n\}$. A natural generalization of Eq. (2)

$$P(\{D_n\}|M) = \frac{1}{(2\pi\sigma^2)^{N/2}} \exp\left[-\frac{1}{2}\sum_{n=1}^{N}\frac{(D_n - E_n)^2}{\sigma^2}\right] \tag{4}$$

shows the problem. Even if each data point is close to the value predicted by the model, a large number of factors on the right-hand side of Eq. (4) can make the exponent much smaller than 1.

Some texts on the data analysis suggest comparison of the probability $P(\{D_n\}|M)$ (instead of judging its magnitude on the absolute scale) to the value of $(2\pi\sigma^2)^{-N/2}\exp[-\sum \xi_n^2/(2\sigma^2)]$, where N numbers $\{\xi_n\}$ are randomly selected from an ensemble characterized by a zero mean and standard deviation σ. The distribution of

$$\chi^2 \equiv \sum_{n=1}^{N} \frac{\xi_n^2}{\sigma^2} \tag{5}$$

is well known in statistics and is called the chi-squared distribution (see Appendix 2). The "goodness criterion" is usually formulated by demanding that a certain percentage of randomly selected sets $\{\xi_n\}$ has their χ^2-values larger than the value of the sum in Eq. (4). This sets up a maximum tolerable value for $\sum (D_n - E_n)^2/\sigma^2$. Motivated by conventional wisdom, one can accept a 5% rejection level and discard hypotheses which have $\sum_{n=1}^{N} (D_n - E_n)^2/\sigma^2 \geq N + 2.3N^{1/2}$ (we assume that N is large and use the Gaussian approximation for the chi-squared distribution; see Appendix 2) and be satisfied with a model if the sum of squares is small. However, this reasoning is a dramatic deviation from the clear logic of the Bayesian approach expressed by Eq. (1). To be convinced that problems are simply swept under the carpet, look at the example in Fig. 1.

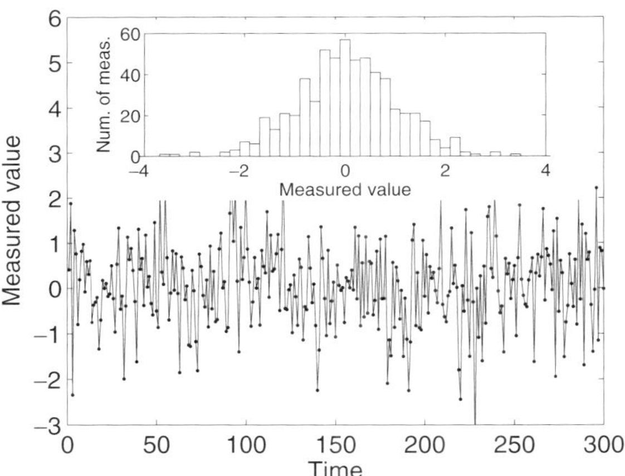

Fig. 1. A computer-simulated time-dependence of a measured variable v (only 300 data points are shown). (The chi-squared value for 600 data points is only 600 (the insert shows a histogram for the complete data). But a model where v is a periodic function gains a strong support if the data are analyzed properly.)

The data in Fig. 1 may represent the frequency of an absorption line of a single molecule as a function of time, which sometimes is called a "frequency trace". This can also be a time dependence of the luminescence intensity of a single molecule. Such data are commonly used to evaluate the number of quasi-stationary states a molecule can occupy when interacting with its environments. But, for the purpose of generality we can think of any variable υ. Let us assume that a working model M_1 predicts the value of υ being constant and time-independent. Do the data agree with this prediction? A visual inspection (which is really not a bad idea) of the plot in Fig. 1 is quite satisfactory. More quantitatively, the standard deviation calculated for the 600 data points is 1.002 and seems consistent with the standard deviation of 1 suggested by the experimental procedure (we do not discuss this procedure but we are assumed to be good experimentalists). The χ^2-value stands at 600. Such a value of χ^2 is within a solid 50% confidence level (as defined above). Can we conclude that M_1 is correct beyond any reasonable doubts? Not at all! The value, $\exp(-\chi^2/2) = \exp(-300)$, on the right-hand side of Eq. (4) leaves an enormously wide room for alternative hypotheses. As an example, consider M_2 saying that the dependence $\upsilon(t)$ is a function whose values change periodically between 0.39, 0.06, and -0.33 starting from 0.39. Somewhat surprisingly, the right-hand side of Eq. (4) for M_2 elevates to $\exp(-286)$. This is also a small number, but it is 10^6 times larger than $\exp(-300)$, a clear indication that the data are much more in favor of M_2 than M_1. Is M_2 the best possible theory? Since $\exp(-286)$ is also very small, there is obviously a room for further improvements. Only the ultimate model stating that the data points should be where they are (and therefore there are 600 different quasi-stationary states in the trace) will have the right-hand side at its absolute maximum.

If you think that a large value of χ^2 can unconditionally justify the rejection of a model, think again. Suppose that a trace shown in Fig. 2 represents here the charge of a single electron as measured in an experiment. In the conventional model of the universe, the charge

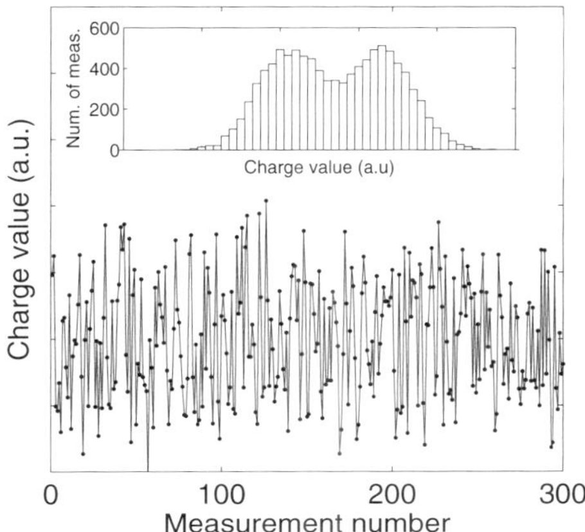

Fig. 2. First 300 results of a numerical experiment where a charge of an electron is measured repeatedly. (The histogram in the insert is the summary of 10,000 measurements which shows a bimodal distribution. If such experimental data were obtained in real life, most physicists would declare the data invalid.)

of an electron is constant. The data points can be grouped in two sets with significantly different means (this becomes obvious if one looks at the histogram shown in the insert of Fig. 2) and the χ^2 calculated for all 10^4 data points is about $1.8 \cdot 10^4$. Should we then conclude that the conventional model is not working? Very few physicists will do this. Most of them instead will reconsider the original estimate for sigma or will declare that the data are not valid (that is they will question the value of $P(D)$). Why? This is because the available background information so strongly supports the model where the electron charge is a constant that no physicist can think at present of any serious alternative. However, a person without any background in physics but with a strong background in statistics may with no hesitation take a different point of view.

2. Hidden parameters

An important difference between single-molecule experiments and ensemble measurements is that microscopic details which affect the outcomes of experiments with single molecules usually are not measurable directly while the ensemble properties depending on macroscopic variables such as temperature, pressure, concentration, etc., are relatively easily accessible. This distinction makes a large difference for hypothesis testing. For the purpose of generality let us consider a microscopic parameter α which is not under experimental control and for which only a probability $P(\alpha|M)$ is known. Using the marginalization and product rules for probabilities (see Appendix 1) one gets

$$P(D|M) = \int P(D|\alpha, M)P(\alpha|M)d\alpha, \tag{6}$$

where the integration runs over all possible values of α. If there is also a factor β which affects the value of $P(D|M)$, this can be taken into account by adding one integration on the right-hand side of Eq. (6),

$$P(D|M) = \iint P(D|\alpha, \beta, M)P(\alpha|\beta, M)P(\beta|M)d\alpha d\beta. \tag{7}$$

The last two terms under the integral have been obtained by applying the product rule to $P(\alpha, \beta|M)$. Simplifications can be achieved if assumptions are made. One, for example, may believe that the likelihood of the data explicitly depends only on α and the model choice. Mathematically, this is expressed as $P(D|\alpha, \beta, M) = P(D|\alpha, M)$. He/she may also write $P(\alpha|\beta, M) = P(\alpha|\beta)$ and $P(\beta|M) = P(\beta)$ thinking that the prior probabilities of α and β are independent on the model selection. These simplifications lead to the expression

$$P(D|M) = \iint P(D|\alpha, M)P(\alpha|\beta)P(\beta)d\alpha d\beta. \tag{8}$$

It is important to recognize that all factors under the integral in Eqs. (7) and (8) contribute to the estimate of the posterior likelihood of M but not just the quality of the fit to the experimental data $P(D|\alpha, M)$ and that all factors should be treated on the same footing. A small value of $P(D|\alpha, M)$ is as unfavorable for M as a small value of $P(\alpha|\beta)$ (which is the conditional probability for α). To see this more clearly, let us assume that $P(\{D_n\}|\alpha, M) = 0$ for all values of α but at a narrow interval around $\alpha = \alpha_0$ where the value of $P(\alpha|\beta)$ does not change significantly and that the value of β is known to be precisely β_0 from a different experiment. Under these common conditions, Eq. (8) reads

$$P(D|M) = P(\alpha_0|\beta_0) \int P(D|\alpha, M) d\alpha. \tag{9}$$

In the view of Eq. (1), the posterior probability for the model to be valid depends on the product of two factors: the prior probability of the fitted value of parameter α and the data likelihood integrated over the valid range of α. Although the data likelihood depends explicitly only on α, parameter β still plays a role because it affects the prior probability of α_0. We will elaborate on this remarkable observation in Sec. 4 discussing an experiment where the outcome predicted by a model explicitly depends on the exact distance between molecules, and implicitly on their concentration.

3. Finding quasi-stationary states

Now we are prepared to tackle the problem of finding the quasi-stationary states and answer the following questions. Is there any reason for believing that the number of different values of $\upsilon(t)$ is not 600? What are the factors that limit the number of parameters in a model? What does the chi-squared test actually prove?

To find the most probable number of states in the frequency trace we begin with the Bayes theorem

$$P(S|\{\upsilon_n\}) \propto P(\{\upsilon_n\}|S) \times P(S), \tag{10}$$

where S is the number of states in a model and the denominator is omitted because it is the same for all the models and therefore does not affect the outcome of the analysis. If we have no prior preference to the total number of states, $P(S)$ is independent of S and can be omitted too. Setting the problem a bit more generally, we will not require $\upsilon(t)$ to be periodic, and we assume that the values of υ are uncorrelated, that is the probability of getting a particular υ depends only on its value. This probability can be related to a set of microscopic parameters $\{v_s, A_s\}$ according to

$$P(\upsilon_n|S, \{v_s, A_s\}) = \sum_{s=1}^{S} \frac{A_s}{(2\pi)^{1/2}\sigma} \exp\left[-\frac{(\upsilon_n - v_s)^2}{2\sigma^2}\right], \tag{11}$$

where A_s is the relative population of state s and v_s is the value of υ in that state. Equation (6), when applied to the situation described above reads

$$P(\{\upsilon_n\}|S) \propto \int \cdots \int P(\{\upsilon_n\}|S, \{v_s, A_s\}) \cdot P(\{v_s, A_s\}|S) d\mathbf{v}^S d\mathbf{A}^S, \tag{12}$$

where $d\mathbf{v}^S \equiv dv_1 dv_2 \cdots dv_S$ and $d\mathbf{A}^S \equiv dA_1 dA_2 \cdots dA_S$ indicate integrations over the corresponding S-dimensional spaces.

We now need to specify the *prior* probability $P(\{v_s, A_s\}|S)$ for the parameters characterizing the S-state model and have to calculate the probability of the data if the model is correct and the parameters are $\{v_s, A_s\}$. For simplicity, let us consider prior uniform probabilities for all v_s within the interval $-\Delta/2 \leq v_s \leq \Delta/2$ and for A_s within the interval $0 \leq A_s \leq 1$. In such a case $P(\{v_s, A_s\}|S) \propto 1/\Delta^S$.

Because the events $\{\upsilon_n\}$ are independent, the probability of observing N events in an experiment equals the product of the probabilities for each event, and therefore,

$$P(\{\upsilon_n\}|S, \{v_s, A_s\}) \propto \prod_{n=1}^{N} \left[\sum_{s=1}^{S} A_s \exp\left[-\frac{(\upsilon_n - v_s)^2}{2\sigma^2}\right]\right] \tag{13}$$

follows from Eq. (11). The final result for the posterior $P(S|\{\upsilon_n\})$ reads

$$P(S|\{\upsilon_n\}) \propto \frac{P(S)}{\Delta^S} \int \cdots \int \prod_{n=1}^{N} \times \left[\sum_{s=1}^{S} A_s \exp\left[-\frac{(\upsilon_n - v_s)^2}{2\sigma^2} \right] \right] d\mathbf{v}^S d\mathbf{A}^S, \quad (14)$$

where σ is considered to have a definitive value. Otherwise, one has to specify a prior distribution for σ and has to add integration over σ to the right-hand side of Eq. (14). A more careful analysis will reveal that the condition $\sum_{s=1}^{S} A_s = 1$ must be satisfied *a priori*, and therefore the integration in Eq. (14) has one dimension of $\mathbf{A}$ less because it should run only over the subspace where the required relation between all A_s holds. By appropriately shaping the integration domain one can include any other restrictions on the values of v_s, A_s, and σ. For example, the information that all the populations are larger than 0.2 will exclude points where any of A_s is smaller than 0.2 from the integration, etc.

In a trivial case, when there is only one data point, Eq. (14) shows that $P(S|\upsilon_1)$ does not depend on S. Not surprisingly, but rather than requiring a spark of intuition this result relies on straightforward calculations. If there are two events in the frequency trace and the prior intervals for v are large enough so that $\Delta \gg \sigma$ and $|\upsilon_1|, |\upsilon_2| < |\Delta - \sigma|$, the integration can be done analytically (see Appendix 3), and the results read

$$P(2|\upsilon_1, \upsilon_2) \propto P(2) \left[\frac{2\pi^{1/2}}{3} \frac{\sigma}{\Delta} \exp\left(-\frac{(\upsilon_1 - \upsilon_2)^2}{4\sigma^2} \right) + \frac{2\pi}{3} \frac{\sigma^2}{\Delta^2} \right] \quad (15)$$

and

$$P(1|\upsilon_1, \upsilon_2) \propto P(1)\pi^{1/2} \frac{\sigma}{\Delta} \exp\left(-\frac{(\upsilon_1 - \upsilon_2)^2}{4\sigma^2} \right). \quad (16)$$

We conclude that the choice between the two hypotheses depends on their prior probabilities, on the values of σ, $|\upsilon_1 - \upsilon_2|$, and on Δ. If $|\upsilon_1 - \upsilon_2| \ll \sigma$, then the ratio $P(1|\upsilon_1, \upsilon_2)/P(2|\upsilon_1, \upsilon_2) \approx (3/2) \times P(1)/P(2)$. More surprisingly, irrespective of the value of $|\upsilon_1 - \upsilon_2|$, the one-state model is also supported by the data if $\Delta \to \infty$. In these two cases, the cost for adding one additional parameter to the model is larger than the achievable gain due to a smaller possible value of the chi-squared. The increase of the conceivable multidimensional space available for the parameters limits the maximum "reasonable" number of parameters in any model (including the one with 600 states). But, if Δ is fixed, then the two-state model quickly gets a clear advantage as the distance $|\upsilon_1 - \upsilon_2|$ increases (this is, perhaps, what most readers expect intuitively). Such dependence of the posterior likelihoods on Δ may look disturbing for some, because it makes the judgment about the two competing hypotheses sensitive to a "subjective" parameter Δ which was not measured but determined exclusively by the personal background of the researcher. However, the approach discussed here is actually more objective than the traditional statistical approach founded on chi-squared statistics and similar techniques where the outcome of the analysis depends on the choice of the statistics. The "subjective" Bayesian approach simply highlights the biases somewhat hidden under the carpet in what some people call orthodox statistical methods. It also takes into account all available information. Notably, the posterior probabilities of the two hypotheses also depend on the corresponding prior probabilities which are also subjective values.

The importance of properly defined priors for the hidden variables can be illustrated by the paradox of two envelopes. The paradox is formulated as follows. You and your friend are shown two envelopes with money. You are told that one envelope has twice the amount of money hidden in the other. You take one envelope, open it, count the money, and may decide to exchange the money for the second envelope. If the amount in the first envelope is, for example, \$1000 you may lose \$500 but you may win \$1000 if you exchange the money with your friend. It seems that taking the envelope chosen by your friend is in average the winning strategy. Apparently your friend thinks that he will do better with your envelope. The reasoning is not logically consistent because it is impossible that both of you will become richer after the exchange. The paradox is easily resolved if you define a prior PDF for the total amount of the money in the two envelopes. For example, you can assume that this PDF is uniform between 0 and X. In this case, you keep the money if the amount is larger than $X/3$ (note that given the prior PDF it is not possible that both envelopes have more than $X/3$). Otherwise, you exchange it for the amount in the second envelope. Of course, the choice of the prior PDF is subjective, but the following conclusions are logically consistent.

Equation (14) is difficult to be analyzed analytically, but the limit of $P(S|\{\upsilon_n\})/P(1|\{\upsilon_n\})$ as $\Delta \rightarrow \infty$ can be obtained easily. It is obvious that $P(1|\{\upsilon_n\}) \propto 1/\Delta$, and to find the limit, we only need to retain the terms in (14) with the slowest decay (these terms are proportional to $1/\Delta$). Such terms appear when the function under the integral depends only on a single variable v_s and therefore integration over all other v_s results in Δ^{S-1}. Therefore, as $\Delta \rightarrow \infty$,

$$\frac{P(S|\{\upsilon_n\})}{P(1|\{\upsilon_n\})} \rightarrow \frac{P(S)}{P(1)} \int \sum_{s=1}^{S} A_s^N d\mathbf{A}^{S-1}, \tag{17}$$

where the region of integration is defined by the relations $0 \leq A_s \leq 1$ and $\sum A_s = 1$. The integral is obviously smaller than 1 and the result obtained for two data points holds. When $\Delta \rightarrow \infty$, the one-state model is preferred irrespective of the data. In a real physical experiment, there is always a finite limit for Δ, the dependence on Δ is not dramatic (see below), and therefore this "paradoxical" result having little practical implications may be disturbing for those who accept only "unbiased" analysis.

In a general case, the analysis of a long frequency trace is computationally involved, and can be done numerically. The multidimensional integration is accelerated if we replace the time trace with a histogram of υ. Because the probability of each value υ is assumed to be uncorrelated with others, such a histogram will not erase any information but only eliminate small fluctuations of the data when these fluctuations are smaller than the width of the bins in the histogram. The probability for observing such a histogram is calculated in Appendix 4.

For illustration purposes, 600 data points were generated by adding normally distributed noise to υ, which was either constant or randomly jumping between two or three equally spaced states. The corresponding histograms of the data sets are shown in Fig. 3. The variance of the noise was used as a normalization factor for the data or, in other words, the variance of the noise was 1. The populations of all states were set equal, and the spacing between the states was 1.5. Note that because the stochastic wandering between the quasi-stationary states is less deterministic than the periodic variation discussed at the beginning of this chapter, a better data quality is needed to make a meaningful conclusion. Therefore, the splitting between the quasi-stationary states in these simulations is larger than that for the data presented in Fig. 1.

Bayesian statistics has been applied to infer the number of quasi-stationary states from the noisy data and the results are presented in Table 1. The analysis has been done using different prior information. The prior limits on the values of $\{v_s\}$ have been taken into account by the two values of Δ. The first column for each set of data is calculated assuming that only $\{v_s\}$ are not known *a priori*. The second column contains the results of analysis when $\{v_s\}$ and σ have been treated as unknowns but the prior values of all $\{A_s\}$ have been matched to those used in the data generation. The prior distribution for σ was assumed to be uniform between 0.5 and 1.5. The results summarized in the third column are evaluated when only the information about σ has been made available *a priori*.

The sensitivity of the analysis to all kinds of prior information is apparent from Table 1. Generally, more accurate prior information makes the distinction between different models easier. Compare, for

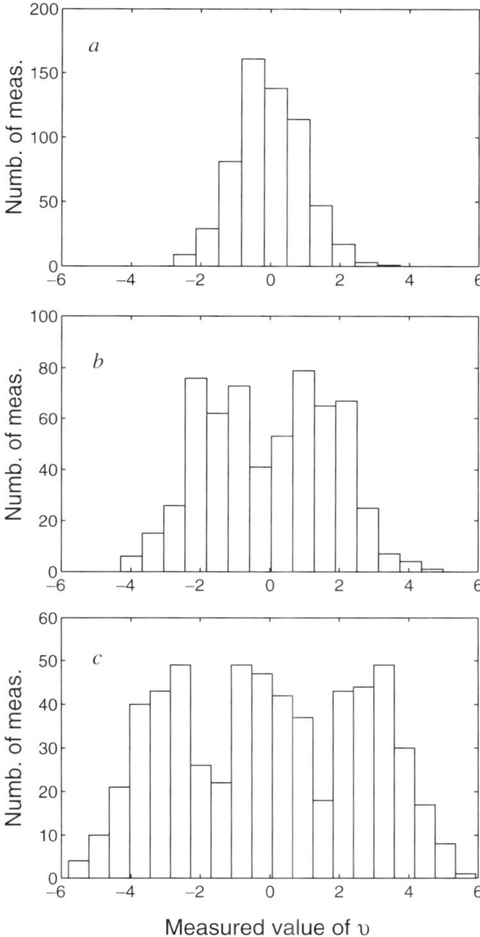

Fig. 3. Histograms of the generated/measured data used for model selection based on the Bayesian analysis. (Panels *a*, *b*, and *c* correspond to the data generated using one, two, and three state models, respectively. The results of the analysis are presented in Table 1.)

example, the results when only $\{v_s\}$ were the variable parameters to the case when neither $\{v_s\}$ nor $\{A_s\}$ are given *a priori*. A wider uncertainty range for $\{v_s\}$ increases the price for adding more states to the model. However, unless the populations of the states are given

Table 1. Results of Bayesian analysis of the simulated data.

	Analysis of data shown in Fig. 3(a)			Analysis of data shown in Fig. 3(b)			Analysis of data shown in Fig. 3(c)		
	$\mathbf{\Delta} = \begin{bmatrix} 8 \\ 20 \end{bmatrix}$			$\mathbf{\Delta} = \begin{bmatrix} 8 \\ 20 \end{bmatrix}$			$\mathbf{\Delta} = \begin{bmatrix} 8 \\ 20 \end{bmatrix}$		
	Variable parameters			Variable parameters			Variable parameters		
	ν_s	ν_s, σ	ν_s, A_s	ν_s	ν_s, σ	ν_s, A_s	ν_s	ν_s, σ	ν_s, A_s
$S = 1$	0.90	0.72	0.85	10^{-154}	10^{-25}	10^{-149}	10^{-592}	10^{-170}	10^{-590}
	0.95	0.85	0.94	10^{-154}	10^{-25}	10^{-149}	10^{-592}	10^{-169}	10^{-592}
$S = 2$	0.085	0.20	0.125	0.92	0.85	0.44	1×10^{-61}	1×10^{-13}	10^{-81}
	0.042	0.14	0.055	0.99	0.975	0.70	2×10^{-62}	3×10^{-12}	10^{-65}
$S = 3$	0.01	0.06	0.02	9×10^{-6}	1×10^{-4}	0.34	1.00	1.00	0.40
	2×10^{-3}	0.01	4×10^{-3}	4×10^{-6}	4×10^{-5}	0.21	1.00	1.00	0.67
$S = 4$	1×10^{-3}	0.02	4×10^{-3}	0.08	0.15	0.22	1×10^{-6}	3×10^{-6}	0.60
	1×10^{-4}	2×10^{-3}	3×10^{-4}	0.01	0.025	0.09	5×10^{-7}	1×10^{-6}	0.33

Notes: The results of the Bayesian analysis of the computer-simulated noisy data (see Fig. 3) representing an observable whose value stochastically jumps between one, two, or three quasi-stationary states. The two rows of numbers for each value of S are the posterior probabilities of the corresponding S-state model calculated when the prior range of ν_s is $-4 \leq \nu_s \leq 4$ (upper number) or $-10 \leq \nu_s \leq 10$ (lower number).

as the background information, the three-state model does not gain a much greater support from the data shown in Fig. 3(c) than the four-state model does.

When $\{A_s\}$ and $\{v_s\}$ are known for every data point and we are interested only in the value of σ which is assumed to be the same for all the data points, one gets

$$P(\sigma|\{\upsilon_n\}) \propto P(\sigma)P(\{\upsilon_n\}|\sigma) \propto P(\sigma)\prod_{n=1}^{N}\frac{1}{\sigma}\exp\left[-\frac{(\upsilon_n - v_n)^2}{2\sigma^2}\right], \tag{18}$$

where $P(\sigma)$ designates the prior distribution of the standard deviation assumed by the model. If $P(\sigma)$ is uniform over a very large region, then probability $P(\sigma|\{\upsilon_n\})$, normalized by the condition $\int P(\sigma|\{\upsilon_n\})d\sigma = 1$, is close to the chi-squared distribution (see Appendix 2). It is easy to see that $P(\{\upsilon_n\}|\sigma)$ has a maximum at $\sigma = \sigma_0$, where $\sigma_0^2 \equiv \sum(\upsilon_n - v_n)^2/N$. If $P(\sigma)$ is not uniform, and has a maximum at $\sigma = \sigma_0$, then the posterior probability will have its maximum still at σ_0. In other words, $\chi^2 \approx N$ indicates that the data are in good agreement with the prior probability $P(\sigma)$. Note that the alternative "periodic" hypothesis discussed earlier actually confirms the same value for σ. Therefore, it is not at all surprising that the chi-squared criterion was not able to tell the difference between the two alternatives.

4. Finding distances between molecules

The Bayesian analysis proves (see Eqs. (8) and (9)) that it is equally important to have a good fit for the experimental data and to have all the fit parameters not too far from their most probable prior values. Of course, if only one model is considered as a possibility, then it does not matter how good the fit is and how likely the values for the fitted parameters are. But, if there is any reason to care about the poor quality of the fit to the data, then a small prior likelihood of the fitted parameters should be equally alarming.

In this section, we analyze an experiment whose outcome depends on the distances between the studied molecules and their closest neighbors. Unfortunately, although the concentration of the molecules can be relatively easy to define by a sample preparation procedure, it is very hard or even impossible to control the exact positions of every single molecule. Therefore, r (the distance between a molecule and its closest neighbor) will be a hidden parameter in the model. The analysis should take into account that a short distance can be an unlikely situation if the concentration of the molecules is low or moderate although interesting effects are most frequently observed when two molecules are separated by 10 nm or less.[9,10] Parameter estimation will be our first goal, followed by a more general analysis.

It has been explained already that parameter estimation (in this case estimation of r) is limited to the consideration of models, which are different only by the values of the parameters. The fundamentals of the models are assumed to be correct *a priori*. Equation (1) and the product and marginalization rules applied to the problem under consideration reveal that

$$P(r|D, M) \propto P(D|r, M) \int P(r|C)P(C)dC. \tag{19}$$

For a non-Bayesian approach, the probability distribution function for the intermolecular distance r depends solely on the value of $P(D|r, M)$, and therefore we only need to know the accuracy of data D and the relation between r and the measured quantity. The Bayesian approach takes into account the prior information about the concentration. The function $P(r|C)$ is derived in Appendix 5. The correction resulting from taking into account the prior information depends on the sharpness of function $P(D|r, M)$. An example is shown in Fig. 4, where the correction is just about a half of the standard deviation. But, the prior information can have a much more profound effect on a more general model selection.

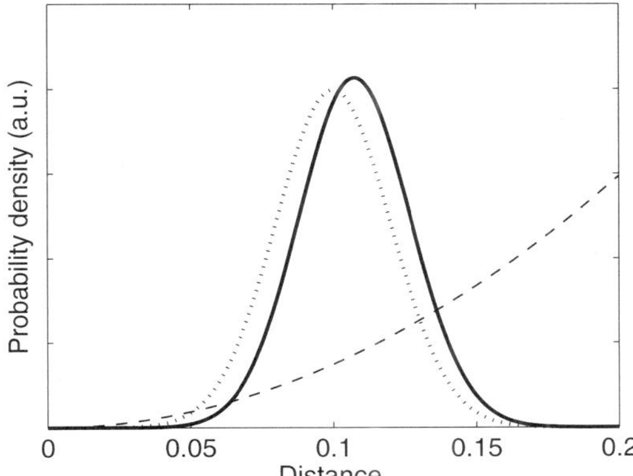

Fig. 4. Probability density for the distance from a single molecule to its closest neighbor. (Dotted line represents the function $P(D|r, M)$ which is assumed to be of a Gaussian shape. Dashed line is the prior probability density for the intermolecular distance given a known concentration of molecules (scaled up vertically for a better visibility). Solid line is the resulting probability density for r.)

Imagine that someone suggests a model saying that all the experimental curves which you have assigned to the presence of two coupled molecules of type A separated by a very short distance can be equally well fitted by assuming the presence of a single impurity molecule of type B. Now it comes down to the comparison of the probabilities for having that unusual impurity and for picking up a pair of molecules with an unusual separation. This comparison can be done, for example, if accurate prior estimates for concentrations of A and B are available. Otherwise, one may enlarge the set of data and look at a histogram for the estimated intermolecular distances. An agreement with the distribution given by Eq. (A15) would favor the "coupled molecules" model.

To approach the problem quantitatively, suppose that we have studied N molecules and have obtained a data set $\{D_n\}$, where n runs from 1 to N. Note that D_n in this example represents all data obtained for the nth molecule and may include more than one data point. We also have model M in place saying that the observed effects depend on r_n, the distances between the molecule and its closest neighbor. A fundamentally different model M' may predict a different set of values $\{r'_n\}$ for the intermolecular distances or may offer a completely different explanation. Which model should be preferred? If the concentration C of the molecules has been measured with an independent trustable method and the probability distribution $P(C)$ has been obtained, then we can use Eq. (8) to find the ratio

$$\frac{P(M|\{D_n\})}{P(M'|\{D_n\})} = \frac{P(M)}{P(M')} \frac{\prod_{n=1}^{N} \iint P(D_n|r, M)P(r|C)P(C)drdC}{P(\{D_n\}|M')}. \tag{20}$$

The following general comment is appropriate before proceeding with mathematical evaluation of this equation. The strength and the most valuable feature of the single-molecule technique is that it allows examination of each molecule. Therefore, it is sometimes possible to find a rare event in a large sample of data. A rare event can also be selected subconsciously because rare events look interesting. However, one should be extremely cautious about making conclusion solely on the basis of such an event even if it fits a theory very well. Keep in mind that if the data were obtained for an ensemble, the rare events would be buried under the noise. The single-molecule approach amplifies the rare events making them a standing out observation. However, the noise of ensemble measurements does not disappear without a trace. It is replaced by the parameter likelihood function. To see this, we will use the same approximations which are described when Eq. (9) is derived from Eq. (8) and obtain

$$\frac{P(M|\{D_n\})}{P(M'|\{D_n\})} = \frac{P(M)}{P(M')} \frac{P(\{r_n\}|C_0) \times \prod_{n=1}^{N} \int P(D_n|r, M)dr}{P(\{D_n\}|M')}, \tag{21}$$

where C_0 is the exact concentration of the molecules. Two factors in the numerator of the right-hand side of Eq. (21) can be evaluated separately.

Let the symbol D_n in Eq. (21) represent J data points valued at $V_{n,j}$ (j runs from 1 to J). If the conventional Gaussian approximation for the noise in the data is valid and if r is close to r_n, then using a

Taylor expansion of $E_j(r)$

$$\int P(D_n|r, M)dr \propto \int_{-\infty}^{\infty} \exp\left[-\sum_{j=1}^{J} \frac{[V_{n,j} - E_j(r_n) - \varphi_j(r - r_n)]^2}{2\sigma^2} \right] dr, \quad (22)$$

where $E_j(r_n)$ is the expectation value predicted by the model and φ_j is the derivative of the function $E_j(r)$ over r at $r = r_n$. Because r_n is the value of r where $\sum_{j=1}^{J} [V_{n,j} - E_j(r)]^2$ has a minimum, $\sum_{j=1}^{J} [V_{n,j} - E_j(r)]\varphi_j = 0$, and therefore the proportionality (22) can be reduced to

$$\int P(D_n|r, M)dr \propto \left(\sum_{j=1}^{J} \varphi_j^2 \right)^{-1/2} \exp\left[-\frac{1}{2} \sum_{j=1}^{J} \frac{[V_{n,j} - E_j(r_n)]^2}{\sigma^2} \right] \quad (23)$$

as shown in Appendix 3. Except for the factor depending on φ_j, this is a conventional chi-squared expression (see Eq. (4) and the following discussion) which is an important contribution to the posterior likelihood but it is not the only contribution. The factor $P(\{r_n\}|C_0)$ should be taken into account as well.

If the molecules are chosen at random, the order in which molecules are picked is irrelevant and the set $\{r_n\}$ can be represented by a histogram with K bins and with m_k events in the kth bin and consequently (see Appendix 4)

$$P(\{r_n\}|C_0) \propto P(\{m_k\}|C_0) \propto \exp\left(-\sum_{k=1}^{K} \frac{(m_k - Np_k)^2}{2Np_k} \right), \quad (24)$$

where p_k is the probability of getting r in the kth bin in each measurement. The maximum of the right-hand side is achieved if all m_k are close to their expectation values given by Np_k. The standard deviation of each m_k is equal to the square root of its expected value. An intuitively appealing requirement that the observed number of events in every bin should not deviate too much from their expected values is now established rigorously. There is no reason why any of the two "chi-squared type" contributions to the posterior probabilities (Eqs. (23) and (24)) should be more important than the other. Once again we arrive to the conclusion that any experimental test of a model depending on a hidden parameter α should include analysis of background information B and the probability $P(\alpha|B)$.

When the probabilities p_k are not known, they should be sufficiently accurately estimated from the experimental data. A failure to do so or a substantial disagreement between the observed bin populations and the prior expectations affects the reliability of the model in the same way as the absence of error bars in the data or large deviations of experimental quantities from their theoretically expected values. For a reason, the curve fitting has got much more attention in papers dealing with single molecules than the prior statistics of the underlying microscopic characteristics used in the models. Probably, this occurs because the two contributions (Eqs. (23) and (24)) do not need special attention in ensemble measurements since they both are included in the overall noise. Contrary to the conventional statistical methods, the Bayesian approach to the data treatment provides a consistent and logical way for scientific inference founded both on the data gathered in an experiment and on the prior information. This approach also shows that single-molecule experiments should cover a representative fraction of the total molecular population to provide sufficient information for the model testing. Even if a currently available set of competing models is limited and the model selection can be done by studying only one molecule (just to verify that $P(D|M)$ is not zero), this may change in future and then the information about the statistics of the representative set will become vital.

Appendix 1: Product and marginalization rules

For any two possibilities A_1 and A_2, the product rule states that

$$P(A_1, A_2|B) = P(A_1|B) \cdot P(A_2|A_1, B) = P(A_2|B) \cdot P(A_1|A_2, B). \tag{A1}$$

The second equality can be expressed in the form

$$P(A_1|A_2, B) = \frac{P(A_2|A_1, B) \cdot P(A_1|B)}{P(A_2|B)}, \tag{A2}$$

which represents the simplest form of Bayes' theorem.

The marginalization rule states that

$$P(A|B) = \sum_{n=1}^{N} P(A, C_n|B), \tag{A3}$$

where $\{C_n\}$ is a set of mutually exclusive but exhaustive possibilities (this means that one of these possibilities but only one is true).

Appendix 2: Chi-squared distribution

Consider a probability density function

$$P(\sigma) \propto \frac{1}{\sigma^{N-1}} \exp\left(-\frac{1}{2\sigma^2}\sum_{n=1}^{N}(x_n - \bar{x})^2\right). \tag{A4}$$

The probability distribution for $z \equiv \chi^2 \equiv \sum_{n=1}^{N}(x_n - \bar{x})^2/\sigma^2$ is

$$P(z) \propto z^{N/2-1} \exp\left(-\frac{z}{2}\right). \tag{A5}$$

The function on the right-hand side of (A5) is a not normalized chi-squared distribution with N degrees of freedom. The normalization constant can be determined using the identity $\int_0^\infty P(z)dz = 1$.

In a more general case when the standard deviation is not the same for all data points, the definition for the chi-squared is

$$\chi^2 \equiv \sum_{n=1}^{N} \frac{(x_n - \bar{x}_n)^2}{\sigma_n^2}. \tag{A6}$$

If the number of data points is large, the χ^2-distribution is well approximated by a normal (Gaussian) distribution with a mean of

$N-2$ and a variance of $2N$. For example, if $N = 300$, then the probability of $\chi^2 > 341$ is only 5%. The probability is 32% for $\chi^2 > 311$ and is 50% to satisfy $\chi^2 > 299$.

Appendix 3: Gaussian integrals

The integration is simplified due to the following identity:

$$\sum_{n=1}^{N} \frac{(x_n - \varphi_n v)^2}{2\sigma_n^2} = \left\{ \sum_{n=1}^{N} \frac{x_n^2}{2\sigma_n^2} - \frac{\left(\sum_{n=1}^{N} \frac{\varphi_n x_n}{\sigma_n^2}\right)^2}{2\sum_{n=1}^{N} \frac{\varphi_n^2}{\sigma_n^2}} + \frac{1}{2}\left[\left(\frac{\sum_{n=1}^{N} \frac{\varphi_n x_n}{\sigma_n^2}}{\sqrt{\sum_{n=1}^{N} \frac{\varphi_n^2}{\sigma_n^2}}}\right) - v\sqrt{\sum_{n=1}^{N} \frac{\varphi_n^2}{\sigma_n^2}}\right]^2 \right\}. \tag{A7}$$

Therefore,

$$\int_{-\infty}^{\infty} \exp\left[-\sum_{n=1}^{N} \frac{(x_n - \varphi_n v)^2}{2\sigma_n^2}\right] dv = \left(\frac{2\pi}{\sum_{n=1}^{N} \frac{\varphi_n^2}{\sigma_n^2}}\right)^{1/2} \exp\left[\frac{\left(\sum_{n=1}^{N} \frac{\varphi_n x_n}{\sigma_n^2}\right)^2}{2\sum_{n=1}^{N} \frac{\varphi_n^2}{\sigma_n^2}} - \frac{1}{2}\sum_{n=1}^{N} \frac{x_n^2}{\sigma_n^2}\right]. \tag{A8}$$

In a special but important case when $\sum_{n=1}^{N} \varphi_n x_n / \sigma_n^2 = 0$, the integral is reduced to a simpler expression

$$\int_{-\infty}^{\infty} \exp\left[-\sum_{n=1}^{N} \frac{(x_n - \varphi_n v)^2}{2\sigma_n^2}\right] dv = \left(\frac{2\pi\sigma^2}{\sum_{n=1}^{N} \frac{\varphi_n^2}{\sigma_n^2}}\right)^{1/2} \exp\left[-\sum_{n=1}^{N} \frac{x_n^2}{2\sigma_n^2}\right]. \tag{A9}$$

Appendix 4: Probability of a histogram

The probability of getting a chronologically ordered sequence of uncorrelated measurements $\{b_n\}$ (n runs from 1 to N and b_n is the bin number to which the result of nth measurement belongs) is

$$P(\{b_n\}|\{p_k\}) = \prod_{k=1}^{K} p_k^{m_k}, \tag{A10}$$

where m_k is the number of results which belong to the kth bin of the histogram (k runs from 1 to K and some of m_k may be zeros) and p_k is the probability of getting a result in k-th bin. However, one should take into account that different sequences $\{b_n\}$ can result in the same histogram, because the order in which the bins are filled with events is of no importance. The total number of permutations between N measurements is $N!$. These $N!$ permutations can be represented as a combination of permutations within bins (there are $\prod m_k!$ such permutations) and permutations where events from different bins change their chronological order (there are apparently $N!/\prod m_k!$ permutations of this kind). Therefore, the probability of a histogram reads

$$P(\{m_k\}|N) = \frac{N!}{\prod m_k!} \prod_{k=1}^{K} p_k^{m_k} = N! \prod_{k=1}^{K} \frac{p_k^{m_k}}{m_k!}. \tag{A11}$$

The right-hand side of this equation can be approximated by a product of Poisson distributions applying the following approximations and identities

$$N! \prod_{k=1}^{K} \frac{p_k^{m_k}}{m_k!} \approx \prod_{k=1}^{K} \frac{(p_k N^{Np_k/m_k})^{m_k} \exp(-Np_k)}{m_k!}$$

$$\approx \prod_{k=1}^{K} \frac{(\mu_k)^{m_k} \exp(-\mu_k)}{m_k!}, \tag{A12}$$

where $\mu_k \equiv Np_k$, $N! \approx N^N \exp(-N)$, $\mu_k \gg |m_k - \mu_k|$, and $N^{\mu_k/m_k} \approx N$. If $\mu \gg |m_k - \mu_k| \ln N$, then the Poisson distribution can be approximated by a Gaussian.[6]

$$P(\{m_k\}|N) \approx \exp\left(-\sum_{k=1}^{K} \frac{(m_k - Np_k)^2}{2Np_k}\right) \times \prod_{k=1}^{K} \frac{1}{\sqrt{2\pi Np_k}}. \quad \text{(A13)}$$

Appendix 5: Probability distribution for the distance to the nearest molecule

The probability to have n molecules at concentration C in a sphere of radius r (volume V) is described by a Poissonian distribution

$$P(n|r, C) = \frac{(CV)^n \exp(-CV)}{n!} = \frac{(4\pi Cr^3/3)^n \exp(-4\pi Cr^3/3)}{n!}. \quad \text{(A14)}$$

The probability to have zero molecules in the sphere is $P(0|r, C) = \exp(-4\pi Cr^3/3)$. Therefore, the probability to have one molecule at the distance between r and $r + dr$ and no molecules within the sphere of radius r is $P(r|C)dr \equiv P(1|r + dr, C) - P(0|r, C)$. With the assistance of Eq. (A14),

$$P(r|C) = 4\pi Cr^2 \exp\left(-\frac{4\pi}{3}Cr^3\right). \quad \text{(A15)}$$

This probability distribution can be "inverted". If the distance to the nearest molecule is a, then the distribution for concentration reads

$$P(C|a) \propto P(a|C)P(C) \propto 4\pi Ca^2 \exp\left(-\frac{4\pi}{3}Ca^3\right) P(C). \quad \text{(A16)}$$

If $P(C)$ is independent on C, then after normalization by the condition $\int P(C|a)dC = 1$, one gets

$$P(C|a) = \frac{16\pi^2 a^6}{9} C \exp\left(-\frac{4\pi Ca^3}{3}\right). \quad \text{(A17)}$$

This function has a maximum at $C = 3/(4\pi a^3)$. If one gets N pairs of molecules with intermolecular distances $\{a_n\}$, then the posterior probability is

$$P(C|\{a_n\}) \propto \left[C \exp\left(-\frac{4\pi}{3} C \frac{1}{N} \sum_{n=1}^{N} a_n^3 \right) \right]^N . \qquad \text{(A18)}$$

References

1. Y. Chen, J. D. Müller, P. T. C. So and E. Gratton, The photon counting histogram in fluorescence fluctuation spectroscopy, *Biophysics Journal* **77** (1999) 553–568.
2. E. A. Donley and T. Plakhotnik, Statistics for single-molecule data, *Single Molecules* **2** (2001) 23–30.
3. K. McHale, A. J. Berglund and H. Mabuchi, Bayesian estimation for species identification in single-molecule fluorescence microscopy, *Biophysics Journal* **86** (2004) 3404–3422.
4. J. B. Witkoskie and J. Cao, Single molecule kinetics II: Numerical Bayesian approach, *Journal of Chemical Physics* **121** (2004) 6373.
5. S. C. Kou, X. S. Xie and J. S. Liu, Bayesian analysis of single-molecule experimental data, *Applied Statistics* **54** (2005) 469–506.
6. D. S. Sivia, *Data Analysis: A Bayesian Tutorial* (Oxford University Press, New York, 1996).
7. E. T. Jaynes, *Probability Theory: The Logic of Science* (Cambridge University Press, Cambridge UK, New York, 2003).
8. A. Gelman, J. B. Carlin, H. S. Stern and D. B. Rubin, *Bayesian Data Analysis*, 2nd edn. (Chapman & Hall/CRC, Boca Raton, Florida, 2004).
9. S. Weiss, Fluorescence spectroscopy of single biomolecules, *Science* **283** (1999) 1676–1683.
10. C. Hettich, C. Schmitt, J. Zitzmann, S. Kühn, I. Gerhardt and V. Sandoghdar, Nanometer resolution and coherent optical dipole coupling of two individual molecules, *Science* **298** (2002) 385–389.

CHAPTER 3

Generating Functions for Single-Molecule Statistics

Frank L. H. Brown

Department of Chemistry and Biochemistry and Department of Physics, University of California, Santa Barbara, CA

1. Introduction

The introduction of single-molecule experimental techniques nearly two decades ago[1,2] has revolutionized the study of condensed phase systems in chemistry, physics, and biology. While certain single-molecule studies focus on large biomolecules and polymers where dynamics are directly observable via microscopy,[3–6] commonly it is necessary to infer single-molecule dynamics from spectroscopic signals (see Reviews 7–11 and references therein). More often than not, interpretation of single-molecule experiments is a complicated affair.

In single-molecule spectroscopy (SMS) experiments, laser irradiation of single chromophoric molecules induces fluorescence of photons. The fluoresced photons are detected; their arrival times, and possibly frequency information as well, serve as the experimental data. SMS spectroscopy, unlike traditional spectroscopies, is inherently the study of a discrete data stream. Every collected photon results from a single spontaneous emission event and it is the time history of these emissions registered in experiment. Even in experiments where single photon resolution is not achieved, it may be possible to interpret data in terms of a discrete set of times associated with underlying molecular transitions (as in on/off blinking and related experiments, see chapters 1 and 11 and Refs. 12 and 13). This chapter is concerned with the modeling of SMS experiments that measure discrete events in time.

SMS experiments are noisy. Some of this noise can be attributed to imperfect detection schemes, however the physical process of spontaneous emission is inherently stochastic. Furthermore, the ubiquitous presence of thermal fluctuations over molecular length scales imparts additional randomness to the physical behavior of single-molecules imbedded in a condensed phase environment. While SMS has been hailed for its ability to probe these fluctuations directly, it remains difficult to extract physical pictures for molecular dynamics based solely on SMS data streams. Some of this difficulty is likely fundamental (current SMS experiments may not collect sufficient data to allow for direct inversion to molecular dynamics), but even if SMS data were sufficient to differentiate between all viable physical hypotheses, it remains an open question as to the best means to simulate such models to allow for comparison with experiment. Indeed, much effort has been expended on the theory of interpreting/modeling SMS trajectories (see Reviews 12 and 14, and references therein) and this subject is a major focus of this book. In particular, chapters 1–4, 6–8, 10 and 11 all consider various aspects of modeling and analyzing SMS data.

The generating function (GF) method is a well-known tool for dealing with discrete random events in time.[15] In chemical physics texts[16,17] the GF is often introduced as an elegant and concise means to derive classical results related to Poisson processes, one-dimensional random walks, and similar toy problems. It is apparently not widely appreciated, however, that these same techniques are easily extended to complex physical models appropriate to SMS and that the resulting equations are well suited to numerical analysis as well as traditional analytical work. In many ways, the GF technique is completely natural for the study of SMS trajectories. GFs were introduced to the field of SMS independently by Brown[18] and Gopich and Szabo[19] in the context of kinetic and stochastic models for dynamics. Zheng and Brown[20,21] and Mukamel[22] extended this picture to quantum dynamics for single chromophores, much in the spirit of earlier work by Cook.[23] Multiple studies in SMS have since applied the generating function approach[24–31] to various physical models and situations relevant to SMS. Please see chapter 6 for additional applications and discussions of the GF formalism beyond the introduction presented here.

This chapter presents an introduction to the GF method with a few simple examples showing how it may be applied to the field of SMS. The following section introduces the formalism of GFs within the context of an idealized Poisson process. For better or worse, Poisson statistics has become the "zeroth order" model, against which the results of photon-counting measurements and related experiments are compared. For this reason, we present an extended discussion of what a Poisson process is and the relation between Poisson statistics in time series measurements and seemingly unrelated physical situations. Section 3 extends this discussion to a more realistic kinetic/stochastic description of single-molecule photon-emission statistics. Although quantum mechanics (Sec. 4) complicates the technical details necessary to perform calculations, all important concepts are clearly displayed within this simple kinetic picture. Sections 5 and 6 wrap up with some applications of the quantum mechanical GF formalism and a brief conclusion. The examples chosen in this work assume individual photon emissions to be the observable stochastic events. Blinking experiments may be described in a mathematically identical fashion, but are not specifically discussed in this chapter. (Chapters 1 and 11 discuss blinking experiments.)

It is important to stress that the GF formalism is just one technique of many that may be fruitfully applied to SMS problems. Furthermore, the perspective on GF's outlined here reflects the personal biases of the author. Different perspectives on how to treat SMS problems similar to those discussed here may be found in several chapters in this book. In particular, the chapters of Sanda and Mukamel (chapter 4), Gopich and Szabo (chapter 6), Cao (chapter 7) and Vlad and Ross (chapter 8) all consider SMS problems very similar to those discussed here. The chapter by Kolomeisky (chapter 9) considers similar theoretical problems in the context of moving motor proteins. The theoretical tools presented in these chapters are complementary to the methods discussed below and readers are strongly encouraged to read these works for different perspectives.

2. The Poisson process and introduction of generating functions

The concept of a "Poisson process" is commonly encountered in the SMS literature. Indeed, some of the statistical measures discussed below are introduced because they allow one to quantify just how different a given system's behavior is from the Poisson limit. In this section, we introduce the idea of a Poisson distribution and its relation to discrete time series as encountered in SMS. The connection between an exponential waiting time distribution for successive events in time and the Poisson distribution is established by introducing the GF formalism.

In chemical physics, the Poisson distribution is often first encountered by students as the probability distribution for the number of ideal gas particles contained within a subvolume v of a much larger container of volume V with particle density ρ. In our eventual connection to time series measurements, it will be convenient to think of space as being one-dimensional so that the ideal gas is confined to a line of total length L with N particles on it ($\rho = N/L$, see Fig. 1). The probability distribution for the number of particles, n, contained within a sublength ℓ of this line can be immediately written down from elementary combinatoric considerations (binomial distribution)

$$P(n) = \left(\frac{\ell}{L}\right)^n \left(\frac{L-\ell}{L}\right)^{(N-n)} \frac{N!}{n!(N-n)!}. \tag{1}$$

This expression reflects the probability that a specific set of n particles are found in ℓ multiplied by the number of specific sets of size n, which may be drawn from N. Since the ideal gas particles do not interact, each one has equal probability to be found anywhere along

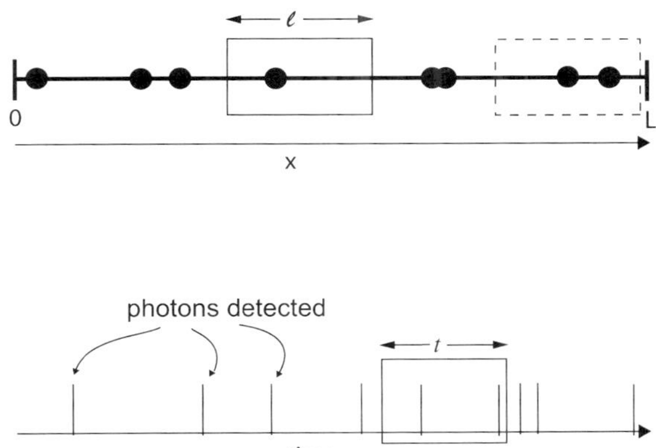

Fig. 1. Top: Particles confined to a line of length L. Total particle number is $N = 8$ with density $\rho = 8/L$. Assuming particle position is completely random over L, the distribution of particle number n within a sublength $\ell < L$ follows Eq. (1). ($n = 1$ and $n = 2$ for the solid and dashed boxed regions shown in this example. The black spheres only serve to indicate the positions of non-interacting particles and may overlap.) As $L, N \to \infty$ while holding ρ constant, this distribution becomes Poissonian (Eq. (2)). Note that the Poisson distribution only makes sense in the context of $N, L \to \infty$ since Eq. (2) predicts a nonvanishing probability to find $n = 9$, for example, which is clearly impossible in the finite L, $N = 8$ example displayed. Similarly, the Poisson distribtuion only applies if the particles behave as an ideal gas with complete absence of any interparticle correlations. Bottom: The set of arrival (detection) times for photons as plotted on a time axis represents a similar statistical problem to the position of particles in one dimension. An idealized model for photon emission (bottom panes of Fig. 2) can lead to a Poisson distribution for the number of photons detected in an interval of duration t.

L independently of the positions of the other particles. If we break our line into a binary description that the particle occupies either ℓ or "not ℓ", then the probability for a single particle to be found in ℓ must be ℓ/L, and the probability to lie outside ℓ is $(L-\ell)/L$. The counting factor is necessary since we only ask "What is the probability that any n particles are found in ℓ?," and never focus on the identity of the particles. This distribution is exact for the problem as presented, but does not yet correspond to a Poisson distribution. The Poisson distribution is the expression taken on by Eq. (1) in the limit $L \to \infty$, $N \to \infty$ while holding ρ constant. This limit considerably simplifies the general expression as follows:

$$\begin{aligned}\lim_{L,N\to\infty} P(n) &= \left(\frac{\ell}{L-\ell}\right)^n \left(1-\frac{\ell}{L}\right)^N \frac{N^n}{n!}\\ &= \left(\frac{\ell\rho}{N}\right)^n \left(1-\frac{\ell\rho}{N}\right)^N \frac{N^n}{n!}\\ &= \frac{(\ell\rho)^n}{n!} e^{-\ell\rho}. \end{aligned} \tag{2}$$

In the first step above we have replaced $N!/(N-n)!$ with N^n; in the second step we have replaced $(L-\ell)$ with L; and in the third line we have replaced $(1-\rho\ell/N)^N$ with $e^{-\rho\ell}$. All three substitutions are justified in the N, $L \to \infty$ limit.

The final line of Eq. (2) is the Poisson distribution. Note that the simple form of this distribution depends upon the assumptions that our particles do not interact and that the sublength ℓ is part of a line of infinite extent. Noninteracting particles enabled us to write down the simple binomial distribution of Eq. (1), while infinitely large L insured that the limiting form of Eq. (2) applies. In this sense, it should be understood that the Poisson distribution is, at best, a good approximation to physical reality. Neither the ideal gas limit nor an infinitely large system can be perfectly realized in physical systems, but the Poisson distribution can be a very effective approximation in the right situations.

There is another useful way to think about the Poisson distribution, which is perhaps more natural in the context of SMS. If we consider Eq. (2) in the limit of small sublengths, $\Delta\ell$, we conclude

$$\begin{aligned} P(0) &= 1-\rho\Delta\ell + O(\Delta\ell^2)\\ P(1) &= \rho\Delta\ell + O(\Delta\ell^2)\\ P(n>1) &= O(\Delta\ell^2). \end{aligned} \tag{3}$$

The above expressions are immediately obtained by expanding Eq. (2) in a Taylor series. This result is interpreted to mean that the probability to find a single particle in the infinitesimal line segment

$[\ell, \ell + d\ell]$ is $\rho d\ell$ with vanishing probability to find multiple particles within this infinitesimal segment. Since there is no position dependence along the line for this result and since finding a particle at one point has no influence on finding other particles elsewhere (so long as we always assume $L, N \to \infty$), we may readily calculate the probability density for the empty space between the two adjacent particles along the line. Starting from a particle, the probability that you do not pass another particle as you walk to the right at a distance x is defined to be $p(x)$. This quantity is calculated by considering a series of small rightward steps.

$$\begin{aligned} p(\Delta x) &= 1 - \rho \Delta x \\ p(2\Delta x) &= (1 - \rho \Delta x)^2 \\ &\vdots \\ p(M\Delta x = x) &= \left(1 - \frac{\rho x}{M}\right)^M \\ p(x) &= e^{-\rho x}. \end{aligned} \tag{4}$$

The first line above follows immediately from Eq. (3). The second line reflects the fact that to make it to $2\Delta x$ without finding any particle you must first make it to Δx without finding any, followed by another empty span of length Δx. To make it to $M\Delta x$ one must find M succesive empty Δx lengths. The final equality follows from the limiting procedure $\Delta x \to dx$ (i.e. $M \to \infty$) and from the application of the identity introduced in Eq. (2). The probability that the first particle encountered is found between x and $x + \Delta x$ is defined to be $W(x)\Delta x$ and is easily calculated as

$$p(x) - p(x + \Delta x) = W(x)\Delta x. \tag{5}$$

The probability density $W(x)$ is then calculated by considering the $\Delta x \to dx$ limit

$$W(x) = -\frac{dp(x)}{dx} = \rho e^{-\rho x}. \tag{6}$$

To connect the above discussion with the SMS time series we consider a simplified model for single-molecule dynamics and

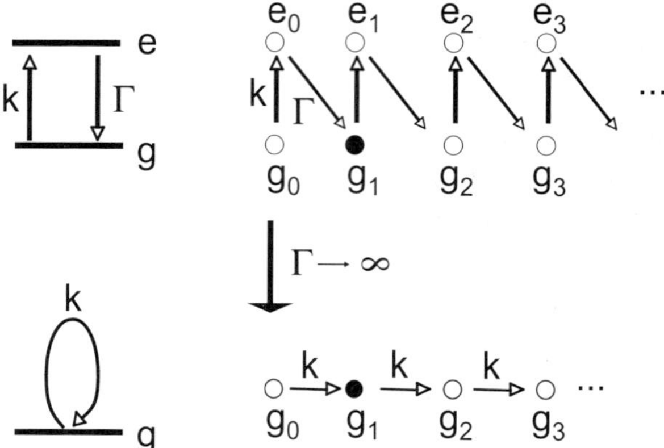

Fig. 2. Kinetic model for photon emission from a two-level chromophore. Top left: The instantaneous electronic state is either ground (g) or excited (e) with Markovian (rate process) transitions allowed between the two. Top right: To study the process of photon emission, the state space is expanded to track the total number of photon emissions that have occurred since the field was turned on. Allowable states are g_n and e_n with n being any nonnegative integer. In addition to the electronic state, the index n counts the number of previously emitted photons. Spontaneous emission via Γ necessarily advances this index by 1. The shaded circle indicates that the system currently resides in the ground electronic state and has previously emitted one photon. Bottom: When $\Gamma \gg k$, we may think of the photon-emission process as being infinitely fast. As soon as the system is excited, emission occurs instantaneously. In this limit, only one rate appears in the problem; the system spends no time in the excited state and the waiting time between successive emissions is exponentially distributed; Poisson statistics are recovered.

photon-counting (Fig. 2). We imagine a two-level chromophore under broadband excitation conditions. To a good approximation,[32] both the excitation of the system caused by the incident radiation as well as the relaxation via spontaneous emission of photons may be treated as simple rate processes in such a picture. We assume that our experimental apparatus is able to measure the photons emitted via spontaneous emission and that this is the only observable available to us.

The arrival times for emitted photons may be plotted on a time axis to create a plot qualitatively similar to the one-dimensional ideal gas problem studied earlier. The distribution of time intervals between photon arrivals (the waiting time distribution between succesive events, $W(t)$) may be inferred from the kinetic scheme implied by Fig. 2:

$$W(t) = \int_0^t d\tau k e^{-k\tau} \Gamma e^{-\Gamma(t-\tau)} = \frac{k\Gamma}{k-\Gamma}(e^{-\Gamma t} - e^{-kt}). \qquad (7)$$

The above expression reflects the neccesary two-step kinetic process involved to emit a photon at time t following the previous emission. Starting from the ground state (where you necessarily end up after the last photon emission at $t = 0$) we can spontaneously emit only by first making it to the excited state at an intermediate time τ with probability $ke^{-k\tau}$ and then dropping back to the ground state at time t ($t - \tau$ from the time we reached the excited state) with probability $\Gamma e^{-\Gamma(t-\tau)}$. The integral accounts for all possible intermediate times at which excitation could possiby occur. The simple exponential probabilites for the individual steps follow from the fact that both upward and downward processes are assumed to be Markovian rate processes. Stated differently, the exponential forms for the two elementary steps may be viewed as the mathematical definition for the rate cartoon in Fig. 2.

It is clear that the spacing of photon emissions in time, as dictated by Eq. (7) does not correspond to the Poissonian form as given in Eq. (6). However, in the limit that either $k \gg \Gamma$ or $\Gamma \gg k$, $W(t)$ does take on the exponental form expected for a Poisson distribution. For concreteness, let us assume that $\Gamma \gg k$ (weak exciting conditions) so that $W(t) \approx ke^{-kt}$. In this limit, the statistical properties of photon detection points on the time axis correspond perfectly with the statistical properties of an ideal gas — both obey Poisson statistics. Actually, while we have shown that a Poissonian distribution of points on the line implies an exponential distribution of interpoint gaps (Eq. (6)), we have not technically demonstrated the reverse. That is, we have not rigorously established the fact that an exponential waiting time distribution between successive events necessarily leads to a Poissonian distribution of events within a time window of a given duration. To establish this fact, we introduce the GF approach.

The lower portion of Fig. 2 displays the kinetic scheme appropriate to a two-level chromophore with $k \gg \Gamma$. To count the number of photon emissions that occur within a specified time interval, we formulate an infinite master equation associated with the indicated kinetics.

$$\frac{d}{dt}P_{g_n}(t) = -kP_{g_n}(t) + kP_{g_{n-1}}(t), \tag{8}$$

where $P_{g_n}(t)$ is the probability that the system has previously emitted n photons. The "g" index indicates that the system is always in the ground state, since the emission rate from the excited state is effectively infinite and the system spends no time there. We include this index for comparison to the more general cases considered in subsequent sections. We assume that the detector was turned on at $t = 0$, so that $P_{g_0}(0) = 1$, with vanishing probabilities for $n > 0$. A convenient way to solve the full set of the P_{g_n} variables is to introduce the auxiliary variable s and the associated GF, defined as

$$G(s,t) \equiv \sum_{n=0}^{\infty} P_{g_n}(t)s^n. \tag{9}$$

Multiplying Eq. (8) by s^n and summing over all non-negative values for n leads to a single equation for the GF

$$\frac{d}{dt}G(s,t) = -k(1-s)G(s,t) \tag{10}$$

with the solution

$$G(s,t) = e^{-k(1-s)t}. \tag{11}$$

Knowledge of the GF is equivalent to knowing the full set of probabilities. Since the above solution for $G(s,t)$ is valid for any value of s that insures the function is well defined ($|s| \leq 1$ is sufficient), it must be the case that the Taylor expansion for $G(s,t)$ in s agrees term by term with definition (9). It is therefore necessarily the case that $P_{g_n}(t) = e^{-kt}\frac{(kt)^n}{n!}$ for the idealized model we have described. In other words, the number of photon emissions occuring within a given time interval of duration t is distributed exactly in accordance with the Poisson distribution (Eq. (2)). The duration t in this example is to be associated with the length ℓ from ideal gas statistics and the rate k with the density ρ.

The GF displays a property, which is extremely useful in the study of photon statistics. Differentiation of $G(s,t)$ with respect to s and evaluation at $s = 1$ provides moments of the $P_{g_n}(t)$ distribution. The "factorial moments"[16] are calculated as

$$\langle n\rangle(t) = \sum_{n=0}^{\infty} P_{g_n}(t)n = \frac{\partial}{\partial s}G(s,t)\bigg|_{s=1}$$

$$\langle n(n-1)\rangle(t) = \sum_{n=0}^{\infty} P_{g_n}(t)n(n-1) = \frac{\partial^2}{\partial s^2}G(s,t)\bigg|_{s=1}$$

$$\vdots$$

$$\begin{aligned}\langle n(n-1)\cdots(n-r+1)\rangle &= \sum_{n=0}^{\infty} P_{g_n}(t)[n(n-1)\cdots(n-r+1)] \\ &= \frac{\partial^r}{\partial s^r}G(s,t)\bigg|_{s=1}. \end{aligned} \tag{12}$$

Applied to the Poisson distribution, these formulae immediately lead to the conclusion that

$$\langle n\rangle(t) = [\langle n^2\rangle - \langle n\rangle^2](t) = kt. \tag{13}$$

Indeed, the equivalence of the average and variance is often taken as the signature of a Poisson process. In fact, in the context of photon statistics it is conventional to define "Mandel's Q parameter"[33]

$$Q(t) = \frac{\langle n^2\rangle - \langle n\rangle^2 - \langle n\rangle}{\langle n\rangle}(t). \tag{14}$$

By construction, this quantity is equal to zero for a Poisson process. Depending upon the physical situation under study its value may range from -1 for a deterministic process up to large positive values. Examples of such deviations from Poisson behavior will be explored in the following sections.

In summary, we have provided three definitions for the Poisson distribution and have demonstrated equivalence between them. For convenience, we state these three definitions using the language of the time series discussed above and presented in the lower panes of Fig. 2.

1. The probability for a photon to be emitted in the interval $[t, t+dt]$ is kdt for all times.
2. The probability distribution for the time between successive photon emissions is $W(t) = ke^{-kt}$. (Equivalently, starting anywhere on the time axis, the waiting time distribution for the next photon emission is $W(t) = ke^{-kt}$. $t = 0$ need not be marked by an emission event, due to the Markovian nature of the emission process.)
3. The probability to emit n photons in a time window of duration t is $p_n(t) = e^{-kt}\frac{(kt)^n}{n!}$.

One will sometimes come across statements to the effect that the Poisson process reflects the random distribution of points on a line. Any one of the three definitions above may be taken as the mathematical definition of this statement. It is important to remember that the Poisson process is not expected to be exact for any real physical phenomena. In the context of photon statistics, this fact should be quite apparent since it is never the case that a physical rate ever becomes truly infinite.

The GF approach introduced above is readily extended to much more complex models for chromophore dynamics. Models including complex kinetic schemes, stochastic modulation of the rates themselves, as well as quantum dynamics for the chromophore are readily treated. Some examples are provided in the following sections. While the Poisson distribution may be a useful reference case to compare against realistic models, we will see that substantial deviations from this limit can and do arise in the context of SMS measurements.

3. Generating functions for photon emission: More complex kinetic models

In the previous section, we confined the mathematical analysis to a kinetic scheme that is overly simplistic for most SMS applications. A more realistic (yet still very simplified) model for photon emission is to consider the full kinetic scheme implied by the upper portions of Fig. 2. If we were interested only in the chromophore (to the exclusion of photon-counting information), we would formulate the dynamics in terms of a simple master equation implied by the top left diagram in Fig. 2:

$$\frac{d}{dt}\begin{pmatrix} P_{\mathrm{g}}(t) \\ P_{\mathrm{e}}(t) \end{pmatrix} = \begin{pmatrix} -k & \Gamma \\ k & -\Gamma \end{pmatrix} \cdot \begin{pmatrix} P_{\mathrm{g}}(t) \\ P_{\mathrm{e}}(t) \end{pmatrix}. \tag{15}$$

Here, P_{g} and P_{e} are the probabilities to find the chromophore in the ground and excited states, respectively. While the solution of these equations does specify the steady state for an ensemble of chromophores under excitation conditions as well as the timescale associated with reaching this state, it does not directly contain any information pertinent to single-molecule photon statistics.

Photon emission information is included by considering the infinite master equation suggested by the extended kinetic scheme in the upper right panel of Fig. 2:

$$\frac{d}{dt}P_{\mathrm{g}_n}(t) = -kP_{\mathrm{g}_n} + \Gamma P_{\mathrm{e}_{n-1}}$$

$$\frac{d}{dt}P_{\mathrm{e}_n}(t) = kP_{\mathrm{g}_n} - \Gamma P_{\mathrm{e}_n}. \tag{16}$$

In this formulation, we have refined our definition of a state to include not only the chromophore's electronic configuration, but also the number of prior photon emissions that have occurred. Accordingly, the probabilities P_{g_n} and P_{e_n} reflect the probabilities to find the chromophore in the ground or excited states with a past history of n photon emissions.

In analogy to the preceding section, we define partial GFs specific to the ground and excited state manifolds of the chromophore

$$G_{\mathrm{g}}(s,t) = \sum_{n=0}^{\infty} P_{\mathrm{g}_n}(t)s^n$$
$$G_{\mathrm{e}}(s,t) = \sum_{n=0}^{\infty} P_{\mathrm{e}_n}(t)s^n. \tag{17}$$

Multiplication of Eqs. (16) by s^n followed by summing over n leads to equations of motion for the GFs

$$\frac{d}{dt}\begin{pmatrix} G_{\mathrm{g}}(s,t) \\ G_{\mathrm{e}}(s,t) \end{pmatrix} = \begin{pmatrix} -k & s\Gamma \\ k & -\Gamma \end{pmatrix} \cdot \begin{pmatrix} G_{\mathrm{g}}(s,t) \\ G_{\mathrm{e}}(s,t) \end{pmatrix} \equiv \mathbb{M}(s) \cdot \tilde{G}(s,t) \tag{18}$$

with the solution

$$\tilde{G}(s,t) = \exp \mathbb{M}(s)t \cdot \tilde{G}(s,0) = \exp \mathbb{M}(s)t \cdot \begin{pmatrix} P_{\mathrm{g}}(0) \\ P_{\mathrm{e}}(0) \end{pmatrix}. \tag{19}$$

The $\tilde{G}$ notation introduced above is a shorthand for the column vector of partial generating functions. Similarly, $\mathbb{M}$ represents the GF analog to the transition matrix. $\mathbb{M}(s = 1)$ is the transition matrix for the master equation itself.

Extraction of photon-counting statistics proceeds as in our previous example. The only difference is that now we have two partial GFs corresponding to the ground and excited state manifolds. Experiment, however, is only sensitive to n and not to the electronic state. The complete GF for photon-emission events is thus $G(s,t) = G_{\mathrm{g}}(s,t) + G_{\mathrm{e}}(s,t)$, which counts all contributions to the photon statistics irrespective of the instantaneous electronic state. The final expression is obtained via diagonalization of $\mathbb{M}$:

$$G(s,t) = \frac{e^{-[(\Gamma+k-f(s))t/2]}(\Gamma + k + f(s)) - e^{-[(\Gamma+k+f(s))t/2]}(\Gamma + k - f(s))}{2f(s)}$$
$$f(s) \equiv \sqrt{(\Gamma - k)^2 + 4\Gamma k s}. \tag{20}$$

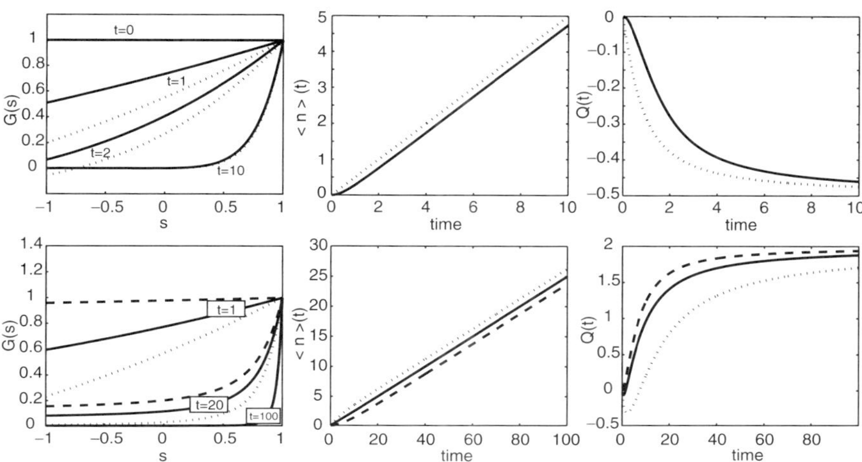

Fig. 3. Top line: Left: Generating function as a function of s at different times. Middle: Average number of emitted photons, $\langle n\rangle(t)$. Right: Mandel's Q parameter, $Q(t)$. We have assumed the model of Fig. 2 with parameter values $\Gamma = k = 1$ in arbitrary time units. In all figures solid lines correspond to expression (20), which assumes that the system begins in the ground state at $t = 0$. Dotted lines assume a steady-state initial condition $P_g(0) = P_e(0) = 1/2$. At short times the initial condition makes a difference, but the effect is quickly washed out. Bottom line: Similar to above, but considering the case of a chromophore coupled to a stochastic two-level system. We have chosen $\omega_{ab} = \omega_{ba} = 0.1$ and the same chromophore parameters as in the top panels for the "a" state, but $k_b = \Gamma_b = 0$. i.e. the a-state corresponds to a system that is fluorescing and absorbing while the b-state is totally inactive. The different lines correspond to different initial conditions. All three reflect equilibrium between electronic states. The solid line additionally reflects equilibrium between a and b, the dotted line reflects full occupation of the a-state at time 0, and the dashed full occupation of b at $t = 0$. (Adapted from Ref. 31.)

(Note that this particular answer assumes that the system began in the ground state at time $t = 0$. See Fig. 3 for the effect of different initial conditions.)

Although this model is very simple, it does contain sufficient physics to explore many of the ideas associated with general photon-counting measurements. k and Γ represent the only physical parameters of this model, reflecting the rate of field-induced electronic excitation and the rate of spontaneous emission. The average rate of photon emission is calculated in a straightforward, if algebraically tedious, manner as (see Eq. (12))

$$\text{Emission rate} \equiv \lim_{t\to\infty} \frac{d}{dt}\langle n\rangle = \lim_{t\to\infty} \frac{d}{dt}\left(\left.\frac{\partial G(s,t)}{\partial s}\right|_{s=1}\right) = \frac{\Gamma k}{(k+\Gamma)}. \tag{21}$$

The steady-state value of the Q parameter is obtained in a similar manner[33]

$$Q \equiv \lim_{t\to\infty} \frac{\langle n^2\rangle - \langle n\rangle^2 - \langle n\rangle}{\langle n\rangle} = \lim_{t\to\infty} \frac{\left(\left.\frac{\partial^2 G(s,t)}{\partial s^2}\right|_{s=1}\right) - \langle n\rangle^2}{\langle n\rangle}$$

$$= -\frac{2\Gamma k}{(\Gamma+k)^2}. \tag{22}$$

The above expressions may also be calculated outside the long time limit, but the answers reflect the initial condition of the system, explicit time dependence as the steady-state is approached, and are algebraically much more complicated. Figure 3 (upper panes) graphs the results of such calculations including time dependence. Note that the Q parameter saturates to a negative value at long times, so that the two-level kinetic scheme does not lead to Poisson statistics (except in the limiting parameter regimes discussed in the previous section). The two-stage cycling from ground to excited and back to ground leads to a diminished variance relative to the Poisson statistics. In the parlance of photon counting or quantum optics, this effect is called "photon antibunching": there is a vanishing probability for two photons to be emitted immediately after one another, which manifests itself through a negative Q value.

From a practical standpoint, it is convenient to calculate low-order photon counting moments directly without having to calculate the full GF. This is especially true for more complex dynamic models that are not amenable to analytical solution. It is easy to derive equations for the factorial moments by differentiating Eq. (18) with respect to s one or more times. The resulting equations for $\partial^m \tilde{G}/\partial s^m$ depend upon the lower order derivatives of $\tilde{G}$ in a simple way.[24] Up to second order, for example, we find

$$\begin{pmatrix} \dot{\tilde{G}}(s,t) \\ \dfrac{\partial \dot{\tilde{G}}(s,t)}{\partial s} \\ \dfrac{\partial^2 \dot{\tilde{G}}(s,t)}{\partial s^2} \end{pmatrix} = \begin{pmatrix} \mathbb{M}(s) & 0 & 0 \\ \mathbb{M}'(s) & \mathbb{M}(s) & 0 \\ 0 & 2\mathbb{M}'(s) & \mathbb{M}(s) \end{pmatrix} \cdot \begin{pmatrix} \tilde{G}(s,t) \\ \dfrac{\partial \tilde{G}(s,t)}{\partial s} \\ \dfrac{\partial^2 \tilde{G}(s,t)}{\partial s^2} \end{pmatrix}. \tag{23}$$

The prime notation indicates differentiation with respect to s. Higher order derivatives on $\mathbb{M}(s)$ do not appear since all these higher derivatives vanish. The elements of $\mathbb{M}(s)$ are always either independent of s or linear in s, so Eq. (23) is completely general. The matrix implied by the block form used above is actually 6×6 for the present model, but is still easily solved numerically by direct exponentiation. Since s derivatives of $G(s,t)$ evaluated at $s = 1$ correspond directly to the photon emission moments by virtue of Eq. (12), solving the above equations at $s = 1$ yields $\langle n \rangle(t)$ and $Q(t)$ immediately. This method was used in the generation of Fig. 3. (If one is only interested in the steady-state values of the various moments, and not on their time dependence, an even simpler numerical scheme may be formulated.[34])

A simple physically motivated extension to the above model comes from considering the case that k and Γ may themselves be fluctuating in time due to stochastic modulation by the environment. A particularly simple form of modulation is two-state jump dynamics.[16] In addition to the ground and excited electronic states, we imagine an additional discrete degree of freedom that jumps in time between two states a and b. Such a model is standard in the study of low-temperature glasses[35] where the two states represent localized conformational states of the glass, the so-called two-level system (TLS) model. This additional degree of freedom evolves with dynamics specified by the master equation,

$$\begin{pmatrix} \dot{P}_a(t) \\ \dot{P}_b(t) \end{pmatrix} = \begin{pmatrix} -\omega_{ab} & \omega_{ba} \\ \omega_{ab} & -\omega_{ba} \end{pmatrix} \cdot \begin{pmatrix} P_a(t) \\ P_b(t) \end{pmatrix} \tag{24}$$

and through the properties of composite Markov processes,[16,18,36] leads to the GF equations for this four-state system

$$\frac{d}{dt}\begin{pmatrix} G_{g;a}(s,t) \\ G_{e;a}(s,t) \\ G_{g;b}(s,t) \\ G_{e;b}(s,t) \end{pmatrix} = \begin{pmatrix} -k_a-\omega_{ab} & s\Gamma_a & \omega_{ba} & 0 \\ k_a & -\Gamma_a-\omega_{ab} & 0 & \omega_{ba} \\ \omega_{ab} & 0 & -k_b-\omega_{ba} & s\Gamma_b \\ 0 & \omega_{ab} & k_b & -\Gamma_b-\omega_{ba} \end{pmatrix} \cdot \begin{pmatrix} G_{g;a}(s,t) \\ G_{e;a}(s,t) \\ G_{g;b}(s,t) \\ G_{e;b}(s,t) \end{pmatrix}$$
$$\equiv \mathbb{M}(s)\cdot \tilde{G}(s,t). \tag{25}$$

$k_{a(b)}$ and $\Gamma_{a(b)}$ are the rates for chromophore excitation and emission when the TLS occupies state $a(b)$. The beautiful thing is that the more complicated system of equations here (relative to Eq. (18)) is, as the notation implies, formally equivalent to the simpler dynamics. In particular, calculation of moments still proceeds via Eq. (23). Once the usual master equation can be written down for system dynamics, it is a simple matter to generalize this expression to $\mathbb{M}(s)$. With $\mathbb{M}(s)$ and $\mathbb{M}'(s)$ in hand, the matrix of Eq. (23) is constructed from these blocks, multiplied against time and exponentiated to arrive at the moments. *If you know the system's dynamics in terms of a master equation, it is a trivial numerical procedure to extend this master equation to a calculation for photon-counting moments.* In fact, using a high-level programming language like Matlab or Mathematica, it takes about five lines of code to generate photon-counting moments once the master equation is specified! Numerical results are illustrated in Fig. 3 for the case of TLS dynamics one order of magnitude slower than the photon emission rates and with extreme coupling to the chromophore such that k_a and Γ_a assume the values introduced previously, but $k_b = \Gamma_b = 0$. The rate of photon emission is halved relative to the earlier model since the system only spends half its time in a photoactive state. Q now saturates to a positive value, reflecting photon bunching. Photons are only emitted when the system is in the a state, which leads to long dark intervals in the photon emission trajectory interspersed with periods of a rapid succession of photon emissions. Saturation to the value 2 reflects this bunching phenomena (which alone would predict $Q \to 5/2$), but tempered a bit by the intrinsic negative values for Q inherent to the emitting state. Also, note that the antibunching effect dominates at short times leading to a negative Q parameter, which transits into bunching over longer timescales.

This section has introduced all the relevant concepts necessary for the calculation of photon-counting moments for models based around stochastic dynamics. Given a master equation for chromophore dynamics, the associated transition matrix can immediately be transformed into the $\mathbb{M}(s)$ matrix needed for calculation of the GF itself or photon-counting moments. As a practical matter, single-molecule measurements contain a limited amount of statistically meaningful information due to finite experimental durations. In most cases, experimental data can be interpreted to provide reliable information on a few low-order moments of the full probability distribution, but not the entire distribution itself. For this reason (and because analytical calculation of the GF is limited to a few simple models), it is useful to have a numerical framework to calculate low-order moments directly, as provided by Eq. (23). The matrix to be exponentiated is of size $(m+1) \cdot N \times (m+1) \cdot N$, where N is the total number of internal states accessible to the chromophore system (i.e $N \times N$ is the size of the corresponding transition matrix for the associated master equation) and m is the order of the highest moment to be calculated.

4. Generating functions for photon emission: Quantum treatment

In the preceding section, the GF route toward photon statistics was presented in the context of completely stochastic models for dynamics. One of the primary appeals of the GF approach is that it may be easily extended to the regime of quantum dynamics with only minor modification. We shall not rigorously extend our derivation to the quantum case here (interested readers could see Refs. 20–22 for the detailed analysis), but rather appeal to simple arguments.

Within the stochastic picture, chromophore dynamics are formulated in terms of a master equation, symbolically written as

$$\dot{\tilde{P}}(t) = \mathbb{M}(s=1) \cdot \tilde{P}(t), \tag{26}$$

where $\tilde{P}(t)$ is the vector of state probabilities and $\mathbb{M}(s)$ is equivalent to the transition matrix when $s = 1$. Generalizing to arbitrary s leads immediately to the GF equations of motion so long as the replacement $\tilde{P}(t) \rightarrow \tilde{G}(s, t)$ is made concurrently. Correspondence with quantum mechanical systems proceeds via the density matrix formalism and relies upon the assumption that photon-emission events are treated as incoherent rate processes. We introduce a linear superoperator $\mathbb{L}$, which acts upon the density matrix to effect time evolution in the quantum mechanical system

$$\dot{\rho}(t) = \mathbb{L}\rho(t). \tag{27}$$

It is most convenient to think of $\rho(t)$ as a vector of dimension N^2 containing all populations and coherences for a system with N quantum states and $\mathbb{L}$ as an $N^2 \times N^2$ matrix for the reduced system dynamics. The radiation field will not be considered quantum-mechanically (hence "reduced dynamics"). Exciting fields are treated classically and spontaneous emission terms (and generalized decay constants) are generated by integrating out the field variables to provide elements within $\mathbb{L}$ corresponding to the emission process(es).[32,37] Within the usual approximations, this leads to spontaneous emission occurring as a Markovian rate process which transfers chromophore population from upper to lower electronic state manifolds.

As in the case of stochastic dynamics, we convert Eq. (27) to a GF picture by appending the variable s to those matrix elements of $\mathbb{L}$ that cause population to appear in lower electronic states as a result of spontaneous emission processes. From a formal perspective, we arrive at equations identical in structure to the stochastic case

$$\begin{aligned} \dot{\tilde{G}}(s, t) &= \mathbb{L}(s) \cdot \tilde{G}(s, t) \\ G_{ab}(s, t) &= \sum_{n=0}^{\infty} \rho_{ab_n}(t) s^n. \end{aligned} \tag{28}$$

Since the elements of ρ (and hence $\tilde{G}$) represent populations and coherences of the density matrix, they are accordingly indexed by two labels. As always, the subscript n refers to the number of prior photon emissions in our expanded description of dynamics. For the actual extraction of moments, it is important to remember that coherences do not contribute to probabilities, so that

$$1 = \sum_{a=1}^{N} \rho_{aa}(t) = \sum_{a=1}^{N} G_{aa}(1, t),$$

$$\langle n \rangle (t) = \sum_{a=1}^{N} \left. \frac{\partial G_{aa}(s, t)}{\partial s} \right|_{s=1}, \tag{29}$$

$$\langle n(n-1) \rangle (t) = \sum_{a=1}^{N} \left. \frac{\partial^2 G_{aa}(s, t)}{\partial s^2} \right|_{s=1}, \quad \text{etc.}$$

Only those portions of $\tilde{G}(s, t)$ and its s derivatives corresponding to populations contribute to the moments in the quantum case. With this single caveat, we can immediately apply the results of the preceding section to the calculation of moments up to order m. Equation (23) still applies provided we use $\mathbb{L}(s)$ in place of $\mathbb{M}(s)$. So, if it is known how to write down the matrix $\mathbb{L}$ for the dynamics of the chromophore, it becomes trivially easy to extract photon-counting moments. In practice, $\mathbb{L}$ may reflect either the Hamiltonian chromophore dynamics with the addition of field interactions[20, 23, 24] or the additional dissipative interactions with an environment treated implicitly as in Redfield dynamics[34, 38] or some combination of Hamiltonian dynamics with stochastic modulation.[20, 24, 29]

5. Quantum dynamics examples

Explicit examples of stochastic dynamics were already detailed in Sec. 3 and we choose here a few examples with various levels of underlying quantum dynamics to demonstrate the wide applicability of the GF formalism. Although many phenomena of interest to SMS may be treated in the context of stochastic models, and indeed most theoretical works to date have focused on such pictures, a quantum treatment allows for studying the effects of excitation frequency on photon statistics and phenomena dependent upon quantum coherence. In addition, quantum models for chromophore dynamics have the ability to predict behaviors based on first principles, whereas purely kinetic treatments typically rely on some level of empiricism to generate the required suite of rate constants.

An elementary and experimentally relevant example of condensed phase chromophore dynamics is the behavior of a two-level chromophore coupled to a stochastically fluctuating TLS. As indicated above, the TLS picture is the standard model for glassy dynamics at low temperatures and many early single-molecule experiments probed exactly these dynamics. The standard mathematical descriptions for a quantum mechanical two-level chromophore under laser irradiation are the optical Bloch equations[32,37]

$$\begin{aligned}
\dot{\rho}_{\mathrm{ee}} &= i\Omega(\rho_{\mathrm{eg}} - \rho_{\mathrm{ge}})\cos(\omega_{\mathrm{L}}t) - \Gamma\rho_{\mathrm{ee}}, \\
\dot{\rho}_{\mathrm{gg}} &= -i\Omega(\rho_{\mathrm{eg}} - \rho_{\mathrm{ge}})\cos(\omega_{\mathrm{L}}t) + \Gamma\rho_{\mathrm{ee}}, \\
\dot{\rho}_{\mathrm{ge}} &= i\omega_0(t)\rho_{\mathrm{ge}} - i\Omega(\rho_{\mathrm{ee}} - \rho_{\mathrm{gg}})\cos(\omega_{\mathrm{L}}t) - \frac{\Gamma}{2}\rho_{\mathrm{ge}}, \\
\dot{\rho}_{\mathrm{eg}} &= -i\omega_0(t)\rho_{\mathrm{eg}} + i\Omega(\rho_{\mathrm{ee}} - \rho_{\mathrm{gg}})\cos(\omega_{\mathrm{L}}t) - \frac{\Gamma}{2}\rho_{\mathrm{eg}}.
\end{aligned} \tag{30}$$

We denote the chromophore-excited and ground states as $|\mathrm{e}\rangle$ and $|\mathrm{g}\rangle$, respectively. $\hbar\omega_0(t)$ is a temporally evolving splitting between these two levels reflecting TLS dynamics, Ω is the Rabi frequency,[37] and ω_{L} is the frequency of the applied laser field. Γ is the spontaneous emission rate. The explicit time dependence imparted by the applied field is conveniently removed by invoking the rotating wave approximation (RWA) and by moving to a rotating reference frame.[37]

$$\begin{aligned}
\dot{\sigma}_{\mathrm{ee}} &= \frac{i\Omega}{2}(\sigma_{\mathrm{eg}} - \sigma_{\mathrm{ge}}) - \Gamma\sigma_{\mathrm{ee}}, \\
\dot{\sigma}_{\mathrm{gg}} &= -\frac{i\Omega}{2}(\sigma_{\mathrm{eg}} - \sigma_{\mathrm{ge}}) + \Gamma\sigma_{\mathrm{ee}}, \\
\dot{\sigma}_{\mathrm{ge}} &= -i\delta_{\mathrm{L}}(t)\sigma_{\mathrm{ge}} - \frac{i\Omega}{2}(\sigma_{\mathrm{ee}} - \sigma_{\mathrm{gg}}) - \frac{\Gamma}{2}\sigma_{\mathrm{ge}}, \\
\dot{\sigma}_{\mathrm{eg}} &= i\delta_{\mathrm{L}}(t)\sigma_{\mathrm{eg}} + \frac{i\Omega}{2}(\sigma_{\mathrm{ee}} - \sigma_{\mathrm{gg}}) - \frac{\Gamma}{2}\sigma_{\mathrm{eg}}.
\end{aligned} \tag{31}$$

Here we have introduced the transformations

$$\begin{aligned} \sigma_{\mathrm{eg}} &= \rho_{\mathrm{eg}} e^{i\omega_{\mathrm{L}} t}, \\ \sigma_{\mathrm{ge}} &= \rho_{\mathrm{ge}} e^{-i\omega_{\mathrm{L}} t}, \\ \sigma_{\mathrm{gg}} &= \rho_{\mathrm{gg}}, \\ \sigma_{\mathrm{ee}} &= \rho_{\mathrm{ee}}, \end{aligned} \tag{32}$$

and discarded all the terms oscillating at frequencies twice that of the applied radiation in accord with the RWA. $\delta_{\mathrm{L}}(t) = \omega_{\mathrm{L}} - \omega_0(t)$ is the time-dependent detuning frequency of the electronic transition. Note that since the rotating frame transformation does not affect the population elements of the density matrix, the σ representation is equivalent to the ρ representation as far as the population evolution is concerned. By extension, both representations are equivalent for the purposes of counting photons, since only the population elements of the GF are used in determining photon-counting moments (see Eq. (29)).

The GF analog to Eqs. (31) are immediately obtained by analogy to the discussion in the preceding sections. The only term involving spontaneous emission that must be appended with the s variable is the one which advances the photon-counting index by putting population into the ground state. The other three terms involving Γ reflect the loss of population from the excited state, and the corresponding decay of associated coherences are not modified:

$$\begin{aligned} \dot{G}_{\mathrm{ee}} &= \frac{i\Omega}{2}(G_{\mathrm{eg}} - G_{\mathrm{ge}}) - \Gamma G_{\mathrm{ee}}, \\ \dot{G}_{\mathrm{gg}} &= -\frac{i\Omega}{2}(G_{\mathrm{eg}} - G_{\mathrm{ge}}) + s\Gamma G_{\mathrm{ee}}, \\ \dot{G}_{\mathrm{ge}} &= -i\delta_{\mathrm{L}}(t) G_{\mathrm{ge}} - \frac{i\Omega}{2}(G_{\mathrm{ee}} - G_{\mathrm{gg}}) - \frac{\Gamma}{2} G_{\mathrm{ge}}, \\ \dot{G}_{\mathrm{eg}} &= i\delta_{\mathrm{L}}(t) G_{\mathrm{eg}} + \frac{i\Omega}{2}(G_{\mathrm{ee}} - G_{\mathrm{gg}}) - \frac{\Gamma}{2} G_{\mathrm{eg}}. \end{aligned} \tag{33}$$

With this specification of $\mathbb{L}(s)$, calculation of the GF or low-order photon-counting moments follow immediately via procedures already outlined. Numerically, solution of these equations may be facilitated by transforming to a "Bloch vector" representation to take advantage of the complex conjugate relationship between σ_{ge} and σ_{eg}, and to make the equations entirely real-valued.[21] For clarity, we do not consider this in the present discussion.

The TLS modulation of dynamics enters by way of $\delta_L(t)$, which is assumed to hop between two splitting energies as the TLS changes state. Internal dynamics of the TLS are assumed to follow Eq. (24) with both rates being equal, denoted by R, for simplicity. The fact that TLS dynamics are assumed to follow a Markovian dynamics allows us to remove the explicit time dependence in the detuning by extending our GF equations of motion twofold. The procedure is completely analogous to the purely stochastic model discussed previously and has been detailed in Refs. 20 and 24. The resulting time-independent $\mathbb{L}(s)$ matrix is 8×8 in size.

In Fig. 4 we illustrate some behaviors associated with this mixed quantum/stochastic model. Here we choose model parameters appropriate to the spectroscopy of a dye molecule in a low-temperature glass, namely $\Gamma = 40\,\text{MHz}$, $\Omega = 4\,\text{MHz}$, and $R = 1\,\text{Hz}$. The laser is assumed on resonance when the system is in one of the possible TLS states ($\delta_L = 0$) and badly off-resonance in the other

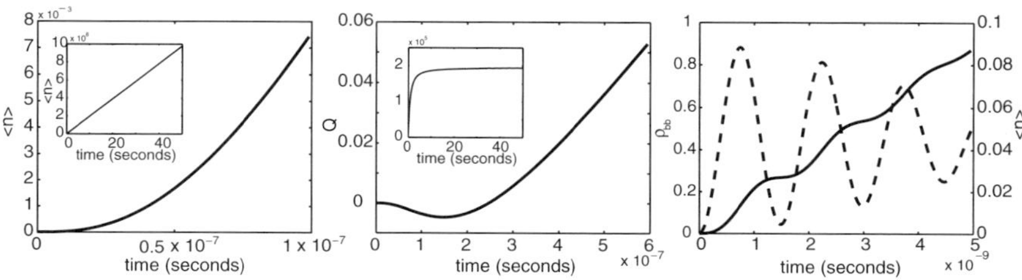

Fig. 4. Photon statistics for a two-level quantum mechanical chromophore coupled to a stochastic TLS. Parameters are as described in the text except for the rightmost panel which incorporates a Rabi frequency of 400 MHz (i.e. 100× that of the other two panes). Inset boxes display behaviors over long time intervals. The right panel superimposes the behavior of the excited state population (dotted line) with average photon number (solid line). The effect of Rabi oscillations on the emission is apparent. (Adapted from Ref. 20.)

state ($\delta_L = 2$ GHz). The initial condition reflects ground electronic state occupation at $t = 0$ and equal occupation of TLS states as dictated by thermal equilibrium. As in the purely stochastic model for chromophore/TLS dynamics, a steady-state emission rate of photons is reached following a brief relaxation to this behavior after the driving field is turned on. Similarly, Mandel's Q parameter eventually settles into a constant bunching behavior, reflecting TLS dynamics. Leading up to this behavior, anti-bunching is observed at short times. One behavior captured with the quantum model, not observable in the classical stochastic treatment is the appearance of Rabi oscillations for sufficiently strong driving fields. These oscillations are due to the coherent excitation of the chromophore, and are completely missed in a purely stochastic treatment.

The influence of Rabi oscillations on photon emission explicitly demonstrates the ability of the GF approach to properly capture coherent dynamics. That said, Rabi oscillations reflect the coherence properties of the applied field and are, perhaps, less indicative of the truly quantum mechanical evolution than effects such as quantum beats and coherent trapping, which explicitly rely upon the coherent superposition of quantum eigenstates. To see the effects of this type of quantum dynamics it is necessary to consider systems with at least three quantum states. We have previously presented GF results for variously configured three-level systems[30] (based on the well-established equations of motion for the density matrix of three-level systems[38–43]), and confine our discussion here to a simple case that displays some quantum effects of interest.

A three-level system in a "V" configuration consists of two quasi-degenerate excited states ($|2\rangle$ and $|3\rangle$) separated from the ground state ($|1\rangle$) by transition frequencies ω_{21} and ω_{31} (see Fig. 5). In general, there are many independent parameters possible with such a model, owing to the various magnitudes and orientations of the two transition dipoles associated with ground–excited transitions. We confine ourselves to situations where these two transition dipoles are either parallel or orthogonal to one another. We further assume that the polarization direction of the applied field (if present) is directed so that

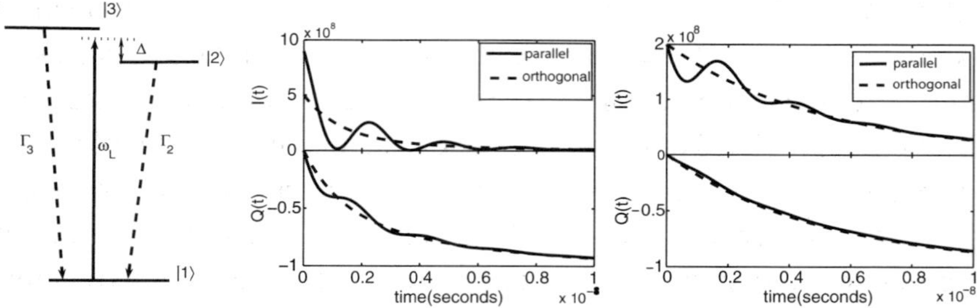

Fig. 5. Left: Schematic diagram for a three-level system in the "V" configuration. Middle and Right: intensity ($\frac{d}{dt}\langle n\rangle$) and Q parameter as a function of time for a system evolving from excited initial conditions in the absence of an applied field. The two panes differ in their imposed initial conditions. The middle pane assumes a coherent superposition of states 2 and 3: $|\psi(0)\rangle = \frac{1}{\sqrt{2}}(|2\rangle + |3\rangle)$, whereas the right pane assumes occupation of state 3: $|\psi(0)\rangle = |3\rangle$. Both figures share the same physical constants: $\Gamma_2/2 = 4 \times 10^8\ \mathrm{s}^{-1}$, $\Gamma_3/2 = 1 \times 10^8\ \mathrm{s}^{-1}$, $\omega_{32} = 25 \times 10^8\ \mathrm{s}^{-1}$. (Adapted from Ref. 30.)

the Rabi frequencies for both transitions are identical. These assumptions restrict the generality of our treatment, but do preserve a fairly simple notation while still allowing us to capture many behaviors of interest. For the "V" system, Eq. (28) reads (RWA and a rotating frame transformation have been applied)[30]

$$\dot{G}_{13} = \left[-i(\Delta - \omega_{32}) - \frac{\Gamma_3}{2}\right] G_{13} + \frac{i\Omega}{2}(G_{11} - G_{33}) - \frac{i\Omega}{2} G_{23} - \eta \frac{1}{2}\sqrt{\Gamma_2 \Gamma_3}\, G_{12}$$

$$\dot{G}_{12} = \left[-i\Delta - \frac{\Gamma_2}{2}\right] G_{12} + \frac{i\Omega}{2}(G_{11} - G_{22}) - \frac{i\Omega}{2} G_{23}^{*} - \eta \frac{1}{2}\sqrt{\Gamma_2 \Gamma_3}\, G_{13}$$

$$\dot{G}_{23} = \left[i\omega_{32} - \frac{\Gamma_2}{2} - \frac{\Gamma_3}{2}\right] G_{23} + \frac{i\Omega}{2} G_{12}^{*} - \frac{i\Omega}{2} G_{13} - \eta \frac{1}{2}\sqrt{\Gamma_2 \Gamma_3}\,(G_{33} + G_{22})$$

$$\dot{G}_{33} = -\Gamma_3 G_{33} - \frac{i\Omega}{2}(G_{13} - G_{13}^*) - \eta\frac{1}{2}\sqrt{\Gamma_2\Gamma_3}(G_{23} + G_{23}^*)$$

$$\dot{G}_{22} = -\Gamma_2 G_{22} - \frac{i\Omega}{2}(G_{12} - G_{12}^*) - \eta\frac{1}{2}\sqrt{\Gamma_2\Gamma_3}(G_{23} + G_{23}^*)$$

$$\dot{G}_{11} = \frac{i\Omega}{2}(G_{13} - G_{13}^*) + \frac{i\Omega}{2}(G_{12} - G_{12}^*) + s\Gamma_3 G_{33} + s\Gamma_2 G_{22}$$

$$+ 2s\eta\frac{1}{2}\sqrt{\Gamma_2\Gamma_3}(G_{23} + G_{23}^*), \tag{34}$$

where $\Gamma_{2(3)}$ is the spontaneous emission rate from state 2(3), Ω is the Rabi frequency for both transitions, and $\Delta \equiv \omega_L - \omega_{21}$ is the laser detuning relative to the $|2\rangle \rightarrow |1\rangle$ transition. The parameter η is equal to unity for parallel transition dipoles or zero for orthogonal dipoles. As always, the calculation of photon-counting moments proceeds immediately from the above equations as outlined in the preceding sections.

In the absence of an explicit exciting field, but assuming that the system begins in an excited configuration at $t = 0$, we readily see the phenomena of quantum beating (Fig. 5). There are at least two interesting points to see in these figures. First, notice that a coherent superposition of excited states is not required to observe the beating phenomena. While this seems inconsistent with the standard qualitative explanation for quantum beats, the effect is well known.[39] Coupling to the quantum radiation field, in addition to providing spontaneous emission pathways to the ground state, causes coupling between the two states of the excited manifold. The original $|2\rangle$, $|3\rangle$ states can no longer be viewed as eigenstates of the system Hamiltonian; so, starting out with all populations in one of these states does not destroy the beating effect. While beating is more pronounced with the naive superposition initial condition (coupling to the radiation field is weak), the characteristic wiggles in $\langle n\rangle$ and Q are observed even for a system with initial occupation of only state $|3\rangle$.

The second effect to be appreciated in Fig. 5 is the loss of the beating phenomena for orthogonal transition dipole moments. The lack of beating in this case is most easily interpreted in terms of quantum measurement theory.[44] In the case of parallel (or at least nonorthogonal) dipole moments there is ambiguity as to which transition an emitted photon originated from. Quantum mechanics tells us that in such a case the emission probability must be computed by adding the two possible emission trajectory amplitudes and squaring — this gives rise to the beating effect. In the case that the two transition dipoles are orthogonal ($\eta = 0$), one could imagine measuring the polarization of the emitted photon. Since an $\hat{x}$-polarized photon could not have originated from a $\hat{y}$-oriented dipole (for example) it is possible to know which transition pathway was followed, and quantum mechanics dictates that we sum the individual trajectory probabilities rather than their amplitudes and then squaring — thus destroying the beating effect.

One of the benefits of a full quantum mechanical description of the chromophore, which we have yet to consider, is the explicit dependence of photon statistics on the frequency of the exciting field. The detuning frequency enters the optical Bloch equations and the related equations (such as Eq. (34)) naturally and thus provides a means to compute the absorption lineshapes (by conservation of energy, $\langle n \rangle$ is proportional to the energy absorbed from the field) as well as the behavior of higher order moments as a function of frequency. In Fig. 6, we consider both the lineshape and the Q parameter spectrum as a function of the laser detuning. The expected power broadening is observed in the peaks corresponding to the two transitions, and the second-order statistics are displayed as well. It is interesting to note the vanishing absorption at frequencies exactly midway between the two resonances for the parallel transition dipole case. By comparison to the orthogonal case, it is clear that one would naively expect strong driving to create significant absorption at all frequencies intermediate between the two resonances. The reason for this "hole" in the absorption profile is well established.[39] Clearly, the two absorption pathways to the excited manifold are not independent of one another as is easily verified mathematically by examining the $\mathbb{L}$ matrix. The 1, 2 coherence is directly coupled to the 1, 3 coherence

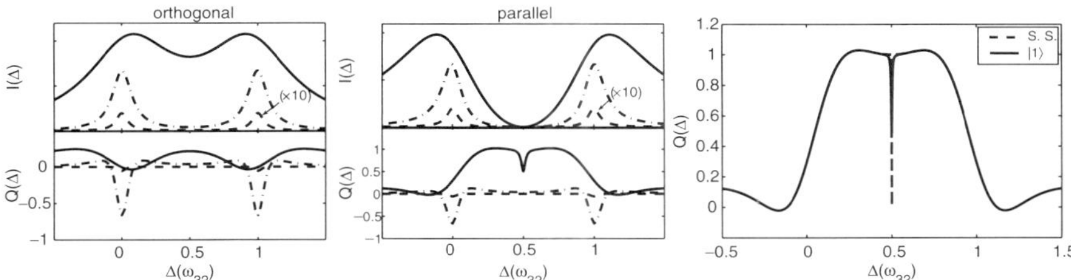

Fig. 6. Left and Middle: Lineshapes and Mandel's Q spectra for the V-type system (transition dipoles for the optical transition are orthogonal on the left or parallel in the middle). Parameters used include: $\Gamma_2 = \Gamma_3 = 0.1\omega_{32}$. $\Omega = 0.5\omega_{32}$ (solid line), $\Omega = 0.1\omega_{32}$ (dash–dot line), $\Omega = 0.01\omega_{32}$ (dashed line). The initial condition is taken as a complete occupation of the ground state, $|1\rangle$. Although the lineshapes are plotted in arbitrary units, the curves are displayed with the correct relative intensities. The lineshape for the weakest excitation is actually shown 10 times larger than it should appear (as indicated) to make its features visible on the same set of axes. Right: Mandel's "steady-state" Q parameter at $\Delta = \omega_{32}/2$ depends upon the initial state of the system. We display the strong excitation case from the middle pane in more detail for two different initial conditions: ground state fully occupied at $t = 0$ and the steady state assumed at $t = 0$. In the case of a steady-state initial condition, photon emission is impossible for $\Delta = \omega_{32}/2$, and both the intensity and Q parameter are identically zero. A different initial condition can allow a few photon emissions as steady state is approached and the observed nonzero Q value. The effect is confined to an extremely narrow range of frequencies. (Adapted from Ref. 30.)

so long as $\eta = 1$ and the dipoles are parallel. Equivalently, these couplings serve to mediate the relative contribution of the two transitions to the overall dipole moment of the system. If system parameters are chosen appropriately, the amplitude of the oscillating dipole may vanish entirely, causing the absorption rate to become zero. With our choice of parameters, the system is poised to exhibit this effect for a detuning of $\Delta = \omega_{32}/2$. This choice guarantees total destructive interference between the two absorption pathways under steady-state conditions, regardless of the strength of the driving field. In the case of orthogonal transition dipoles, such a cancellation is impossible and the two transitions behave very much as if they were decoupled.

We briefly note that the GF formalism is readily extended to multilevel quantum mechanical (and mixed quantum/stochastic) models exceeding the complexity of cases considered in this chapter. More complex models are discussed further in Ref. 34. The primary limitation in extending the present approach is the poor scaling of density matrix equations of motion with the number of quantum states. The $\mathbb{L}$ matrix for evolution of ρ is of size $N^2 \times N^2$ in a model including N quantum states. As with traditional methods for propagating the density matrix itself, this effectively limits the complexity of systems that can be considered to tens of states. We emphasize that this limitation is inherent to quantum mechanical dynamics and not to the GF approach itself, which necessarily shares the same numerical complexity as density matrix time evolution.

6. Conclusion

The continual evolution of experimental single-molecule techniques dictates that new theories and computational algorithms be developed for the purpose of interpreting SMS measurements. The inherently discrete nature of data in a large class of SMS experiments makes the GF approach an appealing theoretical tool for a broad range of SMS problems. We have reviewed here the application of such approaches to experiments that measure photon-counting statistics. Extension to blinking experiments, and related measurements is immediate as the GF approach applies to any linear, memoryless dynamics where the counted events are pure rate processes.

The GF technique is useful as a practical numerical tool since it allows us to translate well-established methods for dynamics of the chromophore (master equations, optical Bloch equations, Redfield equations, etc.) into equations that predict the observables attainable by SMS. Although the equations will become more and more computationally expensive as higher order moments are calculated, the programming effort in extending traditional calculations is negligible. As such, the GF approach is especially appealing for testing hypothetical models against experimental data. Especially in the context of stochastic models for dynamics, it is easy to write down the hypothetical master equation which is immediately translated into GF equations of motion, suitable for simple numerical "experiments".

Acknowledgments

The research summarized herein was supported in part by the Research Corporation and the National Science Foundation. The author thanks Yujun Zheng and Golan Bel for their contributions to the work reviewed here. F. L. H. Brown is an Alfred P. Sloan Research Fellow and a Camille Dreyfus Teacher-Scholar.

References

1. W. E. Moerner and L. Kador, *Physical Review Letters* **62** (1989) 2535.
2. M. Orrit and J. Bernard, *Physical Review Letters* **65** (1990) 2716.
3. S. Gurrieri, E. Rizzarelli, D. Beach and C. Bustamante, *Biochemistry* **29** (1990) 3396.
4. T. Funatsu, Y. Harada, M. Tokunaga, K. Saito and T. Yanagida, *Nature* **374** (1995) 555.
5. T. T. Perkins, D. E. Smith, R. G. Larson and S. Chu, *Science* **268** (1995) 83.
6. H. Noji, R. Yasuda, M. Yoshida and K. Kinosita, *Nature* **386** (1997) 299.
7. T. Plakhotnik, E. A. Donley and U. P. Wild, *Annual Reviews of Physical Chemistry* **49** (1997) 181.
8. W. E. Moerner and M. Orrit, *Science* **283** (1999) 1670.
9. X. S. Xie, *Journal of Chemical Physics* **117** (2002) 11024.
10. S. Weiss, *Science* **283** (1999) 1676.
11. M. Bohmer and J. Enderlein, *ChemPhysChem* **4** (2003) 793.
12. M. Lippitz, F. Kulzer and M. Orrit, *ChemPhysChem* **6** (2005) 770.
13. L. P. Watkins and H. Yang, *Journal of Physical Chemistry B* **109** (2005) 617.
14. E. Barkai, Y. Jung and R. J. Silbey, *Annual Reviews of Physical Chemistry* **55** (2004) 457.
15. W. Feller, *An Introduction to Probability Theory and Its Applications*, 3rd edn. (John Wiley and Sons, New York, 1957).
16. N. G. van Kampen, *Stochastic Processes in Physics and Chemistry* (North-Holland, Amsterdam, 1992).

17. C. W. Gardiner, *Handbook of Stochastic Methods for Physics, Chemistry and the Natural Sciences*, 2nd edn. (Springer-Verlag, Berlin, 1985).
18. F. L. H. Brown, *Physical Review Letters* **90** (2003) Art. No. 028302.
19. I. V. Gopich and A. Szabo, *Journal of Chemical Physics* **118** (2003) 454.
20. Y. Zheng and F. L. H. Brown, *Journal of Chemical Physics* **119** (2003) 11814.
21. Y. Zheng and F. L. H. Brown, *Physical Review Letters* **90** (2003) Art. No. 238305.
22. S. Mukamel, *Physical Review A* **68** (2003) Art. No. 063821.
23. R. J. Cook, *Physical Review A* **23** (1981) 1243.
24. Y. Zheng and F. L. H. Brown, *Journal of Chemical Physics* **121** (2004) 3238.
25. Y. He and E. Barkai, *Physical Review Letters* **93** (2004) Art. No. 068302.
26. Y. He and E. Barkai, *Journal of Chemical Physics* **122** (2005) Art. No. 184703.
27. I. V. Gopich and A. Szabo, *Journal of Chemical Physics* **122** (2005) Art. No. 014707.
28. F. Sanda and S. Mukamel, *Physical Review A* **71** (2005) Art. No. 033807.
29. Y. Zheng and F. L. H. Brown, *Journal of Chemical Physics* **121** (2004) 7914.
30. Y. Peng, Y. Zheng and F. L. H. Brown, *Journal of Chemical Physics* **126** (2007) 104303.
31. F. L. H. Brown, *Accounts of Chemical Research* **39** (2006) 363.
32. R. Loudon, *The Quantum Theory of Light*, 3rd edn. (Oxford, New York, 2000).
33. L. Mandel, *Optics Letters* **4** (1979) 205.
34. G. Bel, Y. Zheng and F. L. H. Brown, *Journal of Physical Chemistry B* **110** (2006) 19066.
35. P. W. Anderson, B. I. Halperin and C. M. Varma, *Philosophical Magazine* **25** (1972) 1.
36. R. Zwanzig, *Accounts of Chemical Research* **23** (1990) 148.
37. C. Cohen-Tannoudji, J. Dupont-Roc and G. Grynberg, *Atom-Photon Interactions* (Wiley-Interscience, New York, 1992).
38. K. Blum, *Density Matrix Theory and Applications*, 2nd edn. (Plenum Press, New York, 1981).
39. D. A. Cardimona, M. G. Raymer and C. R. Stroud, *Journal of Physics B* **15** (1982) 55.
40. L. M. Narducci, M. O. Scully, G.-L. Oppo, P. Ru and J. R. Tredicce, *Physical Review A* **42** (1990) 1630.
41. P. W. Milonni, *Physical Reports C* **25** (1975) 1.
42. M. B. Plenio and P. L. Knight, *Review of Modern Physics* **70** (1998) 101.
43. R. M. Whitley and C. R. Stroud, *Physical Review A* **14** (1976) 1498.
44. M. O. Scully and M. S. Zubairy, *Quantum Optics* (Cambridge University Press, Cambridge, UK, 2006).

CHAPTER 4

Multipoint Correlation Functions for Photon Statistics in Single-Molecule Spectroscopy: Stochastic Dynamics in Liouville Space

František Šanda* and Shaul Mukamel†

*Charles University, Faculty of Mathematics and Physics, Institute of Physics, Ke Karlovu 5, Prague, 121 16 Czech Republic
†Department of Chemistry, University of California, Irvine, CA 92697-2025

1. Photon statistics: Factorial moments vs. correlation functions

Single-molecule spectroscopy (SMS) had opened up new windows into the molecular world by providing a wealth of information about fluctuations of elementary molecular events.[1–4] SMS directly measures statistical distributions of various properties and observes the spectroscopic trace of slowly varying (∼msec) parameter trajectories. Photon arrival trajectories obtained in response to a train of excitation pulses were employed to extract similar information for faster (∼nsec) environmental changes on the radiative lifetime timescale, where the statistical properties of parametric fluctuations affect the photon emission.[5,6] Statistical analysis of SMS signals is usually based on two-point distributions or correlation functions. Higher statistical measures provide independent additional information when the dynamics is nonMarkovian and may not be completely described in terms of two-point Green functions. For example, two-point measures of the fluorescence trace only provide partial information about the underlying internal dynamics, which is usually not sufficient to establish Markovian description of all the relevant degrees of freedom. Multipoint quantities may be instrumental for distinguishing among various possible models.[7–13] Formal analogy between some SMS observables and coherent nonlinear signals[14] will be discussed.

We shall consider two elementary observables in continuous wave (cw) experiments: the number of photons $N(\tau)$ detected during a binning time τ, and the photon emission intensity $I(\tau)$. The two are connected by

$$I(\tau) \equiv \frac{N(\tau+\Delta\tau)-N(\tau)}{\Delta\tau}, \qquad \Delta\tau \to 0. \tag{1}$$

The k-point correlation function $g^{(k)}$ of fluorescence intensities I is defined by

$$g^{(k)}(t_{k-1},\ldots,t_1) \equiv \langle I(t_{k-1}+\cdots+t_1)\cdots I(t_1)I(0)\rangle \tag{2}$$

or in a normalized form,

$$h^{(k)}(t_{k-1},\ldots,t_1) \equiv \frac{\langle I(t_{k-1}+\cdots+t_1)\cdots I(t_1)I(0)\rangle}{\langle I(t_{k-1}+\cdots+t_1)\rangle\cdots\langle I(t_1)\rangle\langle I(0)\rangle}, \tag{3}$$

where $t_j = \tau_j - \tau_{j-1}$ are the intervals between the observation times τ_j. We have set $\tau_0 = 0$ so that $\tau_j = t_j + \cdots + t_1$.

Photon counting distributions may be characterized by the kth factorial moments $F^{(k)}(\tau) \equiv \langle N(N-1)\cdots(N-k+1)\rangle$. Single photon emission is a stochastic event, whose most basic characteristics are its mean and variance. The latter is traditionally expressed in terms of the Mandel parameter

$$M(\tau) \equiv \frac{\langle N(N-1)\rangle - \langle N\rangle^2}{\langle N\rangle}. \tag{4}$$

For Poissonian statistics (independent emission events) we have $M = 0$. Sub-Poissonian $M < 0$ (superPoissonian $M > 0$) statistics shows up as photon antibuching (bunching). For long binning times we expect the distributions of N to approach a Gaussian form, in agreement with the central limit theorem. Some exceptions are the photon distributions of anomalously relaxing systems.[15,16] Such nonergodic systems do not equilibrate and their Mandel parameter grows asymptotically linearly with time[17,18] instead of approaching a limiting value as expected for normal relaxation.[19,20]

A few assumptions are inherent in the modeling of photon counting experiments. The precise emission time is assumed to be sharp, and any coherence between the initial and the final states of the emission process is neglected. The photon emission is then described by a quantum master equation (QME)[21,22]:

$$\left(\frac{d\rho}{dt}\right)_M = \frac{1}{2}\sum_{i>j}\Gamma_{i\to j}(-S_{ij}^{\dagger}S_{ij}\rho + 2S_{ij}\rho S_{ij}^{\dagger} - \rho S_{ij}^{\dagger}S_{ij}), \tag{5}$$

where ρ is the molecular density matrix, $S_{ij}^{\dagger} \equiv |i\rangle\langle j|$ ($S_{ij} \equiv |j\rangle\langle i|$) are raising (lowering) operators connected with the radiative transition from level $|i\rangle$ to a lower level $|j\rangle$, and $\Gamma_{i\to j}$ are the corresponding spontaneous emission rates. Equation (5) may be derived microscopically starting with the quantum Hamiltonian of the radiation field by assuming that the molecular levels are well separated so that coherences may be neglected,[20] i.e. each photon can be associated with a distinct transition between a specific pair of levels. Equation (5) can be generalized to account for cooperative emission.[22,23]

The QME provides a classical trajectory picture for the quantum dynamics of the radiation field. The emission event can be analyzed in classical terms, since observing a photon with a given frequency uniquely defines the state of the system after that event. The same can be said about a quantum measurement. In fact, photon emission is a kind of quantum measurement. It is different from the standard von Neumann measurements[24] since the effect on the molecular wavefunction is not described by a simple projection operator, as the emission is necessarily accompanied by an annihilation of excitation. However, as in the von Neumann measurement, the emission event erases coherences through a "quantum collapse" and defines the future state of the system by providing a statistical prediction as to which photons will be subsequently observed and when.

Single-molecule experiments directly provide time-averaged correlation functions. However, most theoretical approaches focus on ensemble averages, and use the ergodic hypothesis to predict time-averaged observables. Nonergodic, anomalously relaxing, single molecules do not explore the state space uniformly,[17,25] and a description in terms of distributions of time averages is then necessary.[26]

The factorial moments are also introduced for nontrivial time-dependent driving, e.g., to test the quality of single-photon sources, and optimize a given number of emitted photons by manipulating the laser source.[27,28] In these cases the statistical distributions of the emitted photons are more suitable than multipoint correlation functions. For spectroscopic measurements on stationary samples the information content of both measures is similar, as will be shown below.

SMS supplements bulk measurements of biomolecules, quantum dots, and aggregates.[29,30] We assume a stationary cw experiment and the ergodic hypothesis to hold. As a consequence of Eq. (5) the time evolution is not described by a unitary transformation in Hilbert space, and a Liouville space description is preferable.[14,31] We shall adopt tetradic Liouville-space notation. The elements of Liouville space are $|jk\rangle\rangle \equiv |j\rangle\langle k|$. Molecular density ρ, which is a matrix in

Eq. (5) now becomes a vector in this higher space. For instance, Eq. (5) reads

$$\left(\frac{d\rho}{dt}\right)_M = \frac{1}{2}\sum_{i>j}\Gamma_{i\to j}\big(-S_{ij}^{\dagger(L)}S_{ij}^{(L)}\rho + 2S_{ij}^{(L)}S_{ij}^{\dagger(R)}\rho - S_{ij}^{(R)}S_{ij}^{\dagger(R)}\rho\big), \tag{6}$$

where the left (right) superoperators are defined by their action on Hilbert space operator (Liouville space element) X; $A^{(L)}X \equiv AX$ ($A^{(R)}X \equiv XA$). Using this notation we can describe the photon emission by the resetting superoperator R, defined by the positive contributions to the right-hand side of Eq. (5)[32]:

$$R \equiv \sum_{i>j}\Gamma_{i\to j}S_{ij}^{(L)}S_{ij}^{\dagger(R)} = \sum_{i>j}|jj\rangle\rangle\Gamma_{i\to j}\langle\langle ii|. \tag{7}$$

The resetting superoperator R connects the density matrix before and after emission events.

The full dynamics of the system is described by Liouville equation:

$$\frac{d\rho}{dt} = \mathcal{L}\rho + E(t)\mathcal{L}_{\text{int}}\rho + \left(\frac{d\rho}{dt}\right)_M, \tag{8}$$

where $\mathcal{L}_{\text{int}} = (-i/\hbar)[H_{\text{int}}, \ldots]$ describe the interaction with a classical electric field and $\mathcal{L}$ describe the internal dynamics of the system (including a bath), which need not be necessarily induced by Hamiltonian description (e.g., the stochastic Liouville equations).

The superoperator Green function solution $\mathcal{G}(t)$ to Eq. (8) represents the evolution of the optically driven system in Liouville space between emission events. Hereafter in this section we treat all relevant degrees of freedom explicitly so that $\mathcal{G}(t)$ represents a Markovian evolution in that space, and with the rotating wave approximation it only depends on differences between the two times (see Ref. 20). The complete distribution of photons may be conveniently calculated using the generating function technique[33–36] which is explained in

Chap. 3 of this book. The multipoint correlation function (2) can be calculated by[20]

$$g^{(k)}(t_{k-1}, \ldots, t_1) = \mathrm{Tr}[R\mathcal{G}(t_{k-1})R \cdots \mathcal{G}(t_1)R\rho_s], \tag{9}$$

where ρ_s is the steady-state density matrix. The Laplace domain expression ($\mathcal{G}(s) \equiv \int_0^\infty e^{-st}\mathcal{G}(t)dt$) for the kth factorial moment is (Eqs. (C1)–(C5) in Ref. 20)

$$F^{(k)}(s) = k!\frac{1}{s^2}\mathrm{Tr}(R\mathcal{G}(s))^{k-1}R\rho_s. \tag{10}$$

Prefactor $\frac{1}{s^2}$ corresponds to double integration in time domain; and by comparing Eq. (10) with the Laplace transform of Eq. (11) we see that the two quantities are connected by

$$F^{(k)}(t) = k!\int_0^t d\tau' \int_0^{\tau'} d\tau_{k-1} \int_0^{\tau_{k-1}} d\tau_{k-2} \cdots \times \int_0^{\tau_2} d\tau_1 g^{(k)}(\tau_{k-1} - \tau_{k-2}, \ldots, \tau_1 - \tau_0) \tag{11}$$

The lowest $k = 1$ and 2 moments and correlation functions carry identical information and we have

$$N(t) = t\langle I\rangle,$$

where $\langle I\rangle \equiv g^{(1)}$ and

$$\begin{aligned} M(t) &= \frac{2\langle I\rangle}{t}\int_0^t dt' \int_0^{t'} dt_1[h^{(2)}(t_1) - 1] \\ h^{(2)}(t) &= 1 + (2\langle I\rangle)^{-1}\frac{d^2}{dt^2}tM(t). \end{aligned} \tag{12}$$

For $k > 2$, Eq. (11) may not be inverted, since $g^{(k)}$ carry more information than $F^{(k)}$; obviously, one cannot derive the $(k-1)$-parameter multipoint correlation function from the one-parameter factorial moment.

2. Photon statistics in weakly driven systems: Analogy with four wave mixing

We first discuss the basic phenomenology of SMS using a single two-level model system (TLS). The resetting is a nonequilibrium operation and photon counting carries a different information in single molecule and in bulk experiments. Consider the probability $g^{(2)}(t_1)$ to observe two photon emissions at times 0 and t_1. Immediately following an emission event, the steady state is perturbed: The TLS is in the ground state and it cannot emit another photon so that $h^{(2)}(0) = 0$. A single two-level chromophore cannot emit twice in a short time interval. This effect is known as antibunching. In bulk measurements the other molecules may still emit,[37] and antibunching is suppressed since the correlation function $h_V^{(2)}$ for V independent identical chromophores

$$h_V^{(2)} = 1 + \frac{1}{V}(h_1^{(2)} - 1)$$

approaches 1 for $V \to \infty$. Antibunching persists for shorter times than the ground state recovery time, which for a weakly driven TLS coincides with the excited state lifetime Γ^{-1}. If more levels are involved in the optical response, $h^{(2)}$ could become $h^{(2)} \gg 1$. Nevertheless in all cases there is a short time dynamics induced by the deviation from steady state, immediately following photon emission.

The above arguments apply for TLS with fixed parameters. Bath fluctuations on a timescale longer than Γ^{-1} cause another effect observed in SMS experiment. On this timescale the nonequilibrium effect of photon emission may be neglected, and the relaxation to a local steady state (with fixed system parameters) may be considered instantaneous.[38] The fluorescence trace is then a simple classical function of the bath state and may be used to directly probe bath dynamics. Periods of high and low fluorescence intensity are observed, often changing abruptly (blinking) which results in $h^{(2)} \gg 1$ (and superPoissonian photon distributions, bunching).

In Sec. 3 we show how multipoint correlation functions may be used to trace the microscopic origin of the relaxation processes.

Bulk spectroscopy is commonly performed in the weak field $\mu E \ll \Gamma$ (where μ is the dipole moment) regime to avoid saturation effects. In contrast, weak fields are not generally applicable to SMS experiments. Many photons should be recorded from a single-molecule source for good statistics. Many experiments are carried out using strong fields which may not be treated perturbatively. On the other hand, the weak field statistics, if accessible experimentally, may be easier to interpret. All moments simply scale with laser field intensity, e.g., $M \sim E^2$, and the linear regime may be simply controlled. The response can be described in terms of system characteristics and the magnitude of laser field enters as a simple prefactor.

A perturbative treatment of photon counting in the laser intensity should provide good insight into processes which may be probed by photon counting statistics. Consider a two-level system driven by a weak cw laser field

$$H_S = |e\rangle\omega_{eg}\langle e| - Ee^{-i\omega\tau}|e\rangle\mu\langle g| - E^* e^{i\omega\tau}|g\rangle\mu^*\langle e|.$$

The laser detuning from the two-level frequency ω_{eg} will be denoted by $\Delta \equiv \omega - \omega_{eg}$. The system interaction with bath degrees of freedom $(\mathbf{q})$ is described by the Hamiltonian

$$H_{SB} = |g\rangle H_g(\mathbf{q})\langle g| + |e\rangle H_e(\mathbf{q})\langle e|. \tag{13}$$

We further assume that the spontaneous emission rates are independent of the bath variables.

We define the dipole moment operator in the interaction picture

$$\begin{aligned} D_{eg}(\tau) &\equiv \mu \exp(iH_e\tau)\exp(-iH_g\tau); \\ D_{ge}(\tau) &\equiv \mu^* \exp(iH_g\tau)\exp(-iH_e\tau), \end{aligned} \tag{14}$$

and the corresponding superoperators $D_{eg}^{(L)}(\tau)\rho \equiv D_{eg}(\tau)\rho$; acting from the left (ket), and $D_{eg}^{(R)}(\tau)\rho \equiv \rho D_{eg}(\tau)$ acting from the right (bra). The conjugate superoperators $D_{ge}(\tau)$ are defined in a similar manner.

The factorial moments and correlation functions may be described by double-sided Feynman diagrams which depict the evolution of the density matrix and are commonly used in nonlinear spectroscopy (see Figs. 1 and 2 for the lowest two factorial moments). Vertical lines represent the ket (left) or bra (right) of the density matrix, and assume the value e (excited state) or g (ground state). The following factors are assigned to the evolution during the time interval t: 1 for gg, $e^{-(i\Delta+\Gamma/2)t}$ for ge,$e^{(i\Delta-\Gamma/2)t}$ for eg, and $\exp(-\Gamma t)$ for ee. Additional factors are connected with the various interactions. The leading terms in the perturbative expansion for the kth correlation function $g^{(k)}$ are formed by the 2^k diagrams with k-bold horizontal

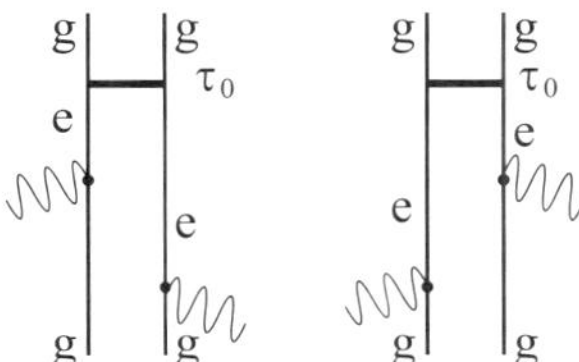

Fig. 1. Double-sided Feynman diagrams for steady-state fluorescence intensity to second (leading) order in the electric field (Eq. (15)).

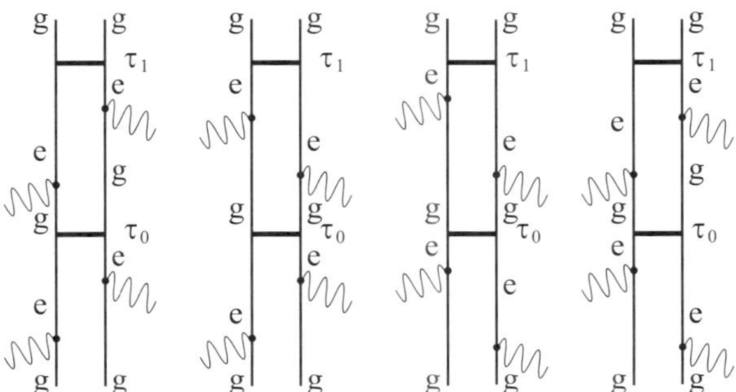

Fig. 2. Double-sided Feynman diagrams for the second factorial moment or two-point correlation function at steady state to fourth (leading) order in the electric field (Eq. (18)).

lines representing open photon detector and two (left and right) wavy lines between each pair of consecutive emissions representing interactions with the optical field, which induce the $g \to e$ transition. The interaction factors are $-iED_{eg}^{(L)}(\tau)$ for exciting ket, and $iE^* D_{ge}^{(R)}(\tau)$ for exciting bra. Bold horizontal line represents photon detection; change $ee \to gg$ and carries a factor $(\Gamma/|\mu|^2)D_{ge}^{(L)}(\tau)D_{eg}^{(R)}(\tau)$. (Due to the final tracing, the factors for the last emission are canceled since Tr $D_{ge}^{(L)}(\tau)D_{eg}^{(R)}(\tau)\rho = |\mu|^2$ Tr ρ.) Rules for constructing the complete perturbative expansion may be found in Ref. 39. Strong field $\mu E \gg \Gamma$ counting statistics, nevertheless, should be rather calculated nonperturbatively starting from Eq. (9) (see discussion below).

For the correlation function $g^{(k)}$, the emission events are fixed at times $\tau_0, \ldots, \tau_k$, and the interactions are integrated over all times consistent with time ordering, as suggested by the respective diagrams. To calculate the factorial moments, the emission times need to be integrated as well over the binning times. The first two integrations must run from $-\infty$ (adiabatic switching) to build the steady state at τ_0.

We shall focus on the lowest two moments of photon counting. The fluorescence intensity and the average photon counting rate calculated to second order in the laser field assume the form (Fig. 1)

$$I=\Gamma E^2 \int_{-\infty}^{0} d\tau_2' \int_{-\infty}^{\tau_2'} d\tau_1' e^{\Gamma\tau_2'} e^{(i\Delta-\Gamma/2)t_2'} \langle\langle D_{ge}^{(R)}(\tau_2') D_{eg}^{(L)}(\tau_1')\rangle\rangle_g + c.c., \tag{15}$$

and

$$\langle N\rangle(t) = \Gamma E^2 \int_{0}^{t} d\tau_3' \int_{-\infty}^{\tau_3'} d\tau_2' \int_{-\infty}^{\tau_2'} d\tau_1' e^{-\Gamma t_3'} e^{(i\Delta-\Gamma/2)t_2'} \times \langle\langle D_{ge}^{(R)}(\tau_2') D_{eg}^{(L)}(\tau_1')\rangle\rangle_g + c.c., \tag{16}$$

with $t_j' \equiv \tau_j' - \tau_{j-1}'$. Brackets $\langle\langle \cdots \rangle\rangle \equiv \mathrm{Tr} \cdots \rho$ denote averaging where (superoperator) actions are in Liouville space. Equations (15) and (16) were obtained by direct application of the above rules. One integration in Eq. (15) and two of the integrations in Eq. (16) can be carried out to yield the classical formula for photon absorption[14]

$$\langle N\rangle(t) = It$$

with

$$I = E^2\int_0^\infty d\tau' e^{(i\Delta-\Gamma/2)\tau'}\langle\langle D_{ge}^{(R)}(\tau')D_{eg}^{(L)}(0)\rangle\rangle_g + c.c. \qquad (17)$$

The linear absorption lineshape is thus determined by the autocorrelation function of the dipole operator, which depends on the dephasing on between two interactions with the laser field. The two-point correlation function is (Fig. 2)

$$\begin{aligned} g^{(2)}(t_1) = \frac{\Gamma^2E^4}{|\mu|^2}&\int_0^{t_1} d\tau_5'\int_0^{\tau_5'} d\tau_4'\int_{-\infty}^0 d\tau_2'\int_{-\infty}^{\tau_2'} d\tau_1' e^{-\Gamma(t_1-\tau_5'-\tau_2')} \\ &\times\big[e^{(i\Delta-\Gamma/2)t_5'}e^{(i\Delta-\Gamma/2)t_2'}\langle\langle D_{ge}^{(R)}(\tau_5')D_{eg}^{(L)}(\tau_4')D_{ge}^{(L)}(0) \\ &\quad\times D_{eg}^{(R)}(0)D_{ge}^{(R)}(\tau_2')D_{eg}^{(L)}(\tau_1')\rangle\rangle_g \\ &\quad+e^{(-i\Delta-\Gamma/2)t_5'}e^{(i\Delta-\Gamma/2)t_2'}\langle\langle D_{eg}^{(L)}(\tau_5')D_{ge}^{(R)}(\tau_4')D_{ge}^{(L)}(0) \\ &\quad\times D_{eg}^{(R)}(0)D_{ge}^{(R)}(\tau_2')D_{eg}^{(L)}(\tau_1')\rangle\rangle_g\big] + c.c. \end{aligned} \qquad (18)$$

The second factorial moment is

$$\begin{aligned} F^{(2)}(t) = 2\frac{\Gamma^2E^4}{|\mu|^2}&\int_0^t d\tau_6'\int_0^{\tau_6'} d\tau_5'\int_0^{\tau_5'} d\tau_4' \\ &\times\int_0^{\tau_4'} d\tau_3\int_{-\infty}^{\tau_3'} d\tau_2'\int_{-\infty}^{\tau_2'} d\tau_1' e^{-\Gamma(t_6'+t_3')} \\ &\times\big[e^{(i\Delta-\Gamma/2)t_5'}e^{(i\Delta-\Gamma/2)t_2'}\langle\langle D_{ge}^{(R)}(\tau_5')D_{eg}^{(L)}(\tau_4')D_{ge}^{(L)}(\tau_3') \\ &\quad\times D_{eg}^{(R)}(\tau_3')D_{ge}^{(R)}(\tau_2')D_{eg}^{(L)}(\tau_1')\rangle\rangle_g \\ &\quad+e^{(-i\Delta-\Gamma/2)t_5'}e^{(i\Delta-\Gamma/2)t_2'}\langle\langle D_{eg}^{(L)}(\tau_5')D_{ge}^{(R)}(\tau_4')D_{ge}^{(L)}(\tau_3') \\ &\quad\times D_{eg}^{(R)}(\tau_3')D_{ge}^{(R)}(\tau_2')D_{eg}^{(L)}(\tau_1')\rangle\rangle_g\big] + c.c. \end{aligned} \qquad (19)$$

Higher order correlation functions may be constructed in a similar manner. The Liouville-space correlation functions Eqs. (16)–(18) may be recast in terms of ordinary (Hilbert space) dipole correlation functions (brackets $\langle\cdots\rangle$ denote averaging where the operators act in Hilbert space). We get

$$\langle\langle D_{ge}^{(R)}(\tau_2)D_{eg}^{(L)}(\tau_1)\rangle\rangle_g = J(\tau_2,\tau_1),$$

where

$$J(\tau_2,\tau_1) \equiv \langle D_{ge}(\tau_2)D_{eg}(\tau_1)\rangle_g.$$

Similarly, we introduce the six-point correlation function

$$\begin{aligned}&F(\tau_1,\tau_2,\tau_3,\tau_4,\tau_5,\tau_6)\\&\quad\equiv \langle D_{ge}(\tau_1)D_{eg}(\tau_2)D_{ge}(\tau_3)D_{eg}(\tau_4)D_{ge}(\tau_5)D_{eg}(\tau_6)\rangle_g.\end{aligned}$$

We then have

$$\begin{aligned}&\langle\langle D_{ge}^{(R)}(\tau_5)D_{eg}^{(L)}(\tau_4)D_{ge}^{(L)}(\tau_3)D_{eg}^{(R)}(\tau_3)D_{ge}^{(R)}(\tau_2)D_{eg}^{(L)}(\tau_1)\rangle\rangle_g\\&\quad= F(\tau_2,\tau_3,\tau_5,\tau_4,\tau_3,\tau_1),\\&\langle\langle D_{eg}^{(L)}(\tau_5)D_{ge}^{(R)}(\tau_4)D_{ge}^{(L)}(\tau_3)D_{eg}^{(R)}(\tau_3)D_{ge}^{(R)}(\tau_2)D_{eg}^{(L)}(\tau_1)\rangle\rangle_g\\&\quad= F(\tau_2,\tau_3,\tau_4,\tau_5,\tau_3,\tau_1).\end{aligned}$$

In Liouville space we only require forward time ordering. The related Hilbert space correlation functions have a more complex structure for time orderings, because they mix the forward time orderings for the ket and backward orderings for the bra. For the spin boson model[14] these correlation functions may be exactly calculated in a closed form using the second-order cumulant expansion.[39]

We now explain the physical significance of Eq. (18). The molecule, initially ($\tau = -\infty$) at equilibrium in the ground state, interacts with the electric field, via the bra (from the right) $D_{ge}^{(R)}$, and the ket (left) $D_{eg}^{(L)}$ at times τ_1' and τ_2', respectively. The propagation in $|ge\rangle\rangle$ or $|eg\rangle\rangle$ between τ_1' and τ_2' (i.e. the interval t_2') includes the decay $e^{-\Gamma t_2'/2}$, and the propagation in $|ee\rangle\rangle$ between τ_2' and $\tau_0 = 0$ includes $e^{-\Gamma\tau_2'}$ representing the excited state lifetime. The density matrix at

$\tau_0 = 0$ represents the steady state, and the average photon count (Eq. (16)) is obtained at this point. At $\tau_0 = 0$ we observe the emitted photon and the molecule moves to the ground state described by the factors $D_{ge}^{(L)}(\tau_0)D_{eg}^{(R)}(\tau_0)$. The excitation at times τ_4' and τ_5', and the second emission at time τ_1 can be described in a similar way. Following the emission at τ_0, the molecule is no longer at equilibrium in the ground state. Its state depends on the spontaneous emission rate, the relaxation rate, and laser detuning. Thus, correlations build up and the probability of subsequent absorption from the initial state may be different.

The fully microscopic expressions given above include a proper description of bath reorganization and the effects of resetting. The connection between nonlinear spectroscopy and SMS for classical stochastic ω_{eg} fluctuations applied to the weak field counting formula Eq. (32) was discussed in Refs. 40 and 41. This approach neglects antibunching (valid for the long time limit) induced by the resetting, some unphysical Hilbert space time orderings of D_{eg} operators contribute, which never show up in the full microscopic derivation, and the Stokes shift is neglected. The stochastic Liouville equations provide a different approach to SMS photon counting statistics.[36] They rigorously incorporate Markovian stochastic ω_{eg} fluctuations into the full Liouville Equations Eq. (8) and resolve the first two difficulties. It has two important additional advantages: it allows a nonperturbative treatment in the electric field because it may be often expanded in finite dimensional linear space and is therefore applicable to strong fields. Photon counting observables for non-Gaussian fluctuations may then be calculated exactly. On the other hand, the SLE misses finite temperature effects. Bath reorganization and the Stokes shift are neglected since the bath evolution is independent of the state of the system.[42] For the spin-boson model of bath fluctuations, Eq. (13) may be expanded as a sum of stochastic coordinates, and should include some temperature correction terms [42, 61]. The stochastic Liouville equations are recovered in the high temperature limit. The first correction introduce a Stokes shift between the ground state and excited state harmonic surface. Further corrections may systematically bring in additional oscillators, whose Matsubara frequencies depend on temperature. This connects the SLE with the microscopic spin-boson bath model, and suggests a practical nonperturbative way for calculating strong field effects.

To compare the microscopic expressions with the SLE approach we note that for classical stochastic models, the left (ket) and the right (bra) bath density matrix variables are identical, and may be considered as stochastic c-number variables ($D_{ge}(\tau) = D^*_{eg}(\tau)$) rather than operators in the bath Hilbert space. The factor corresponding to emission is thus eliminated,

$$(1/|\mu|^2)D^{(L)}_{ge}(\tau)D^{(R)}_{eg}(\tau) \to (1/|\mu|^2)D^*_{eg}(\tau)D_{eg}(\tau) = 1, \quad (20)$$

and the six-point correlation function reduces to a four-point function. The two-point correlation function becomes

$$\begin{aligned} g^{(2)}(t_1) = \frac{\Gamma^2 E^4}{|\mu|^2} \int_0^{t_1} d\tau_5' \int_0^{\tau_5'} d\tau_4' \int_{-\infty}^{0} d\tau_2' \int_{-\infty}^{\tau_2'} d\tau_1' e^{-\Gamma(t_1-\tau_5'-\tau_2')} \\ \times \big[e^{(i\Delta-\Gamma/2)t_5'} e^{(i\Delta-\Gamma/2)t_2'} \langle\langle D_{ge}(\tau_5') D_{eg}(\tau_4') \\ \times D_{ge}(\tau_2') D_{eg}(\tau_1')\rangle\rangle_g + e^{(-i\Delta-\Gamma/2)t_5'} e^{(i\Delta-\Gamma/2)t_2'} \\ \times \langle\langle D_{eg}(\tau_5') D_{ge}(\tau_4') D_{ge}(\tau_2') D_{eg}(\tau_1')\rangle\rangle_g\big] + c.c. \end{aligned} \quad (21)$$

The time variable associated with the emission event in Eq. (19) can be integrated out, giving

$$\begin{aligned} F^{(2)}(t) = 2E^4(\Gamma/|\mu|)^2 \int_0^t d\tau_4' \int_0^{\tau_4'} d\tau_3 \int_{-\infty}^{\tau_3'} d\tau_2' \int_{-\infty}^{\tau_2'} d\tau_1 \\ \times (\xi(\tau_2') - e^{-\Gamma t_3'})(1 - e^{-\Gamma(t-\tau_4')}) e^{(i\Delta-\Gamma/2)t_2'} \\ \times \big[e^{(i\Delta-\Gamma/2)t_4'} \langle D_{ge}(\tau_4') D_{eg}(\tau_3') D_{ge}(\tau_2') D_{eg}(\tau_1')\rangle \\ + e^{(-i\Delta-\Gamma/2)t_4'} \langle D_{ge}(\tau_4') D_{eg}(\tau_2') D_{ge}(\tau_4') D_{eg}(\tau_3')\rangle\big] + c.c., \end{aligned} \quad (22)$$

where $\xi(t) = 1$ for $t > 0$ and $\exp(\Gamma t)$ for $t < 0$.

We next compare these results with the coherent four-wave mixing response which depends on similar correlation functions. In ultrafast nonlinear spectroscopy three laser pulses (with wavevectors $\mathbf{k_1}, \mathbf{k_2}, \mathbf{k_3}$) interact with the probed molecule at prescribed times. Spatial coherence is retained, and the response to laser pulses is observed in specific phase-matching directions. Combining signals from various directions allows to select a specific Liouville space pathway which represents the evolution of the molecule density nature during the course of the experiment. The signal generated in the $\mathbf{k_I} = \mathbf{k_1} + \mathbf{k_2} + \mathbf{k_3}$ (photon echo) phase matching direction is[14]

$$S_{\mathrm{I}}(t_3, t_2, t_1) = \left(\frac{i}{\hbar}\right)^3 e^{-i\omega_{eg}t_3} e^{i\omega_{eg}t_1} \times \Big(\big\langle\big\langle D_{ge}^{(L)}(\tau_3) D_{eg}^{(L)}(\tau_2) D_{eg}^{(R)}(\tau_1) D_{ge}^{(R)}(\tau_0) \big\rangle\big\rangle_g + \big\langle\big\langle D_{ge}^{(L)}(\tau_3) D_{eg}^{(R)}(\tau_2) D_{eg}^{(L)}(\tau_1) D_{ge}^{(R)}(\tau_0) \big\rangle\big\rangle_g \Big) + c.c., \tag{23}$$

and for the $\mathbf{k_{II}} = \mathbf{k_1} - \mathbf{k_2} + \mathbf{k_3}$ direction we get

$$S_{\mathrm{II}}(t_3, t_2, t_1) = \left(\frac{i}{\hbar}\right)^3 e^{-i\omega_{eg}t_3} e^{-i\omega_{eg}t_1} \times \Big(\big\langle\big\langle D_{ge}^{(L)}(\tau_3) D_{eg}^{(L)}(\tau_2) D_{ge}^{(L)}(\tau_1) D_{eg}^{(L)}(\tau_0) \big\rangle\big\rangle_g + \big\langle\big\langle D_{ge}^{(L)}(\tau_3) D_{eg}^{(R)}(\tau_2) D_{ge}^{(R)}(\tau_1) D_{eg}^{(L)}(\tau_0) \big\rangle\big\rangle_g \Big) + c.c. \tag{24}$$

Photon counting and nonlinear spectroscopy are thus related to the same type of correlation functions, but on very different timescales: miliseconds vs femtoseconds, respectively. We assume $\Gamma = 0$ in Eqs. (23) and (24) because spontaneous emission is less important at femtosecond timescale.

Finally we note that similar correlation functions also appear in the description of time- and frequency-resolved fluorescence[43]

$$S(\omega_L, \omega_S, t) = \frac{2}{\hbar^2} \mathrm{Re}[S_1(\omega_L, \omega_S, t) + S_2(\omega_L, \omega_S, t) + S_3(\omega_L, \omega_S, t)], \tag{25}$$

where ω_L is the laser frequency, ω_S the fluorescence frequency, and

$$S_1(\omega_L, \omega_S, t) = \int_0^\infty d\tau \int_{-\infty}^{t-\tau} d\tau_1 \int_{-\infty}^{\tau_1} d\tau_2 e^{-i\omega_S \tau} e^{-i\omega_L(\tau_1-\tau_2)} \times e^{-\Gamma(\tau+\tau_1-\tau_2)/2} \langle\langle D_{ge}^{(L)}(t) D_{eg}^{(R)}(t-\tau) D_{ge}^{(R)}(\tau_1) \times D_{eg}^{(L)}(\tau_2)\rangle\rangle_g E(\tau_1) E^*(\tau_2) \tag{26}$$

$$S_2(\omega_L, \omega_S, t) = \int_0^\infty d\tau \int_{-\infty}^{t-\tau} d\tau_2 \int_{-\infty}^{\tau_2} d\tau_1 e^{-i\omega_S \tau} e^{-i\omega_L(\tau_1-\tau_2)} \times e^{-\Gamma(\tau+\tau_2-\tau_1)/2} \langle\langle D_{ge}^{(L)}(t) D_{eg}^{(R)}(t-\tau) D_{eg}^{(L)}(\tau_2) \times D_{ge}^{(R)}(\tau_1)\rangle\rangle_g E(\tau_1) E^*(\tau_2) \tag{27}$$

$$S_3(\omega_L, \omega_S, t) = \int_{-\infty}^{t} d\tau_2 \int_{t-\tau_2}^{\infty} d\tau \int_{-\infty}^{t-\tau} d\tau_1 e^{-i\omega_S \tau} e^{-i\omega_L(\tau_1-\tau_2)} \times e^{-\Gamma(2t-\tau-\tau_1-\tau_2)/2} \langle\langle D_{ge}^{(L)}(t) D_{eg}^{(L)}(\tau_2) D_{eg}^{(R)}(t-\tau) \times D_{ge}^{(R)}(\tau_1)\rangle\rangle_g E(\tau_1) E^*(\tau_2). \tag{28}$$

Unlike coherent nonlinear signals, fluorescence emissions are nondirectional (i.e. not phase-matched).

3. Multipoint correlation functions for slow fluctuations

Slow environment fluctuations in many SMS experiments may be analyzed using multipoint correlation functions of some classical stochastic coordinate x.[6,11–13] Various microscopic models which share the same two-point correlation functions may be distinguished through their higher order correlation functions. We consider an SMS observable given by a multipoint correlation function of a function $f(x)$ of such coordinate x

$$g_f^{(k+1)}(t_k, \ldots, t_1) = \langle f(x(\tau_k)) \cdots f(x(\tau_0))\rangle, \tag{29}$$

where $t_j \equiv \tau_j - \tau_{j-1}$ are the time intervals. By expanding $f(x)$ in a Taylor series $f(x) = \sum_l f_l x^l$, we can recast it in terms of multipoint correlation functions of x

$$g_f^{(k+1)}(t_k, \cdots, t_1) = \sum_{l_0} \cdots \sum_{l_k} f_{l_0} \cdots f_{l_k} \langle x^{l_k}(\tau_k) \cdots x^{l_0}(\tau_0) \rangle.$$

The most detailed description of a k-point measurement is given by the probability density function (PDF)

$$P^{(k+1)}(\tau_k y_k, \ldots, \tau_0 y_0) \equiv \langle \delta(y_k - x(\tau_k)) \cdots \delta(y_0 - x(\tau_0)) \rangle. \quad (30)$$

Multipoint correlation functions are given by moments of the PDFs

$$g_f^{(k+1)}(t_k, \ldots, t_1) = \sum_{y_k} \cdots \sum_{y_0} f(y_k) \cdots f(y_0) \times P^{(k+1)}(\tau_k y_k, \ldots, \tau_0 y_0). \quad (31)$$

When the moment-generating function

$$\zeta(\xi_k, \ldots, \xi_0) \equiv \sum_{l_0=0}^{\infty} \cdots \sum_{l_k=0}^{\infty} \frac{(\xi_k)^{l_k}}{l_k!} \cdots \frac{(\xi_0)^{l_0}}{l_0!} \langle x^{l_k}(\tau_k) \cdots x^{l_0}(\tau_0) \rangle = \left\langle \exp\left(\sum_j \xi_j x(\tau_j) \right) \right\rangle$$

converges around $\zeta = 0$, the PDFs can be obtained by inverse Laplace transform

$$P^{(k+1)}(\tau_k y_k, \ldots, \tau_0 y_0) = \int_{-i\infty}^{i\infty} d\xi_0 \cdots \int_{-i\infty}^{i\infty} d\xi_k e^{-(\xi_0 y_0 + \cdots + \xi_k y_k)} \zeta(\xi_k, \ldots, \xi_0).$$

For a certain class of dynamic models, the PDFs can thus be obtained from the complete set of correlation functions. However, the moments do not always exist and the moment-generating function may not converge. In this sense PDFs are more general than moments. Lower moments are more accessible experimentally, because measuring higher moments or the PDF itself requires more extensive sampling.

We shall examine two examples of SMS observables described by correlation functions of a classical coordinate.

1. Multipoint correlation functions $g_I^{(k)}$ of fluorescence intensities. We consider a resonantly pumped two-level chromophore, whose transition frequency undergoes slow stochastic fluctuations $\delta\omega$. The time-dependent fluorescence intensity is

$$I(\delta\omega) = \frac{(\mu E)^2}{(\mu E)^2 + \Gamma^2 + \delta\omega^2},$$

where μE is the Rabi frequency. We assume that the fluorescence quantum yield does not fluctuate so that the fluorescence intensity is proportional to the absorption. $\delta\omega$ is a classical coordinate, whose dynamics and correlations may be probed through multipoint correlations of time- and frequency-resolved fluorescence intensities

$$g_I^{(k+1)}(t_k, \ldots, t_1) = \langle I(\delta\omega(\tau_k)) \cdots I(\delta\omega(\tau_0))\rangle. \tag{32}$$

2. Photon arrival trajectories (PAT).[44,45] The molecule is excited by a train of laser pulses and after each pulse one or no photon is emitted. The delays of emitted photons with respect to each pulse are recorded (Fig. 3). These delays depend on the competition between photon emission and quenching. We consider a chromophore and a quencher attached to a polymer (e.g., an enzyme).

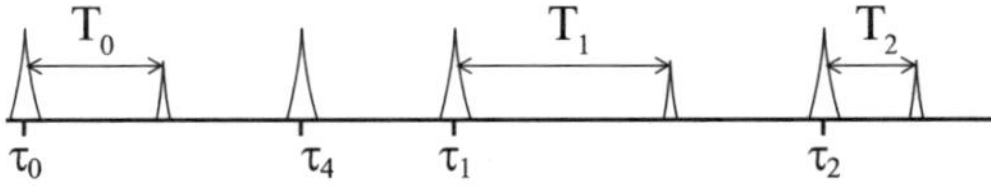

Fig. 3. Photon arrival experiment. Train of pulses at times τ_j excites the molecule. The time delays T_j between the photon arrival time and the exciting pulse at τ_j are recorded. Some excitations are not followed by a response, such as the pulse τ_4. According to our definition this path does not contribute to $P^{(2)}(\tau_4, \tau_1)$ etc.

Conformational dynamics thus affects the quenching rate. Increasing the quenching rate affects the photon arrival statistics in two ways: faster arrival times and lower fluorescence yield. The $(k+1)$-time photon arrival joint function $P^{(k+1)}(T_k\tau_k, \ldots, T_0, \tau_0)_T$ considered here may be calculated solely from the experimentally accessible photon trajectories where a photon is observed after each of the incoming pulses at $\tau_0, \ldots, \tau_k$ (T ensemble). The trajectories where photon was not recorded after some of the pump pulses $\tau_0, \ldots, \tau_k$ are not included. The observable is the correlation function

$$g_T^{(k+1)}(t_k, \ldots, t_1) \equiv \langle T(\tau_k) \cdots T(\tau_0)\rangle_T, \tag{33}$$

where the photon arrival time $T(\tau_k)$ is the delay between the kth excitation pulse and the detected photon, and $t_j \equiv \tau_j - \tau_{j-1}$ are intervals between pulses. Excitation quenching shortens its lifetime and the average photon arrival time. When the quenching rate is determined by a donor–acceptor distance Q, the photon arrival time may be mapped into the statistics of Q. The two common quenching mechanisms are fluorescence resonance energy transfer (FRET) $\gamma_{\mathrm{FRET}} = k_{\mathrm{FRET}}(R_0/Q)^6$, where R_0 is the Förster radius,[46] and electron transfer (ET) $\gamma_{\mathrm{ET}} = k_{\mathrm{ET}}\exp(-\beta Q)$.[44] The total fluorescence decay rate is $\gamma(t) = [\gamma_0 + \gamma_j(Q(t))]$, where γ_0 is the radiative rate and γ_j (j = FRET or ET) is the quenching rate.

We next turn to the photon arrival time statistics. The probability density function for the photon arrival trajectory is generally given by a path integral

$$P^{(k+1)}(\tau_k T_k, \ldots, \tau_0 T_0)_T = C_{k+1}^{-1}\langle e^{-\int_0^{T_k}\gamma(Q(\tau_k+t'))dt'} \cdots e^{-\int_0^{T_0}\gamma(Q(\tau_0+t'))dt'}\rangle_Q, \tag{34}$$

where

$$C_{k+1} = \int \cdots \int \langle e^{-\int_0^{T_k}\gamma(Q(\tau_k+t'))dt'} \cdots e^{-\int_0^{T_0}\gamma(Q(\tau_0+t'))dt'}\rangle_Q \times dT_k \cdots dT_0$$

is the normalization factor. $\langle\,\rangle_Q$ denotes ensemble averaging over all trajectories of the stochastic coordinate. It differs from the T ensemble since the quantum yield decreases with quenching rate, and trajectories with high quenching are underrepresented in the T ensemble.

Evaluating the right-hand side of Eq. (34) requires a path integral over trajectories $Q(t)$. In Sec. 3.2, we shall evaluate it for small Gaussian fluctuations where γ depends linearly on Q. This model is exactly solvable using the second-order cumulant expansion. The calculation is considerably simplified for slow fluctuations whereby Q does not change during the decay time of single excitation and we can set $Q(\tau_k + t') \approx Q(\tau_k)$. The path integral equation (34) then reduces to a k-point correlation function

$$P^{(k+1)}(\tau_k T_k, \ldots, \tau_0 T_0)_T = C_{k+1}^{-1} \left\langle e^{-\gamma(Q(\tau_k))T_k} \cdots e^{-\gamma(Q(\tau_0))T_0} \right\rangle_Q \tag{35}$$

and

$$C_{k+1} = \langle \gamma^{-1}(Q(\tau_k)) \cdots \gamma^{-1}(Q(\tau_0)) \rangle_Q.$$

Since the summations (integrations) over T in Eq. (31) lead to various inverse powers of γ, we define $\varphi \equiv \gamma^{-1}$. The correlation function (Eq. (33)) is given by

$$g_T^{(k+1)}(t_k, \ldots, t_1) = \frac{\langle \varphi^2(Q(\tau_k)) \cdots \varphi^2(Q(\tau_0)) \rangle_Q}{\langle \varphi(Q(\tau_k)) \cdots \varphi(Q(\tau_0))) \rangle_Q}.$$

In the slow fluctuation limit we can thus recast the averaging over photon arrivals in terms of quenching rate fluctuations. When these fluctuations are small compared to the mean quenching rate, the variation of the quantum yield with quenching rate is commonly ignored and the correlation of arrival times $\delta T \equiv T - \bar{T}$; $\bar{T} \equiv \langle T \rangle_T$ is by the inverse decay rates $\delta\varphi \equiv \varphi - \langle \varphi \rangle_Q$:

$$\begin{aligned} g_{\delta T}^{(k+1)}(t_k, \ldots, t_1) &= \langle \delta T(\tau_k) \cdots \delta T(\tau_0) \rangle_T \sim \langle \delta\varphi(\tau_k) \cdots \delta\varphi(\tau_0) \rangle_Q \\ &= g_{\delta\varphi}^{(k+1)}(t_k, \ldots, t_1). \end{aligned} \tag{36}$$

In this approximation correlations of photon arrival times coincide with those of the lifetime. Equation (36) has been used in the modeling of single-molecule FRET and ET experiments.[6,11,47]

A formal relation may be derived by perturbative expansion in $\delta\varphi$ of $\bar{T} = \langle\varphi^2(Q)\rangle_Q / \langle\varphi(Q)\rangle_Q$ and

$$\begin{aligned}\langle\delta T(\tau_1)\delta T(\tau_0)\rangle_T = C_2^{-1}\big(&\langle\varphi^2(Q(\tau_1))\varphi^2(Q(\tau_0))\rangle_Q \\ &- \bar{T}\langle\varphi^2(Q(\tau_1))\varphi(Q(\tau_0))\rangle_Q \\ &- \bar{T}\langle\varphi(Q(\tau_1))\varphi^2(Q(\tau_0))\rangle_Q\big) + \bar{T}^2. \end{aligned} \quad (37)$$

Note that the average $C_2^{-1}\langle\varphi^2(Q(\tau_1))\varphi(Q(\tau_0))\rangle_Q \neq T$ may not be simply reduced to $\bar{T}$ even though it represents an average arrival time at τ_1, since the averaging is performed over paths where a photon is detected both at τ_0 and τ_1. This is different from the averaging over paths with a photon detected at τ_1, which is required for $\bar{T}$. Even though the leading terms in the expansion $\delta\varphi$ for two-point correlations recover Eq. (36), for higher order quantities it can be violated as we show for Gaussian modulation of ET quenching case in Sec. 3.2. Therefore, Eq. (36) needs to be used with some caution.

Note that PDFs of Q are less accessible experimentally than correlation functions. Because of the stochastic nature of photon emission a large number of photon arrival events must be recorded before Q is changed, for a proper sampling of the quenching rate at given Q. This is a much stronger condition than the stationarity of Q during a single emission event, assumed in Eq. (35).

When γ_0 is small, the correlation functions (Eq. (33)) reduce to simple functional forms such as $\langle e^{-\beta Q(\tau_k)} \ldots e^{-\beta Q(\tau_0)}\rangle$ (ET) and $\langle Q^6(\tau_k)\cdots Q^6(\tau_0)\rangle$ (FRET). In either case we can relate the multipoint correlation functions of the photon arrival trajectory to correlation functions of a stochastic coordinate Q. In some special cases the two-point Green function completely characterizes the process. The higher, multipoint, quantities provide additional information that may be used to test the validity of various models. Below we discuss how these multipoint correlation functions may be used to characterize the nature of bath relaxation. Three approaches for modeling the dynamics will be surveyed.

3.1. *Markovian dynamics*

A Markovian description is possible when the state of the system is fully characterized by a few collective variables, and the future evolution of the system only depends on their current values. The probability for a path $x(\tau)$ for $\tau > 0$ starting at $x(0)$ is independent of its past history $\tau < 0$. The dynamics is then described by simple rate equations and all necessary information is contained in the two-point Green function $G(\tau_j x_j, \tau_{j-1} x_{j-1})$. The joint PDF may be calculated as a product of two-point Green functions

$$\begin{aligned} P^{(k+1)}&(\tau_k x_k, \ldots, \tau_0 x_0) \\ &= G(\tau_k x_k, \tau_{k-1} x_{k-1}) \cdots G(\tau_1 x_1, \tau_0 x_0) \rho(x_0, \tau_0). \end{aligned} \tag{38}$$

Two-point Green functions of a Markovian process must satisfy the Chapman–Kolmogorov equation

$$G(\tau_2 x_2, \tau_0 x_0) = \int dx_1 G(\tau_2 x_2, \tau_1 x_1) G(\tau_1 x_1, \tau_0 x_0). \tag{39}$$

For discrete variables, G may be viewed as a matrix with indices x_n and x_{n-1}, and Eq. (39) should be interpreted as a matrix product. The Markovian property is built-in: the probabilities of the $x_0 \to x_1$ and the $x_1 \to x_2$ moves are independent; the probability of the entire trajectory is simply the product of probabilities of its two segments. The integration over x_1 in Eq. (39) indicates going from x_0 to x_2 through all possible values x_1. If the process is stationary, Green's function depends only on the time intervals $t_j = \tau_j - \tau_{j-1}$ and defines a one-parameter semigroup. It can be expressed via operator exponential correlation functions of a Markovian process, characterized by multi-exponential decays.

It should be noted that Markovian dynamics is not merely an intrinsic property of a given physical process, but rather depends on the chosen level of description. Some processes may become Markovian when the system's description is expanded by including additional degrees of freedom, which otherwise cause memory effects. When reducing the amount of information kept on the system, i.e. projecting-off some bath degrees of freedom, the description becomes nonMarkovian and memory effects build up. Formal extension to a Markovian description is often a reasonable strategy for theoretical modeling. The evolution of a fluctuating quantum system is then fully described by the two-point Green function. Similar strategy was used in Sec. 1 of this chapter. In practice it makes sense to denote the process "Markovian" only when a Markovian description is possible using a few, preferably experimentally accessible, degrees of freedom.

One common Markovian model which involves discrete multistate jumps (known from lineshapes theory) has been utilized to describe blinking between several fluorescence intensity levels.

Consider the simplest model of two states d and u, given its master equation

$$\frac{d\rho}{dt} = \hat{W}\rho; \quad \hat{W} = \begin{pmatrix} -\Lambda_u & \Lambda_d \\ \Lambda_u & -\Lambda_d \end{pmatrix};$$

and its Green function solution

$$\hat{G}(t) = \exp[\hat{W}t] = \hat{1} + \frac{1 - \exp[-(\Lambda_u + \Lambda_d)t]}{\Lambda_u + \Lambda_d}\hat{W}. \tag{40}$$

We shall next denote φ in state u and d as φ_u and φ_d, respectively, e.g., for ET, $\varphi_u = \exp(\beta Q_u)$.

Below we present the first three PAT correlation functions in the slow limit

$$g_T^{(1)} = \langle T \rangle_T = \frac{\Lambda_u \varphi_u^2 + \Lambda_d \varphi_d^2}{\Lambda_u \varphi_u + \Lambda_d \varphi_d}$$

$$g_T^{(2)}(t_1) = \frac{\Lambda_d \varphi_d^4 + \Lambda_d \varphi_d^4 - s(t_1)\Lambda_u \Lambda_d (\varphi_u^2 - \varphi_d^2)^2}{\Lambda_d \varphi_d^2 + \Lambda_d \varphi_d^2 - s(t_1)\Lambda_u \Lambda_d (\varphi_u - \varphi_d)^2},$$

where we denoted $s(t) \equiv (1 - \exp[-(\Lambda_u + \Lambda_d)t])/(\Lambda_d + \Lambda_u)$, and

$$g_T^{(3)}(t_2, t_1) = \frac{\begin{array}{l}\Lambda_d\varphi_d^6 + \Lambda_u\varphi_u^6 + \Lambda_u\Lambda_d[(s(t_1) + s(t_2))(\varphi_d^4\varphi_u^2 + \varphi_u^4\varphi_d^2 - \varphi_u^6 \\ \quad - \varphi_d^6) + s(t_1)s(t_2)(\varphi_u^2 - \varphi_d^2)^2(\Lambda_u\varphi_d^2 + \Lambda_d\varphi_u^2)]\end{array}}{\begin{array}{l}\Lambda_d\varphi_d^3 + \Lambda_u\varphi_u^3 + \Lambda_u\Lambda_d[(s(t_1) + s(t_2))(\varphi_d^2\varphi_u + \varphi_u^2\varphi_d \\ \quad - \varphi_u^3 - \varphi_d^3) + s(t_1)s(t_2)(\varphi_u - \varphi_d)^2(\Lambda_u\varphi_d + \Lambda_d\varphi_u)]\end{array}}.$$

Beyond the slow limit the path integral equation (34) may be readily calculated using the stochastic Liouville equations.[42] To do this the time evolution need to be separated for the two types of intervals. Between $\tau_{j-1} + T_{j-1}$ and τ_j the generating function is given by Eq. (40). For the intervals between τ_j and T_j, generating function $\hat{G}^\gamma$ must account for the quenching

$$\hat{G}^\gamma(t) = \exp\left[\begin{pmatrix} -k_u - \gamma_d & k_d \\ k_u & -k_d - \gamma_u \end{pmatrix} t\right],$$

and the PDFs are given by their ordered product

$$\begin{aligned} &P^{(k+1)}(T_k\tau_k, \ldots, T_1\tau_1, T_0\tau_0) \\ &\quad = \mathrm{Tr}\, \hat{G}^\gamma(T_j)\hat{G}(t_k - T_{k-1}) \cdots \hat{G}(t_1 - T_0)\hat{G}^\gamma(T_0)\rho(0). \end{aligned}$$

PDFs and correlation functions may be readily calculated using elementary algebra, however, the final expressions are rather lengthy.

Another common model is a random walk in a potential described by the Fokker–Planck equation[48]

$$\frac{d\rho(r,t)}{dt} = \Lambda\left[\frac{\partial^2}{\partial x^2} - \frac{\partial}{\partial x}\frac{F(x)}{\mathbf{kT}}\right]\rho(r,t), \qquad (41)$$

where Λ is the relaxation rate, $\mathbf{T}$ is the temperature, and F is the force. For a harmonic force $F(x) = -\mathcal{M}\Omega^2 x$, the Green function of the Fokker–Planck equation (41) is

$$G(\tau_{j+1}x_{j+1};\tau_j x_j) \equiv \sqrt{\frac{\mathcal{M}\Omega^2}{2\mathbf{kT}\pi(1-e^{-2\Lambda(\tau_{j+1}-\tau_j)})}} \times \exp\left[\frac{-\mathcal{M}\Omega^2(x_{j+1}-e^{-\Lambda(\tau_{j+1}-\tau_j)}x_j)^2}{2\mathbf{kT}(1-e^{-2\Lambda(\tau_{j+1}-\tau_j)})}\right] \tag{42}$$

and the equilibrium distribution

$$\rho(x_0) = \sqrt{\frac{\mathcal{M}\Omega^2}{2\mathbf{kT}\pi}} \exp\left[\frac{-\mathcal{M}\Omega^2}{2\mathbf{kT}}x_0^2\right]. \tag{43}$$

Equations (42) and (43) show that this is special case of Gaussian process, which will be described in the next section, where we also provide the correlation functions for PAT with ET quenching.

3.2. *Gaussian dynamics*

Consider a harmonic oscillator x coupled to a harmonic bath with coordinates q_j. The system is described by the Hamiltonian

$$H = \frac{p^2}{2\mathcal{M}} + \frac{\mathcal{M}\Omega^2 x^2}{2} + \sum_j \left[\frac{p_j^2}{2m_j} + \frac{m_j\omega_j^2}{2}\left(q_j - \frac{c_j}{m_j\omega_j^2}x\right)^2\right]. \tag{44}$$

x may be viewed as a collective coordinate given by a linear combination of the normal modes of H.[49] All correlation functions for this model may be obtained in a single step using the generating functional technique described below.

Quantum evolution may not be described by trajectories; each value ascribed to some observable should be established by a measurement, which necessarily changes the state of the system.[49] In general, quantum correlation functions of x are not directly relevant for the description of multipoint measurements since they do not include the effect of wavefunction collapse accompanying a quantum measurement. The quantum-classical correspondence may be established by adopting a semiclassical description through the Wigner distribution function, and the measurement is represented by superoperator action in Liouville space $X_+\rho = (x\rho + \rho x)/2$, i.e. $X_+ = (X^{(L)} + X^{(R)})/2$. A classical measurement is thus related to $\delta(X_+ - x)$.

All quantities of interest may be derived from the generating functional

$$\mathcal{S}(J) \equiv \left\langle \mathcal{T} \exp\left[-i \int_0^t d\tau (J(\tau) X_+(\tau)) \right] \right\rangle, \tag{45}$$

where $\mathcal{T}$ is a time ordering operation of superoperators. In the classical limit it reads

$$\mathcal{S}(J) \equiv \left\langle \exp\left[-i \int_0^t d\tau (J(\tau) x(\tau) d\tau) \right] \right\rangle. \tag{46}$$

The model defined by Eq. (44) is exactly solvable. The exact generating functional is obtained by the second-order cumulant expansion

$$\mathcal{S}(J) = \exp\left[\int_0^t d\tau'' \int_0^t d\tau' \left(-\frac{1}{2} g_x^{(2)}(\tau'' - \tau') J(\tau'') J(\tau') \right) \right], \tag{47}$$

where T is the temperature, and the correlation function

$$g_x^{(2)}(t) = \hbar \int_{-\infty}^{\infty} \frac{d\omega}{2\pi} \cos(\omega t) \coth\left(\frac{\hbar\omega}{2\mathbf{kT}} \right) C(\omega).$$

All relevant information is contained in the spectral density $C(\omega)$ of the collective coordinate which determines the two-point correlation functions $g_x^{(2)}$[14,50,51]:

$$C(\omega) = \frac{1}{\mathcal{M}} \frac{\omega \upsilon(\omega)}{(\Omega^2 + \omega\Sigma(\omega) - \omega^2)^2 + \omega^2 \upsilon^2(\omega)}, \tag{48}$$

where

$$\upsilon(\omega) = \frac{\pi}{\mathcal{M}} \sum_j \frac{c_j^2}{2m_j\omega_j^2} \left[\delta(\omega - \omega_j) + \delta(\omega + \omega_j)\right]. \tag{49}$$

Σ and υ are related by the Kramers–Kronig relation

$$\Sigma(\omega) = -\frac{1}{\pi} pp \int_{-\infty}^{\infty} d\omega' \frac{\upsilon(\omega')}{\omega' - \omega}. \tag{50}$$

The classical model is recovered in the high-temperature limit by retaining the first term in the expansion

$$\hbar \coth\left(\frac{\hbar\omega}{2\mathbf{kT}}\right) = \frac{2\mathbf{kT}}{\omega} + \frac{\hbar^2\omega}{6\mathbf{kT}} - \frac{\hbar^4\omega^3}{360(\mathbf{kT})^3} + \cdots \tag{51}$$

This gives

$$g_x^{(2)}(t) = \int_{-\infty}^{\infty} \frac{d\omega}{2\pi} \cos(\omega t) \left(\frac{2\mathbf{kT}}{\omega}\right) C(\omega). \tag{52}$$

Correlation functions are obtained by functional differentiation with respect to $J(t)$, and setting $J = 0$

$$\langle x(\tau_k) \cdots x(\tau_0)\rangle = i^k \left.\frac{\delta \mathcal{S}(J)}{\delta J(\tau_k) \cdots \delta J(\tau_0)}\right|_{J=0}. \tag{53}$$

The joint distribution (Eq. (30)) for multipoint measurements at times τ_j is

$$\begin{aligned} &P^{(k+1)}(\tau_k x_k, \ldots, \tau_0 x_0) \\ &\quad = \int \prod_{j=0}^{k} \frac{ds_j}{2\pi} \exp\left(-i \sum_{j=0}^{k} s_j x_j\right) \left\langle T \exp\left(\sum_j i s_j x(\tau_j)\right)\right\rangle. \end{aligned} \tag{54}$$

The last factor in Eq. (54) may thus be obtained as a generating functional by taking

$$J(\tau) = -\sum_j s_j \delta(\tau - \tau_j), \tag{55}$$

where s_j is the Fourier variable conjugate to x_j.

We further introduce matrix notation along the trajectory for the two-point correlation functions:

$$\bar{M}_{jk} \equiv g_x^{(2)}(\tau_j - \tau_k).$$

Higher correlation functions may be computed using Eq. (53) yielding the factorized forms

$$\begin{aligned} \langle x(\tau_j)x(\tau_l)\rangle &= \bar{M}_{jl}, \\ \langle x(\tau_j)x(\tau_l)x(\tau_m)\rangle &= 0, \\ \langle x(\tau_j)x(\tau_l)x(\tau_m)x(\tau_n)\rangle &= \bar{M}_{jl}\bar{M}_{mn} + \bar{M}_{jm}\bar{M}_{ln} + \bar{M}_{jn}\bar{M}_{lm}. \end{aligned} \tag{56}$$

These are typical for Gaussian fluctuations. Multipoint correlation functions are given by the sums of all possible pairings.

The joint probability distribution, Eq. (54), finally reads

$$\begin{aligned} P^{(k+1)}&(\tau_k x_k, \ldots, \tau_0 x_0) \\ &= \int_{-\infty}^{\infty} \frac{\prod_j ds_j}{(2\pi)^{k+1}} \exp\left(-i\sum_j s_j x_j - \frac{1}{2}\sum_{jl} \bar{M}_{jl} s_j s_l\right) \\ &= \frac{1}{\sqrt{(2\pi)^{k+1}\det \bar{M}}} \exp\left(-\frac{1}{2}\sum_{jl} (\bar{M})^{-1}_{jl} x_j x_l\right). \end{aligned} \tag{57}$$

Gaussian dynamics may or may not be Markovian. The Gaussian–Markovian case, known as the Uhlenbeck–Ornstein process, is obtained by choosing the overdamped oscillator spectral density setting $\gamma \gg \Omega$ in Eq. (48):

$$C(\omega) = \frac{1}{\mathcal{M}\Omega^2}\frac{\omega\Lambda}{\omega^2 + \Lambda^2}. \tag{58}$$

Using Eq. (52) this yields

$$g^{(2)}(t) = \frac{\mathbf{kT}}{\mathcal{M}\Omega^2}\exp(-\Lambda|t|). \tag{59}$$

The correlation matrix elements can now be factorized as $(\bar{M})_{ij} = M_i M_{i-1} \cdots M_{j+1} M_j$, where M_j depend on the time intervals between successive measurements, $M_j = \exp[(-\Lambda(\tau_{j+1} - \tau_j))]$. The inverse matrix is tridiagonal, and the joint distribution equation (57) is finally factorized into a product of Green's function:

$$P^{(k+1)}(\tau_k x_k, \ldots, \tau_0 x_0) = \rho(x_0) \prod_{j=0}^{k-1} G(\tau_{j+1} x_{j+1}; \tau_j x_j), \qquad (60)$$

where the Green function equation (42) and the equilibrium distribution equation (43) of the Fokker–Planck equation (41) are recovered.[49] The factorization (Eq. (60)) is a manifestation of the Markovian dynamics. It implies that ρ at a given time is sufficient to determine the future dynamics without the knowledge of the past history. The present derivation of Eq. (60) did not make any explicit assumption about Markovian property of the master equation.

Note however that Eq. (57) may not be generally recast in the form of Eq. (60). Memory effects show up for the spectral density other than Eq. (58). Gaussian processes with long algebraic tailed correlations were used to interpret some single-molecule experiments.[11,47]

When the oscillator spectral density can be represented by a sum of a few (K) terms of the type of Eq. (58) the dynamics is equivalent to that of K-uncorrelated overdamped harmonic degrees of freedom. The dynamics may be still considered Gaussian–Markovian, but in the higher K-dimensional space. This illustrates the point made previously. It might often be possible to develop a Markovian description by retaining a large number of degrees of freedom. However, if the only accessible observable is Q and the necessary number of degrees of freedom is too large, the Markovian description may not be very practical.

We shall apply the Gaussian model to calculate the joint probability of photon arrivals (Eq. (34)) for an arbitrary fluctuation timescale without invoking the slow fluctuation limit. Assuming small fluctuations, the decay rate is linear in the coordinate $\gamma = \bar{\gamma} + \gamma' Q$. We can then express the PDF of arrival times exactly using the second-order cumulant expansion

$$
\begin{aligned}
P^{(k+1)}&(\tau_k T_k, \ldots, \tau_0 T_0) \\
&= C_{k+1}^{-1} e^{-\sum_{l=0}^{k} \bar{\gamma} T_l} \\
&\quad \times \exp\left[\frac{\gamma'^2}{2} \sum_{i,j=0}^{k} \int_0^{T_i} dt_i' \int_0^{T_j} dt_j' g_x^2(\tau_i + t_i' - \tau_j - t_j')\right].
\end{aligned}
$$

For the Gaussian–Markovian process we substitute $g^{(2)} = \frac{\mathbf{kT}}{\mathcal{M}\Omega^2} e^{-\Lambda t}$ and get

$$
\begin{aligned}
P(\tau_k T_k, \ldots, \tau_0 T_0) = C_{k+1}^{-1} e^{-\sum_{l=0}^{k} \bar{\gamma} T_l}& \\
\times \exp\Bigg[\frac{\mathbf{kT}\gamma'^2}{\mathcal{M}\Omega^2\Lambda^2} \Bigg(&\sum_{l=0}^{k} (\Lambda T_l + e^{-\Lambda T_l} - 1) \\
&+ \sum_{ij,\tau_i > \tau_j}^{k} D(\tau_i, \tau_j, T_i, T_j)\Bigg)\Bigg].
\end{aligned}
$$

For $T_j < |\tau_i - \tau_j|$, which implies that the excitation had decayed before the next laser pulse, we get

$$
D(\tau_i, \tau_j, T_i, T_j) = e^{-\Lambda(\tau_i - \tau_j)}(1 - e^{-\Lambda T_i})(e^{\Lambda T_j} - 1).
$$

Most generally, we have

$$
\begin{aligned}
D(\tau_i, \tau_j, T_i, T_j) = 2\Lambda(T_j + \tau_j - \tau_i)& \\
+ e^{-\Lambda(\tau_i - \tau_j)}&(1 - e^{-\Lambda T_i})(e^{\Lambda T_j} - 1) \\
- 2 \sinh \Lambda&(\tau_j + T_j - \tau_i)
\end{aligned}
$$

for $|\tau_i - \tau_j| < T_j < |\tau_i - \tau_j + T_i|$, and

$$
\begin{aligned}
D(\tau_i, \tau_j, T_i, T_j) = 2\Lambda T_i &- e^{-\Lambda(T_j - \tau_i + \tau_j)}(e^{\Lambda T_i} - 1) \\
&- e^{-\Lambda(\tau_i - \tau_j)}(1 - e^{-\Lambda T_i})
\end{aligned}
$$

when $T_j > |\tau_i - \tau_j + T_i|$. This limit should be avoided since photons from different pulses may not be distinguished in this case, which complicates the interpretation.

PDFs may be integrated numerically to get multipoint correlation functions. When $\gamma(Q)$ is not linear, the higher cumulants must be calculated.

The cumulant expansions and factorization equation (56) may be readily used to calculate the PAT correlation functions in slow limit. For instance, multipoint correlation function for ET quenching may be readily obtained by the second-order cumulant

$$g_T^{(1)} = k_{\mathrm{ET}}^{-1} \exp\left[\frac{3\mathbf{kT}\beta^2}{2\mathcal{M}\Omega^2}\right],$$

$$g_T^{(2)}(t_1) = k_{\mathrm{ET}}^{-2} \exp\left[\frac{3\mathbf{kT}\beta^2}{\mathcal{M}\Omega^2}\right] \exp[3\beta^2 g_x^{(2)}(t_1)],$$

$$\begin{aligned} g_T^{(3)}(t_2, t_1) = k_{\mathrm{ET}}^{-3} \exp\left[\frac{9\mathbf{kT}\beta^2}{2\mathcal{M}\Omega^2}\right] &\exp[3\beta^2(g_x^{(2)}(t_1) \\ &+ g_x^{(2)}(t_1 + t_2) + g_x^{(2)}(t_2))]. \end{aligned} \tag{61}$$

Equation (61) may be directly applied to the Gaussian–Markovian process (Eq. (41)) by choosing $g_x^{(2)}$ according to Eq. (52). The second-order cumulant expansion may be similarly used to derive closed expressions for lineshapes and factorial moments for a two-level system modulated by Gaussian fluctuations either for stochastic or fully quantum Gaussian noise.[39]

We next illustrate the discussion of Eq. (36) and compare correlation functions of δT and $\delta\varphi$.

$$g_{\delta\varphi}^{(2)}(t_1) = \exp[\beta^2 g_x^{(2)}(0)]\left(\exp[\beta^2 g_x^{(2)}(t_1)] - 1\right), \tag{62}$$

$$\begin{aligned} g_{\delta T}^{(2)}(t_1) = \exp[3\beta^2 g_x^{(2)}(0)](&\exp[3\beta^2 g_x^{(2)}(t_1)] \\ &- 2\exp[\beta^2 g_x^{(2)}(t_1)] + 1). \end{aligned} \tag{63}$$

Leading δx^2 terms of Eqs. (62) and (63) coincide. For three-point functions

$$g_{\delta\varphi}^{(3)}(t_2, t_1) = \exp[(3/2)\beta^2 g_x^{(2)}(0)][\Theta - \exp[\beta^2 g_x^{(2)}(t_1)] - \exp[\beta^2 g_x^{(2)}(t_2)] - \exp[\beta^2 g_x^{(2)}(t_1 + t_2)] + 2],$$

where

$$\Theta = \exp[\beta^2 (g_x^{(2)}(t_1) + g_x^{(2)}(t_2) + g_x^{(2)}(t_1 + t_2))], \quad \text{and}$$

$$\begin{aligned} g_{\delta T}^{(3)}(t_2, t_1) = \exp[(9/2)\beta^2 g_x^{(2)}(0)]\big[&\Theta^3 - \Theta(\exp 2\beta^2 g_x^{(2)}(t_1) \\ &+ \exp 2\beta^2 g_x^{(2)}(t_2) - \exp 2\beta^2 g_x^{(2)}(t_1 + t_2)) \\ &+ \exp \beta^2 g_x^{(2)}(t_1) + \exp \beta^2 g_x^{(2)}(t_2) \\ &+ \exp \beta^2 g_x^{(2)}(t_1 + t_2) - 1\big], \end{aligned}$$

however, δx^3 vanishes and the leading δx^4 terms differ significantly. Approximation equation (36) is thus not suitable for modeling higher (odd) moments.

3.3. *NonMarkovian dynamics: Renewal processes, continuous time random walks*

A nonMarkovian process is not fully characterized by its probability density at a given time; the history of the system does make a difference. In some cases reduced master equations for the system may be still derived. For example, a master equation which includes a time-convolution obtained using the Zwanzig projection operator technique[52] shows explicitly the memory functions and may be used for calculating two-point Green functions. However, it does not offer a prescription for calculating multipoint quantities. One cannot assume statistical independence of paths between points $x_0 \to x_1$ and $x_1 \to x_2$, and the resulting two-point PDFs may not generally be used to compute multipoint PDF using factorization into two-point Green function (Eq. (38)). The same is true for time-convolutionless master equations.[53]

In order to compute multipoint quantities one must introduce a more microscopic description of the process. The continuous time random walk (CTRW) provides an example of a solvable nonMarkovian model which nevertheless allows to compute multipoint quantities.[54] It generalizes ordinary random walks by introducing waiting time distribution functions (WTDF) for jumps. This model is suitable for modeling long-time memory effects.

We consider a particle moving stochastically on lattice points x (time is continuous, space is discrete). We define the transition probability W_{xy} for the jump from y to x

$$p(x; i+1) = \sum_y W_{xy} p(y; i), \tag{64}$$

where $p(x; i)$ is the probability to be at x after the ith jump and $\sum_x W_{xy} = 1$ ($W_{xx} = 0$ as jump necessarily implies change of position). This model represents a random walk.

The CTRW supplements this model by introducing the waiting time probability distribution for successive jumps, $\Psi_{xy}(t)$, normalized as $\int_0^\infty \Psi_{xy}(t)dt = W_{xy}$.

A fundamental property of the CTRW model is that the memory effects enter solely through the time elapsed from the last jump. At a given jump event the WTDF for the next jump Ψ is independent of the history. This *renewal* (resetting) property makes it possible to compute all statistical averages, even in the absence of a Markovian description for the probability distribution, and provides a convenient formalism for describing long-time memory effects.

To proceed, we introduce the matrix of survival probabilities $\Phi(t)$, that no jump had occurred prior to t. It is connected to the waiting time distribution by

$$\Phi_{xy}(t) = \delta_{xy} \int_t^\infty \sum_z \Psi_{zy}(t')dt'. \tag{65}$$

We denote the initial probability density to find the particle at x at time τ_0 by $\rho_x(\tau_0)$. The WTDF for the first jump after τ_0 ($\Psi'(t)$) may be different from $\Psi(t)$ since it depends on the past evolution of the system, including the time elapsed from the last jump. For long-tailed WTDF, the choice of $\Psi'(t)$ may substantially change the nature of the ensemble. Two important types of ensembles, which differ by the initial conditions, are commonly used.

(i) Assuming that all particles had arrived at their sites exactly at the initial time we simply set $\Psi'(t) = \Psi(t)$. This defines a nonstationary process and implies that specific preparation was made at the initial time. For models with short memory, this does not affect the long-time behavior. WTDFs with long algebraic tails where the CTRW may never be equilibrated are widely used to model subdiffusion and aging effects.[54] However, the lack of ergodicity makes it difficult to connect the calculated ensemble correlation functions with SMS experiments based on time averages.

(ii) Stationary processes require WTDF for the first jump

$$\Psi'_{xy}(t) = \int_t^\infty \Psi_{xy}(t')dt'/\bar{t}_y, \tag{66}$$

where $\bar{t}_y \equiv \int_0^\infty \tau \sum_x \Psi_{xy}(\tau)d\tau$ is the mean waiting time. Equation (66) is closely connected with microscopic reversibility, since it may be interpreted as connecting for first jump WTDF forward and backward (survival function).[55–57]

CTRW with infinite $\bar{t}$ does not represent a stationary random walk, since it asymptotically approaches a static sample and the mobility vanishes. A stationary CTRW is thus only possible if $\bar{t}$ is finite. This model does not require any special initial preparation; the initial time τ_0 is an arbitrarily chosen point in an ongoing random walk. This model is relevant for SMS since the time and ensemble averages are connected by the ergodic hypothesis. In addition we require the initial density ρ to be invariant to the jump event

$$\hat{W}\rho(\tau_0) = \rho(\tau_0). \tag{67}$$

Two simple models for $\hat{\Psi}$ are the discrete two-state jump model

$$\hat{\Psi} = \begin{pmatrix} 0 & \psi_d \\ \psi_u & 0 \end{pmatrix}$$

and the continuous variable dynamics describing diffusion in a harmonic potential[48]

$$\hat{\Psi} = \psi(t)\left(1 + D\bar{t}\left[\frac{\partial^2}{\partial x^2} + \frac{\partial}{\partial x}\frac{M\Omega^2 x}{\mathbf{kT}}\right]\right), \tag{68}$$

where D is the generalized diffusion constant.

The evolution of the density may be described by a master equation with fractional time derivatives.[58] However, the probability densities do not provide a full description of the system dynamics. Even though the (time-dependent) ratio connecting the density with jump rate can be defined,[59] the tendency to jump varies among various past trajectories, depending on the time elapsed from the last jump, and such rates cannot be used to compute multipoint correlation functions, as is done for Markovian processes. Instead, the density of arrival times plays a fundamental role in the theory. The total densities are finally obtained by convoluting the arrival densities with the survival function $\hat{\Phi}$.

We have developed a diagrammatic Green function method for calculating multipoint functions for this model. The stochastic path should be divided into segments between the first (at $\tau_{k-1} + \zeta_{2k-1}$) and the last ($\tau_k - \zeta_{2k}$) jump (renewal) in each interval t_k and connecting segments. Since some intervals contain no jump, they should be represented separately, and the complete k-point PDF consists of 2^k contributions which are represented by the diagrams shown in

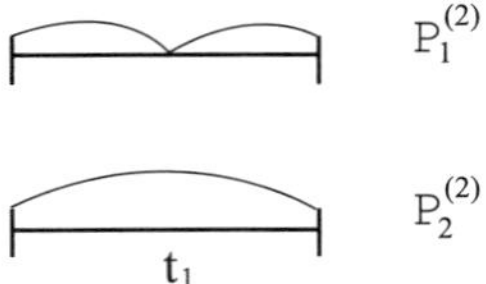

Fig. 4. Two contributions to the two-point PDF. Each diagram represents a path with (line meets the axis) or without (line does not meet the axis) some jump in each time interval.

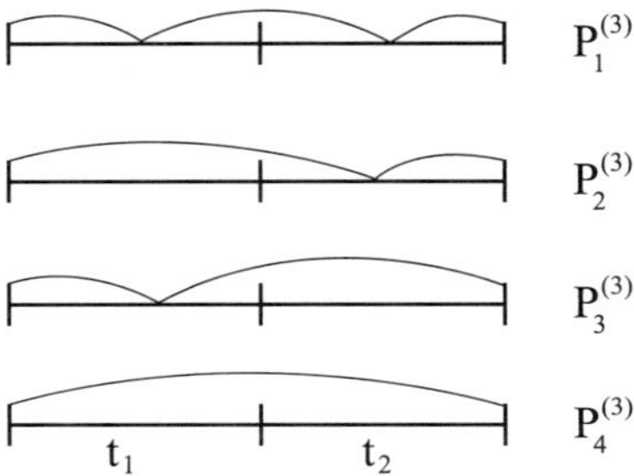

Fig. 5. The four contributions to the three-point PDFs.

Figs. 4–6. The presence of any ($\geq$1) jump in a given time interval is depicted by a trajectory that touches the time axis. The probability $\Sigma_{xy}(t_k - \zeta_{2k-1} - \zeta_{2k})$ for jumping at $\tau_{k-1} + \zeta_{2k-1}$ to x if the path arrived at $\tau_k - \zeta_{2k}$ to y is given by solving the integral equation

$$\hat{\Sigma}(\tau) = \int_0^t \hat{\Psi}(\tau - \tau')\hat{\Sigma}(\tau')d\tau'; \quad \Sigma_{xy}(0) = \delta(\tau)\delta_{xy}. \tag{69}$$

This factor must be multiplied with probability for the connecting segments (between t_l and t_m interval), including the information about the x_j value carried by the trajectory at the boundary point τ_s projecting to the x_k at the interval boundaries τ, which is simply $\Psi(\tau_{l-1} - \tau_m + \zeta_{2l-1} + \zeta_{2m})|x_m\rangle\langle x_m|\delta_{l-1,l-2}\cdots\delta_{m+1,m})$. Similar factors must be included for the first and last segments with $\Psi \to \Psi'$, Φ.

Finally, we must integrate over all possible values of ζ, which are compatible with a diagram, e.g., for the first contribution to the

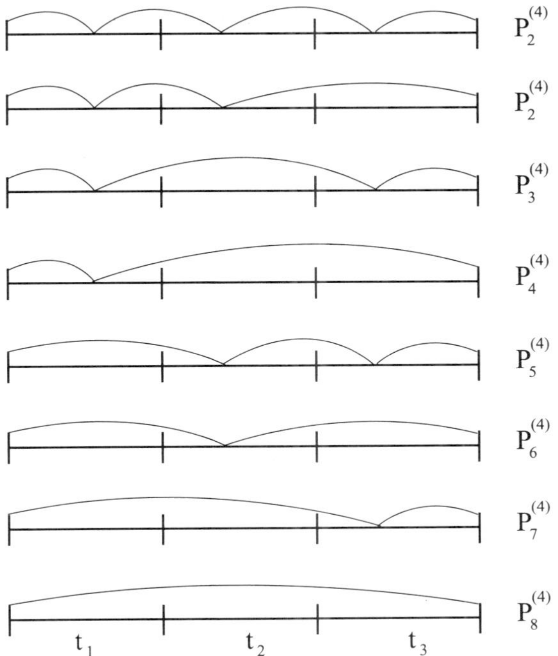

Fig. 6. The eight contributions to the four-point PDFs.

two-interval PDF, $P_1^{(3)}$

$$
\begin{aligned}
P_1^{(3)}&(t_2 + t_1 x_2, t_1 x_1, 0 x_0) \\
&= \int_0^{t_2} d\zeta_4 \int_0^{t_2-\zeta_4} d\zeta_3 \int_0^{t_1} d\zeta_2 \int_0^{t_1-\zeta_2} d\zeta_1 \\
&\quad \times \langle x_2 | \hat{\Phi}(\zeta_4) \hat{\Sigma}(t_2 - \zeta_4 - \zeta_3) \hat{\Psi}(\zeta_3 + \zeta_2) | x_1 \rangle \\
&\quad \times \langle x_1 | \hat{\Sigma}(t_1 - \zeta_2 - \zeta_1) \hat{\Psi}'(\zeta_1) | x_0 \rangle.
\end{aligned} \tag{70}
$$

These equations may be conveniently solved in Laplace space by defining $\hat{\Psi}(s) \equiv \int_0^\infty e^{-st} \hat{\Psi}(t) dt$, where s_k is the conjugate variable to the interval time t_k. Equation (69) is solved by $\Sigma(s) = [1 - \hat{\Psi}(s)]^{-1}$, and Eq. (70) is solved by the convolution theorem. Formal derivation of this algorithm may be found in Ref. 10. In the rest of this section

Table 1. Two contributions to the two-point PDF. Each column corresponds to one diagram of Fig. 4. + means interval with and − without a jump.

	$P_1^{(2)}$	$P_2^{(2)}$
t_1	+	−

we thus switch to Laplace space and set $P^{(k+1)}(s_k x_k, \ldots, s_1 x_1, x_0) = \int_{\tau_{k-1}}^{\infty} d\tau_k e^{s_k(\tau_k - \tau_k - 1} \cdots \int_0^{\infty} d\tau_1 e^{s_1(\tau_1-\tau_0)} P^{(k+1)}(\tau_k x_k, \ldots, \tau_1 x_1, \tau_0 x_0)$

The two diagrams in Fig. 4 (Table 1) represent the contribution to the two-point PDFs, which are given by

$$P^{(2)}(s_1 x_1, x_0) = \sum_{j=1}^{2} P_j^{(2)} \rho_{x_0}(\tau_0), \tag{71}$$

$$P_1^{(2)} = \langle x_1|\hat{\Phi}(s_1)[1 - \hat{\Psi}(s_1)]^{-1}\hat{\Psi}'(s_1)|x_0\rangle,$$

$$P_2^{(2)} = \langle x_1|\hat{\Phi}'(s_1)|x_0\rangle.$$

The four diagrams in Fig. 5 (Table 2) represent the contribution to the three-point PDFs, which are given by

$$P^{(3)}(s_2 x_2, s_1 x_1, x_0) = \sum_{j=1}^{4} P_j^{(3)} \rho_{x_0}(\tau_0) \tag{72}$$

$$P_1^{(3)} = \left\langle x_2 \left| \hat{\Phi}(s_2)[1 - \hat{\Psi}(s_2)]^{-1} \frac{(\hat{\Psi}(s_2) - \hat{\Psi}(s_1))}{s_1 - s_2} \right| x_1 \right\rangle \times \langle x_1|[1 - \hat{\Psi}(s_1)]^{-1}\hat{\Psi}'(s_1)|x_0\rangle$$

$$P_2^{(3)} = \left\langle x_2 \left| \hat{\Phi}(s_2)[1 - \hat{\Psi}(s_2)]^{-1} \frac{(\hat{\Psi}'(s_1) - \hat{\Psi}'(s_2))}{s_2 - s_1} \right| x_1 \right\rangle \langle x_1|x_0\rangle$$

$$P_3^{(3)} = \left\langle x_2 \left| \frac{(\hat{\Phi}(s_2) - \hat{\Phi}(s_1))}{s_1 - s_2} \right| x_1 \right\rangle \langle x_1|[1 - \hat{\Psi}(s_1)]^{-1}\hat{\Psi}'(s_1)|x_0\rangle$$

Table 2. The four contributions to the three-point PDF corresponding to Fig. 5.

	$P_1^{(3)}$	$P_2^{(3)}$	$P_3^{(3)}$	$P_4^{(3)}$
t_2	+	+	−	−
t_1	+	−	+	−

$$P_4^{(3)} = \left\langle x_2 \left| \frac{(\hat{\Phi}'(s_2) - \hat{\Phi}'(s_1))}{s_1 - s_2} \right| x_1 \right\rangle \langle x_1 | x_0 \rangle.$$

Higher order correlation functions are constructed in a similar manner. Figure 6 (Table 3) shows eight diagrams representing all contributions to the four-point PDFs.

We shall demonstrate how Eqs. (71) and (72) work for the simplest PDFs: symmetric two-state ($\psi_u = \psi_d = \psi$, $\Phi_{xy} = \delta_{xy}\phi(t)$) with states $|-a_0\rangle$ and $|a_0\rangle$ undergoing CTRW from equilibrium $\rho_{a_0}(t_0) = \rho_{-a_0}(t_0) = 1/2$. The two-point PDF becomes

$$P^{(2)}(s_1 a_0, a_0) = P^{(2)}(s_1(-a_0), -a_0) = \frac{1}{2s_1} - \frac{\psi'(s_1)}{2s_1(1 + \psi(s_1))},$$

$$P^{(2)}(sa_0, -a_0) = P^{(2)}(s_1(-a_0), a_0) = \frac{\psi'(s_1)}{2s_1(1 + \psi(s_1))},$$

Table 3. The eight contributions to the four-point PDF corresponding to Fig. 6.

	$P_1^{(4)}$	$P_2^{(4)}$	$P_3^{(4)}$	$P_4^{(4)}$	$P_5^{(4)}$	$P_6^{(4)}$	$P_7^{(4)}$	$P_8^{(4)}$
t_3	+	+	+	+	−	−	−	−
t_2	+	+	−	−	+	+	−	−
t_1	+	−	+	−	+	−	+	−

and the two-point PAT correlation function is

$$g_T^{(2))}(t_1) = \frac{\varphi_u^4 + \varphi_d^4 - (\varphi_u^2 - \varphi_d^2)^2 \int_{-i\infty}^{i\infty} ds_1 e^{s_1 t_1} \frac{\psi'(s_1)}{s_1(1+\psi(s_1))}}{\varphi_u^2 + \varphi_d^2 - (\varphi_u - \varphi_d)^2 \int_{-i\infty}^{i\infty} ds_1 e^{s_1 t_1} \frac{\psi'(s_1)}{s_1(1+\psi(s_1))}}.$$

For the three-point PDF we obtain

$$\begin{aligned}
&P^{(3)}(s_2 x_2, s_1 x_1, x_0)\\
&= \frac{1}{(s_1 - s_2)} \frac{\phi(s_2)\psi'(s_1)(\psi(s_2) - \psi(s_1))}{(1 - \psi^2(s_1))(1 - \psi^2(s_2))}\\
&\quad \times \left[\psi(s_1)\psi(s_2)(\delta_{a_0 x_0}\delta_{a_0 x_1}\delta_{a_0 x_2} + \delta_{-a_0 x_0}\delta_{-a_0 x_1}\delta_{-a_0 x_2})\right.\\
&\qquad + \psi(s_1)(\delta_{a_0 x_0}\delta_{a_0 x_1}\delta_{-a_0 x_2} + \delta_{-a_0 x_0}\delta_{-a_0 x_1}\delta_{a_0 x_2})\\
&\qquad + \psi(s_2)(\delta_{-a_0 x_0}\delta_{a_0 x_1}\delta_{a_0 x_2} + \delta_{a_0 x_0}\delta_{-a_0 x_1}\delta_{-a_0 x_2})\\
&\qquad \left. + (\delta_{-a_0 x_0}\delta_{a_0 x_1}\delta_{-a_0 x_2} + \delta_{a_0 x_0}\delta_{-a_0 x_1}\delta_{a_0 x_2})\right]\\
&\quad + \frac{1}{s_1 - s_2} \frac{(\psi'(s_2) - \psi'(s_1))\phi(s_2)}{1 - \psi^2(s_2)}\\
&\quad \times \left[\psi(s_2)(\delta_{a_0 x_0}\delta_{a_0 x_1}\delta_{a_0 x_2} + \delta_{-a_0 x_0}\delta_{-a_0 x_1}\delta_{-a_0 x_2})\right.\\
&\qquad \left. + (\delta_{a_0 x_0}\delta_{a_0 x_1}\delta_{-a_0 x_2} + \delta_{-a_0 x_0}\delta_{-a_0 x_1}\delta_{a_0 x_2})\right]\\
&\quad + \frac{1}{(s_1 - s_2)} \frac{\psi'(s_1)(\phi(s_2) - \phi(s_1))}{1 - \psi^2(s_1)}\\
&\quad \times \left[\psi(s_1)(\delta_{a_0 x_0}\delta_{a_0 x_1}\delta_{a_0 x_2} + \delta_{-a_0 x_0}\delta_{-a_0 x_1}\delta_{-a_0 x_2})\right.\\
&\qquad \left. + (\delta_{-a_0 x_0}\delta_{a_0 x_1}\delta_{a_0 x_2} + \delta_{a_0 x_0}\delta_{-a_0 x_1}\delta_{-a_0 x_2})\right]\\
&\quad + \frac{\phi'(s_2) - \phi'(s_1)}{s_1 - s_2} \left(\delta_{a_0 x_0}\delta_{a_0 x_1}\delta_{a_0 x_2} + \delta_{-a_0 x_0}\delta_{-a_0 x_1}\delta_{-a_0 x_2}\right).
\end{aligned} \tag{73}$$

Beyond the slow limit PAT statistics can be calculated by a modification of the present algorithm as described in Refs. 55, 56, and 60, where the same path integral as Eq. (34) was calculated for nonlinear spectroscopies.

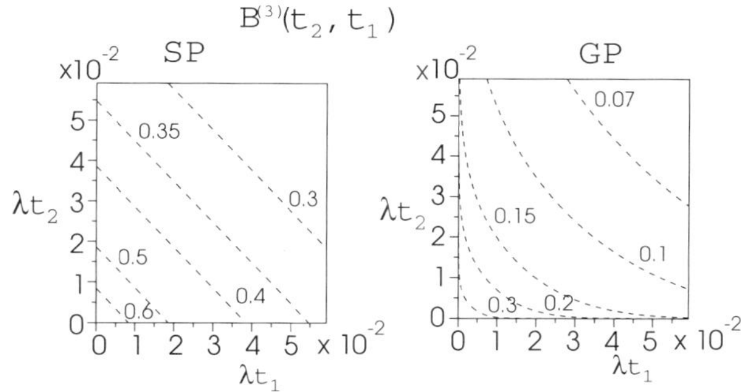

Fig. 7. Left: Contour plot for the three-point correlation function $\langle \delta \exp[\beta Q(\tau_2)]\delta \exp[\beta Q(\tau_1)]\delta \exp[\beta Q(\tau_0)]\rangle$, where $\delta \exp \beta \equiv \exp[\beta Q] - \langle \exp[\beta Q]\rangle$ for a stationary CTRW in a harmonic potential with WTDF (Laplace space) $\psi(s) = \frac{1+(\kappa\lambda)^{\alpha}}{1+(\kappa(s+\lambda))^{\alpha}}$ with $\alpha = 0.8$, $\frac{\lambda \mathbf{kT}}{DM\Omega^2} = 10^{-5}$, $(\beta\Omega^2/\mathbf{kT})^2 = 0.372$. Right: The same plot for Gaussian model with two-point correlation function identical to the left panel. Adapted from Ref. 10.

We next discuss the application of the CTRW models to slow dynamics and point out the differences from Gaussian fluctuations with the same algebraic long-tail correlation functions. In CTRW with algebraic WTDF tails a substantial fraction of particles remains for a long time at the same position (no jump). These fractions are responsible for the long tails of the correlation functions. Based on this picture, we may draw some conclusions. In Fig. 7, we display the three-point correlation function $B^{(3)}$ of $\delta \exp(\beta x)$ (which corresponds to ET quenching correlation function) for a CTRW in a harmonic potential[10] Eq. 68 with algebraic tails $\psi(t) \sim t^{\alpha-1}e^{-\lambda t}$. λ is introduced as a cutoff parameter to represent a stationary ensemble (SP) with a finite $\bar{t}$. Three-point correlation plots are also shown for the Gaussian model with the same two-point correlation function. The linear contours of the CTRW model suggest that correlations depend on the total time $t_1 + t_2$, and not on $t_1 - t_2$ alone. It reflects the dominant asymptotic contribution of particles which survive the entire history and stay near the starting point. Gaussian dynamics shows more curved contours which depend on both t_1 and t_2 in a more complex way.

We see substantial differences between the asymptotic behavior of the two models. We further recast these observations into a simple analytic argument. We shall consider Gaussian and CTRW processes with power law two point correlation $g_x^{(2)} \sim t^{-\alpha}$; for CTRW see Appendix B of Ref. 10 for the construction of the model. We shall compare the four-point correlation function $g_x^{(4)} = \langle x(3t)x(2t)x(t)x(0)\rangle$ at long times for both models. For the Gaussian model $g^{(4)} \sim [g_x^{(2)}(t)]^2 + [g_x^{(2)}(2t)]^2 + g_x^{(2)}(3t)g_x^{(2)}(t)$, i.e. for power law $g^{(2)}(t) \sim t^{-\alpha}$ it gives $g^{(4)} \sim t^{-2\alpha}$ while CTRW is asymptotically given by fraction of long-surviving particles and so $g^{(4)}(t) \sim g^{(2)}(t) \sim t^{-\alpha}$.

4. Summary

We have introduced factorial moments of photon statistics and related them to the multipoint correlation functions. We have also given the weak field perturbation theory of photon counting and drew the connection to nonlinear spectroscopies, recasting them in terms of the same multipoint correlation functions of dipole moments. We have shown two limits of single-molecule experiments, where the correlation function of observed quantities may be related to the correlation functions of classical stochastic coordinate. We have discussed three examples of nonequivalent solvable stochastic models which are particularly suitable for phenomenological modeling of single experiments.

Acknowledgments

The authors acknowledge the support of the National Science Foundation (Grant CHE-0745892) and NIRT (Grant EEC-0303389) and the National Institutes of Health (Grant GM 59230). F. Š. acknowledges the support of the Ministry of Education, Youth and Sports of the Czech Republic (project MSM 0021620835) and the Grant Agency of the Czech Republic (Grant. No. 202/07/P245).

References

1. W. E. Moerner and L. Kador, *Physical Review Letters* **62** (1989) 2535.
2. M. Orrit and J. Bernard, *Physical Review Letters* **65** (1990) 2716.
3. F. Kulzer and M. Orrit, *Annual Reviews Physical Chemistry* **55** (2004) 585.
4. I. S. Osad'ko, *Selective Spectroscopy of Single Molecules*, Springer Series in Chemical Physics, Vol. 69 (Springer, Berlin, 2002).
5. C. W. Gardiner and P. Zoller, *Quantum Noise: A Handbook of Markovian and Non-Markovian Quantum Stochastic Methods with Applications to Quantum Optics* (Springer, 2004).
6. S. C. Kou, X. S. Xie and J. S. Liu, *Applied Statistics* **54** (2005) 469.
7. V. Barsegov and S. Mukamel, *Journal of Physical Chemistry* **108** (2004) 15.
8. V. Barsegov and S. Mukamel, *Journal of Chemistry Physics* **117** (2002) 9465.
9. V. Barsegov, V. Chernyak and S. Mukamel, *Journal of Chemical Physics* **116** (2002) 4240.
10. F. Šanda and S. Mukamel, *Physical Review E* **72** (2005) 031108.
11. S. C. Kou and X. S. Xie, *Physical Review Letters* **93** (2004) 180603.
12. X. Brokmann, J.-P. Hermier, G. Messin, P. Desbiolles, J.-P. Bouchaud and M. Dahan, *Physical Review Letters* **90** (2003) 120601.
13. O. Flomenbom, K. Velonia, D. Loos, S. Masuo, M. Cotlet, Y. Engelborghs, J. Hofkens, A. E. Rowan, R. J. M. Nolte, F. C. de Schryver and J. Klafter, *Proceedings of the National Academy of Sciences* (USA) **102** (2005) 2368.
14. S. Mukamel, *Principles of Nonlinear Optical Spectroscopy* (Oxford University Press, New York, 1995).
15. M. Kuno, D. P. Fromm, H. F. Hamman, A. Gallagher and D. J. Nesbitt, *Journal of Chemical Physics* **112** (2000) 3117.
16. K. T. Shimizu, R. G. Neuhauser, C. A. Leatherdale, S. A. Empedocles, W. K. Woo and M. G. Bawendi, *Physical Review B* **63** (2001) 205316.
17. Y. Jung, E. Barkai and R. Silbey, *Chemical Physics* **284** (2002) 181.
18. G. Margolin, V. Protasenko, M. Kuno and E. Barkai, *Journal of Physical Chemistry B* **110** (2006) 19053.
19. Y. He and E. Barkai, *Physical Review Letters* **93** (2004) 068302.
20. F. Šanda and S. Mukamel, *Journal of Chemical Physics* **125** (2006) 014507.
21. G. Lindblad, *Communications in Mathematical Physics* **48** (1976) 199.
22. G. S. Agarwal, in *Springer Tracts in Modern Physics* 70, ed. G. Höhler (Springer, New York, 1974).
23. R. H. Dicke, *Physical Review* **93** (1954) 99.
24. J. von Neumann, *Mathematical Foundations of Quantum Theory* (Princeton University Press, 1955).
25. G. Bel and E. Barkai, *Physical Review Letters* **94** (2005) 240602.
26. G. Margolin and E. Barkai, *Physical Review Letters* **94** (2005) 080601.
27. D. E. Makarov and H. Metiu, *Journal of Chemical Physics* **115** (2001) 5989.

28. S. A. Rice and M. Zhao, *Optimal Control of Molecular Dynamics* (Willey, New York, 2000).
29. E. Barkai, Y. Jung and R. Silbey, *Annual Reviews of Physical Chemistry* **55** (2004) 457.
30. S. Mukamel, *Physical Review A* **68** (2003) 063821.
31. R. Zwanzig, *Nonequilibrium Statistical Mechanics* (Oxford University Press, New York, 2001).
32. G. C. Hegerfeldt, *Physical Review A* **47** (1993) 449.
33. R. J. Cook, *Physical Review A* **23** (1981) 1243.
34. D. Lenstra, *Physical Review A* **26** (1982) 3369.
35. R. J. Glauber, in *Quantum Optics and Electronics*, eds. C. DeWitt, A. Blandin and C. Cohen-Tannoudji (Gordon and Breach, New York, 1964).
36. Y. Zheng and F. L. H. Brown, *Journal of Chemical Physics* **119** (2003) 11814.
37. T. Basché, W. E. Moerner, M. Orrit and H. Talon, *Physical Review Letters* **69** (1992) 1516.
38. Y. Zheng and F. L. H. Brown, *Journal of Chemical Physics* **121** (2004) 7914.
39. F. Šanda and S. Mukamel, *Physical Review A* **71** (2005) 033807.
40. Y. Jung, E. Barkai and R. Silbey, *Advances in Chemical Physics* **123** (2002) 199.
41. E. Barkai, Y. Jung and R. Silbey, *Physical Review Letters* **87** (2001) 207403.
42. Y. Tanimura, *Journal of the Physical Society of Japan* **75** (2006) 082001.
43. U. Harbola, J. B. Maddox and S. Mukamel, *Physical Review B* **73** (2006) 075211.
44. H. Yang, G. Luo, P. Karnchanaphanurach, T.-M. Louie, I. Rech, S. Cova, L. Xun and X. S. Xie, *Science* **302** (2003) 262.
45. V. Barsegov and S. Mukamel, *Journal of Chemical Physics* **116** (2002) 9802.
46. T. Förster, in *Modern Quantum Chemistry, Part III: Action of Light and Organic Molecules*, ed. O. Sinanoglu (Academic, NewYork, 1965), p. 63.
47. W. Min, G. Luo, B. J. Cherayil, S. C. Kou and X. S. Xie, *Physical Review Letters* **94** (2005) 198302.
48. H. Risken, *The Fokker-Plank Equation* (Springer, Berlin, 1989).
49. V. Chernyak, F. Šanda and S. Mukamel, *Physical Review E* **73** (2006) 036119.
50. A. O. Caldeira and A. J. Leggett, *Physical A* **121** (1983) 587.
51. T. l. C. Jansen and S. Mukamel, *Journal of Chemical Physics* **119** (2003) 7979.
52. R. Zwanzig, *Lectures in Theoretical Physics (Boulder)* **3** (1960) 106.
53. N. Hashitsume, F. Shibata and M. Shingu, *Journal of Statistical Physics* **17** (1977) 155.
54. J. Klafter, M. F. Shlesinger and G. Zumofen, *Physics Today* **49** (1996) 33.
55. F. Šanda and S. Mukamel, *Physical Review E* **73** (2006) 011103.
56. F. Šanda and S. Mukamel, *J. Chem. Phys.* **127** (2007) 154107.
57. H. Qian and H. Wang, *Europhysical Letters* **76** (2006) 15.

58. R. Metzler and J. Klafter, *Physics Reports* **339** (2000) 1.
59. F. Barbi, M. Bologna and P. Grigolini, *Physical Review Letters* **95** (2005) 220601.
60. F. Šanda and S. Mukamel, *Physical Review Letters* **98** (2007) 080603.
61. V. Chernyak and S. Mukamel, *J. Chem. Phys.* **105** (1996) 4565.

CHAPTER 5

Thermodynamics and Kinetics from Single-Molecule Force Spectroscopy

Gerhard Hummer and Attila Szabo

Laboratory of Chemical Physics, National Institute of Diabetes and Digestive and Kidney Diseases, National Institutes of Health, Bethesda, MD 20892-0520, USA

1. Introduction

The mechanical manipulation of single molecules with atomic force microscopes (AFM), laser optical tweezers, magnetic tweezers, etc., provides unprecedented insights into their structure, dynamics, and interactions.[1–16] In such single-molecule force-spectroscopy experiments, a molecule or molecular assembly is subjected to mechanical forces exerted by the pulling apparatus, possibly via intervening molecular linkers (Fig. 1). Pulling causes mechanical stress in the molecular system that, eventually, induces a molecular transition such as the unfolding of a nucleic acid or protein, or the dissociation of a molecular complex. Soon after the first experimental demonstrations of these powerful techniques, analogous computer simulation formalisms have been developed[9,17–20] that yield atomistically detailed pictures of molecular rupture processes.

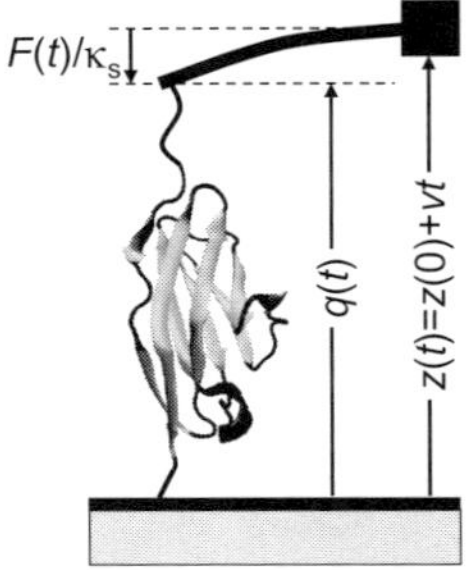

Fig. 1. Schematic of single-molecule force spectroscopy. The anchored sample, as indicated by the protein cartoon, moves at a speed v relative to the pulling apparatus. The vertical arrows indicate the externally *controlled* distance $z(t) = z(0) + vt$ between the attachment surface and the pulling-spring anchor, and the *fluctuating* molecular extension $q(t) = z(t) - F(t)/\kappa_S$, where $F(t)$ is the instantaneous force, and κ_S the spring constant of the pulling apparatus.

In the experiments, forces and molecular extensions are measured with pico-Newton and Ångstrom accuracies. With such resolutions, it may at first sight seem easy to extract useful quantitative information from the pulling experiments, such as equilibrium thermodynamic properties (binding constants, folding free energies, etc.) and kinetic properties (dissociation and unfolding rates, etc.). However, the quantitative analysis of the experiments is complicated by the fact that they are often carried out under nonequilibrium conditions. In typical pulling experiments, a time-varying external force actively perturbs the molecular system, resulting in nonequilibrium effects and hysteresis.

The objective of this chapter is to develop the necessary theoretical framework to extract reliable thermodynamic and kinetic information about microscopic molecular processes from pulling experiments. In the first part of the chapter (Sec. 2), we will describe the theory that permits rigorous thermodynamic measurements from repeated nonequilibrium pulling experiments. In the second part of the chapter (Sec. 3), we will show how kinetic information can be extracted from such experiments.

2. Thermodynamics from single-molecule pulling experiments

We have previously discussed how Jarzynski's identity[21,22] and its extension[23] required to get free energy surfaces generalize standard (equilibrium) free energy perturbation theory, and how they are related to the Feynman–Kac theorem for path integrals in quantum mechanics, the Kubo–Anderson theory of spectral line shapes, and kinetics with fluctuating rates.[24] Here, we adopt a different strategy and simply present the key results in a concise and self-contained way.

In the simplest representation of a pulling experiment, a molecular system is anchored at one end (e.g., at a surface or suction pipette) and attached to a harmonic spring at the other (such as an AFM cantilever or laser tweezer; Fig. 1). The spring and surface are then gradually moved apart, building up tension in the molecular system, and ultimately causing unfolding or dissociation. During the measurement, the distance $z(t)$ between the two anchoring points follows a prescribed protocol (for instance, by pulling at a constant velocity v such that $z(t) = vt$). In contrast, the molecular extension $q(t)$ fluctuates as a function of time t, with $q(t)$ determined by the distance between the two attachment points of the molecular construct.

To obtain the free energy along the controlled distance z, one can directly use Jarzynski's identity[21,22] between nonequilibrium work averages and free energies. In this way one can find free energy differences between thermodynamic states of the entire system, including the pulling spring, defined by different $z(t)$. However, the desired free energy is normally that of the molecular system alone, defined by different molecular extensions q. The free energy profile $G_0(q)$ along the molecular coordinate q alone can be determined using the procedure we have developed previously.[23,24]

2.1. *Theory*

To provide a framework for analyzing nonequilibrium pulling experiments, we define a Hamiltonian energy function that couples the

Hamiltonian H_0 of the molecular system to the pulling spring,

$$H(\mathbf{x}, t) = H_0(\mathbf{x}) + V(\mathbf{x}, t), \tag{1}$$

where $\mathbf{x}$ is the phase space coordinate of the system, and $V(\mathbf{x}, t)$ is the coupling between the spring and the molecule. Note that V depends parametrically on time because the spring moves according to the protocol $z(t)$. Typically, we can assume that the coupling is harmonic,

$$V(\mathbf{x}, t) = V[q(\mathbf{x}), t] = \frac{\kappa_S}{2}[q(\mathbf{x}) - z(t)]^2, \tag{2}$$

where $q = q(\mathbf{x})$ is the molecular extension. In its general form, the extended form of Jarzynski's identity[23,24] is:

$$\begin{aligned} \frac{e^{-\beta[H_0(\mathbf{x})+V(\mathbf{x},t)]}}{\int e^{-\beta[H_0(\mathbf{x}')+V(\mathbf{x}',0)]}d\mathbf{x}'} &= \left\langle \delta[\mathbf{x} - \mathbf{x}(t)]e^{-\beta\int_0^t \frac{\partial V[\mathbf{x}(\tau),\tau]}{\partial \tau}d\tau} \right\rangle \\ &= \left\langle \delta[\mathbf{x} - \mathbf{x}(t)]e^{-\beta W(t)} \right\rangle, \end{aligned} \tag{3}$$

where the left-hand side is the equilibrium Boltzmann distribution in phase space for the Hamiltonian $H(\mathbf{x}, t)$ corresponding to $z = z(t)$. $\beta = 1/(k_B T)$ is the inverse temperature T with k_B being Boltzmann's constant, $\delta(\mathbf{x})$ is Dirac's delta function, and the right-hand side is an average over nonequilibrium pulling trajectories. These trajectories must be initiated at time $t = 0$ from an *equilibrium* distribution corresponding to Hamiltonian $H(\mathbf{x}, 0)$ and evolve according to the time-dependent Hamiltonian $H(\mathbf{x}, t)$. Along each of these trajectories, the work performed between times 0 and t is accumulated,

$$W(t) = \int_0^t \frac{\partial V[\mathbf{x}(\tau), \tau]}{\partial \tau} d\tau. \tag{4}$$

The Boltzmann factor of the work is the weight factor assigned to that particular trajectory. By integrating Eq. (3) over $\mathbf{x}$, we obtain

Jarzynski's identity[21,22]:

$$e^{-\beta G(t)} = \langle e^{-\beta W(t)} \rangle, \tag{5}$$

where

$$e^{-\beta G(t)} = \frac{Q(t)}{Q(0)} = \frac{\int e^{-\beta H(\mathbf{x},t)} d\mathbf{x}}{\int e^{-\beta H(\mathbf{x}',0)} d\mathbf{x}'} \tag{6}$$

is the free energy difference for the entire system between times t and 0. $Q(t)$ is the canonical partition function for Hamiltonian $H(\mathbf{x}, t)$. Technically, $G(t)$ is a Helmholtz, not Gibbs free energy. The latter is usually denoted as F which we reserve here for forces.

We are interested in the free energy surface along the molecular extension q for the system in the absence of the pulling spring. This surface is defined by the Boltzmann average

$$e^{-\beta G_0(q)} = \langle \delta[q - q(\mathbf{x})] \rangle_0 \equiv \frac{\int \delta[q - q(\mathbf{x})] e^{-\beta H_0(\mathbf{x})} d\mathbf{x}}{\int e^{-\beta H_0(\mathbf{x})} d\mathbf{x}} \tag{7}$$

up to a constant factor that is usually not relevant. If we now multiply both sides of Eq. (3) by $e^{\beta V(q,t)} \delta[q - q(\mathbf{x})]$ and integrate with respect to $\mathbf{x}$, using the definition of $G_0(q)$, we obtain[23,24]

$$e^{-\beta G_0(q)} = \left\langle \delta[q - q(\mathbf{x}(t))] e^{-\beta \left(\int_0^t \frac{\partial V\{q[\mathbf{x}(\tau)],\tau\}}{\partial \tau} d\tau - V\{q[\mathbf{x}(t)],t\} \right)} \right\rangle. \tag{8}$$

$\langle \cdots \rangle$ is an average over nonequilibrium trajectories that start from a Boltzmann distribution corresponding to the Hamiltonian $H(\mathbf{x}, 0) = H_0(\mathbf{x}) + V[q(\mathbf{x}), 0]$ and evolve according to the time-dependent Hamiltonian $H(\mathbf{x}, t)$.

A more conventional definition of the work than that given in Eq. (4) involves the mechanical work performed on the pulling spring, $\int F dq$. To obtain such an expression, we integrate the identity $dV = (\partial V/\partial q) dq + (\partial V/\partial t) dt$ along the trajectory from $[t = 0, q = q(0)]$ to $[t = t, q = q(t)]$. This procedure gives a relation between the work

$W(t)$ of Eq. (4) and the mechanical work,

$$W(t) = \int_0^t \frac{\partial V[q(\tau), \tau]}{\partial \tau} d\tau = \int_{q(0)}^{q(t)} F\, dq + V[q(t), t] - V[q(0), 0], \tag{9}$$

where we used the fact that the restoring force is $F(q, t) = -\partial V/\partial q$. The integral over q is along the position-versus-time contour[24] connecting $q(0)$ and $q(t)$. We can thus write the free energy along q as[23]

$$e^{-\beta G_0(q)} = \left\langle \delta[q - q(t)] e^{-\beta(\int_{q(0)}^{q(t)} F\, dq - V[q(0),0])} \right\rangle, \tag{10}$$

which constitutes the central result of this section.

2.2. *Free energy surfaces from a quasi-harmonic approximation*

The implementation of the above rigorous formalism requires many pulling trajectories. Since it is not always practical to obtain such data sets, it is of interest to introduce approximate ways of obtaining free energy surfaces that require less extensive data. We now show how estimates of the molecular free energy surface $G_0(q)$ can be obtained from averages of the Jarzynski work $W(t)$. The partition function $Q(t)$ in Eq. (6) for a harmonic pulling apparatus with spring constant κ_S is,

$$\begin{aligned} Q(t) &= \int d\mathbf{x}\, e^{-\beta H_0(\mathbf{x}) - \beta\kappa_S[q(\mathbf{x}) - z(t)]^2/2}, \\ &= \int dq \int d\mathbf{x}\, \delta[q - q(\mathbf{x})] e^{-\beta H_0(\mathbf{x}) - \beta\kappa_S[q(\mathbf{x}) - z(t)]^2/2}, \\ &= \int dq\, e^{-\beta G_0(q) - \beta\kappa_S[q - z(t)]^2/2}, \end{aligned} \tag{11}$$

where we expressed the partition function in terms of the free energy surface $G_0(q)$, as defined up to a constant in Eq. (7). We now evaluate the q integral using the method of steepest descents. We assume that the integrand is continuous and strongly peaked around a value $q^* = q^*(t)$ which is the (single) maximum of the exponent,

$$G_0'(q^*) + \kappa_S[q^* - z(t)] = 0, \tag{12}$$

where the prime denotes the derivative with respect to q. By approximating the integrand as a Gaussian centered at $q^*(t)$, we obtain

$$e^{-\beta G(t)} = \frac{Q(t)}{Q(0)} \approx \frac{(2\pi)^{1/2} e^{-\beta G_0(q^*) - \beta\kappa_S[q^* - z(t)]^2/2}}{[\beta(\kappa_S + G_0''(q^*))]^{1/2} Q(0)}. \tag{13}$$

At the same level of approximation, we find that the derivative of the total free energy with respect to time (denoted by a dot) is:

$$\begin{aligned} \dot{G}(t) &= -\frac{1}{\beta}\frac{\partial}{\partial t} \ln Q(t), \\ &= \kappa_S \dot{z}(t)[z(t) - q^*]. \end{aligned} \tag{14}$$

Differentiating this equation and Eq. (12) with respect to time and assuming, for the sake of simplicity, that $\ddot{z} = 0$ we find that the second derivative of the free energy is

$$\ddot{G}(t) = \frac{\kappa_S \dot{z}^2 G_0''(q^*)}{\kappa_S + G_0''(q^*)}. \tag{15}$$

By using Eqs. (14) and (15) in Eq. (13), we obtain an approximate expression for the free energy surface,

$$G_0\left[q = z(t) - \frac{\dot{G}(t)}{\kappa_S \dot{z}}\right] \approx G(t) - \frac{\dot{G}^2(t)}{2\kappa_S \dot{z}^2} + \frac{1}{2\beta} \ln\left[1 - \frac{\ddot{G}(t)}{\kappa_S \dot{z}^2}\right], \tag{16}$$

that determines $G_0(q)$ up to a constant. Thus, the free energy profile G_0 at a position q shifted from $z(t)$ is determined by the total free energy $G(t)$ and its time derivatives.

This approximate relation reduces to the stiff-spring approximation[25] if one assumes in addition that κ_S is large, and is exactly equivalent to the one we introduced previously[24] based on moments. In the limit of a stiff spring, $\kappa_S \to \infty$, we can expand the terms in Eq. (16) in powers of $1/\kappa_S$. Truncated at first order, this expansion produces the so-called stiff-spring approximation of Park *et al.*,[25]

$$G_0[q = z(t)] \approx G(t) - \frac{1}{2\kappa_S \dot{z}^2}\left[\frac{\ddot{G}(t)}{\beta} - \dot{G}(t)^2\right] \tag{17}$$

(see also Ref. 26). Equation (16) can thus be viewed as a *resummed* version of the stiff-spring approximation with a potentially higher accuracy and larger range of validity, being useful even when the pulling spring is soft. One can consider Eq. (16) as a "quasi-harmonic approximation" because it is exact when all fluctuations are Gaussian.

Interestingly, Eq. (16) corresponds *exactly* to the cumulant approximation to the free energy we have obtained previously in a different way.[24] To show this, we first express the time derivatives of $G(t)$ in terms of work-weighted averages of the force. For a harmonic spring, the work is $W(t) = \int_0^t d\tau \kappa_S \dot{z}(z(\tau) - q[\mathbf{x}(\tau)])$. By differentiating Jarzynski's identity, Eq. (5), with respect to time we find that

$$\begin{aligned}\dot{G}(t) &= \kappa_S \dot{z}\left[z(t) - \frac{\langle q[\mathbf{x}(t)]e^{-\beta W(t)}\rangle}{\langle e^{-\beta W(t)}\rangle}\right] \\ &= \kappa_S \dot{z}[z(t) - \overline{q(t)}]\end{aligned} \tag{18}$$

or, expressed in terms of the measured forces,

$$\begin{aligned}\dot{G}(t) &= \frac{\dot{z}\langle F(t)\exp -\beta W(t)\rangle}{\langle \exp -\beta W(t)\rangle} \\ &= \dot{z}\overline{F(t)},\end{aligned} \tag{19}$$

where $\overline{q(t)}$ and $\overline{F(t)}$ are the work-reweighted averages of the molecular extension and force at time t, respectively. Similarly, the second derivative can be expressed in terms of the variance of the position or, equivalently, the force

$$\begin{aligned}\ddot{G}(t) &= \kappa_S \dot{z}^2 - \beta \dot{z}^2 \kappa_S [\overline{q^2(t)} - \overline{q(t)}^2] \\ &= \kappa_S \dot{z}^2 - \beta \dot{z}^2 [\overline{F^2(t)} - \overline{F(t)}^2],\end{aligned} \tag{20}$$

where we again assumed that $\ddot{z} = 0$. Using these identities, we can rewrite Eq. (16) in terms of reweighted force averages

$$G_0\left[q = z(t) - \frac{\overline{F(t)}}{\kappa_S}\right] \approx G(t) - \frac{\overline{F(t)}^2}{2\kappa_S} + \frac{1}{2\beta} \ln \frac{\beta[\overline{F^2(t)} - \overline{F(t)}^2]}{\kappa_S}. \tag{21}$$

The logarithmic term will typically be small. Equation (21) is the same as Eq. (37) of Ref. 24 evaluated at $q = \overline{q(t)}$. Similarly, Eq. (40) of Ref. 24, $G_0'[z(t) - \overline{F(t)}/\kappa_S] \approx \overline{F(t)}$ can be rewritten as $G_0'[\overline{q(t)}] = \kappa_S[z(t) - \overline{q(t)}]$ which is identical to Eq. (12) here. The above quasi-harmonic formalism is thus indeed exactly equivalent to the approximate formalism of Ref. 24.

2.3. *Force-extension integrals*

The force-extension integral in Eq. (10) can be evaluated as a Riemann sum,

$$\omega(t) = \int_{q(0)}^{q(t)} F\, dq \approx \sum_{i=1}^{N} \frac{(q_i - q_{i-1})(F_i + F_{i-1})}{2}, \tag{22}$$

where i labels positions consecutive in time, $q_i = q(t_i)$ and $F_i = F(q_i, t_i)$ are the corresponding forces, with $t_0 = 0$ and $t_N = t$. Note that in general, F is a multivalued function of q, such that the sum contains both positive and negative contributions.

2.4. *Analysis using weighted histograms*

Histogram methods are suitable if a sufficiently large number of trajectories have been collected. Formally (i.e., for an infinite number of trajectories), the entire free energy surface could be obtained from the data at any time t by using Eq. (10). However, because at time t the trajectories will likely be clustered in the vicinity of $z(t)$, it is best to combine data from different times t. This can be accomplished by adapting the weighted histogram approach of Ferrenberg and Swendsen.[27] For *equilibrium* umbrella sampling[28] corresponding to a harmonic biasing potential $V(q, t)$ held steady at $z = z(t) = \text{const.}$, their histogram reweighting formula is:

$$e^{-\beta G_0(q)} = \frac{\sum_t \langle \delta[q - q(t)] \rangle}{\sum_t \exp\{-\beta[V(q,t) - G(t)]\}}. \tag{23}$$

In this expression, $G(t)$ must be found self-consistently. In nonequilibrium sampling, in contrast, we can use Jarzynski's identity for $G(t)$. However, we have to unbias each observation with respect to the nonequilibrium work, $W(t)$. In Eq. (23), we correspondingly replace $\langle \delta[q - q(t)] \rangle \rightarrow \langle \delta[q - q(t)] \exp[-\beta W(t)] \rangle / \langle \exp[-\beta W(t)] \rangle$ in the numerator, and $\exp[\beta G(t)] \rightarrow 1/\langle \exp[-\beta W(t)] \rangle$ in the denominator. This procedure results in the following expression for the free energy profile[23]:

$$e^{-\beta G_0(q)} = \frac{\sum_t \frac{\langle \delta[q - q(t)] \exp[-\beta W(t)] \rangle}{\langle \exp[-\beta W(t)] \rangle}}{\sum_t \frac{\exp[-\beta V(q,t)]}{\langle \exp[-\beta W(t)] \rangle}}, \tag{24}$$

where the sums are over the histograms collected at different times t.

To illustrate the use of this formula, consider the following experiment. The molecular extensions $q_k(t_i)$ and accumulated work values $W_k(t_i)$ (from Eq. (4)) are recorded for trajectories $k = 1, \ldots, K$ at discrete times $t = t_i$ ($i = 0, \ldots, N$). At every time slice t_i, we calculate the average

$$\langle \exp[-\beta W(t_i)] \rangle \approx \eta_i \equiv K^{-1} \sum_{k=1}^{K} \exp[-\beta W_k(t_i)]. \tag{25}$$

To calculate $G_0(q_l)$ at discrete values $q_l = l\Delta q$, we collect histograms $h_i(l)$ at times t_i:

$$h_i(l) = K^{-1} \sum_{k=1}^{K} e^{-\beta W_k(t_i)} \theta_l[q_k(t_i)], \tag{26}$$

where $\theta_l(q)$ is one if $(l - 1/2)\Delta q \leq q < (l + 1/2)\Delta q$ and zero otherwise. We then estimate the free energy profile by averaging over all time slices t_i:

$$G_0(q_l) = -\beta^{-1} \ln \frac{\sum_{i=0}^{N} h_i(l)/\eta_i}{\sum_{i=0}^{N} \exp[-\beta V(q_l, t_i)]/\eta_i}. \tag{27}$$

Note that this procedure becomes problematic if the work distributions are very broad relative to $k_B T$ because the exponential estimators are biased and will be dominated by just a few trajectories.[24,29–33]

2.5. *Crooks relation*

Crooks[34,35] derived a powerful relation between the distributions of work values obtained from forward and backward transformations. Up to now, we considered only the forward case (i.e., start at equilibrium at $z = z(0)$ and run trajectories for a time τ until $z = z(\tau)$). The normalized distribution of the work accumulated along such trajectories is denoted by $p_f[w = W(\tau)]$. In the backward case, we start trajectories from an equilibrium distribution corresponding to $z = z(\tau)$ and again run them for time τ, but with the pulling apparatus moving according to a time-reversed protocol, $\underline{z}(t) = z(\tau - t)$, so that at the end of the process $\underline{z}(\tau) = z(0)$. The resulting distribution of the work is denoted by $p_b[w = -W(\tau)]$. The Crooks relation then is

$$\frac{p_f[w = W(\tau)]}{p_b[w = -W(\tau)]} = e^{\beta[w - G(\tau)]}. \tag{28}$$

This relation follows from detailed balance and microscopic time reversibility. Jarzynski's identity can be obtained from it by integrating over all w and using the fact that p_f and p_b are normalized. If one ignores fluctuations in q (see below), one can apply Eq. (28) to experiments to achieve substantially greater accuracy than by using Jarzynski's identity alone.[36] In particular, biases in the free energy estimators are reduced,[30,37,38] and broadening of work distributions from instrument noise is expected to largely cancel.

2.6. *Alternative approaches*

It should be pointed out that instead of the nonequilibrium methods described above, the free energy surface $G_0(q)$ can also be obtained by other methods. One such approach is umbrella sampling,[28] with the spring held steady at different anchor positions z_i. In another approach,[39] the propagator (Green's function) $p(q', t_1|q, t_0)$ is estimated through repeated "clamp-and-release" steps, in which the "bead" has to be captured, positioned, released, and recaptured. The free energy surface can then be estimated from the detailed-balance relation satisfied by the propagators.[39] Alternatively, the Chapman–Kolmogorov relation can be used to obtain $G_0(q)$, since $\exp(-\beta G_0(q))$ is invariant under convolution with $p(q', t_1|q, t_0)$.[40]

2.7. *Practical implementation*

Initial condition. In applications of the nonequilibrium free-energy reconstruction formalism, a number of requirements must be satisfied.[24] The trajectories should be initiated from the proper equilibrium ensemble. This procedure requires that before pulling starts, the position of the anchor $z(t)$ should be held steady long enough for the combined system of molecule, linker, and spring to relax to equilibrium. However, we showed[24] that in many cases, this may not be necessary. In particular, if the path integral in Eq. (10) can be factorized into an initial part (for $q(0) < q_0$) and a final part (for $q > q_0$), and if the two factors are uncorrelated, then the formalism simplifies substantially because (1) the correction factor $V[q(0), 0]$ accounting for the energy in the spring at time $t = 0$ drops out and (2) the initial equilibration can be avoided:

$$e^{-\beta G_0(q)} \approx \langle e^{-\beta \int_{q_0}^{q} F\, dq} \rangle. \tag{29}$$

This simplification has been made first in our free-energy analysis of nonequilibrium extraction of bacteriorhodopsin from a membrane[23] and then in later experimental studies.[15,36] We note here that the energy $V[q(0), 0]$ stored in the spring at time 0 can actually be large relative to $k_B T$ because the system equilibrates on the *combined* surface of spring and molecule.

Averages and trajectory alignment. To evaluate the average in Eq. (10), one needs to know the absolute molecular extension q. This can be achieved by aligning force extension curves, for instance by using fits to worm-like chain models.

Instrument noise. A potentially severe problem arises from artificially broadened work distributions. To give the individual trajectories their proper weight, the accumulated work should be accurate to within $\sim k_B T$. If the actual work distribution denoted by $p_f(w)$ is broadened by uncorrelated Gaussian noise, then the observed work distribution is a convolution of p_f and a Gaussian of zero mean and standard deviation σ_W. By deconvolution, it can be shown that the free energy estimated from Jarzynski's identity will be too low by $\beta\sigma_W{}^2/2$,

$$G_{obs}(t) = -k_B T \ln\langle \exp(-\beta W)\rangle = G(t) - \beta\sigma_W{}^2/2 \tag{30}$$

where $G(t)$ is the actual free energy difference. Thus as the noise in the work distribution increases beyond $k_B T$, the quality of the estimate rapidly deteriorates. One possible way to correct for instrument-noise induced broadening of the work distribution is to perform reference measurements without load (by pulling with and without an attached molecule) and then subtracting the free energy profiles.[24]

Measurement protocol $z(t)$. The path average in Eq. (10) corresponds to an *expectation value*. Therefore, it is not essential to follow identical protocols $z(t)$ in the different measurements,[41] as long as changes between the protocols are without bias.

Separation of linker contributions. An important challenge in the analysis is that the $G_0(q)$ will normally contain contributions from intervening molecular linkers as well as the system of interest. The contributions of the linkers can be subtracted, at least approximately, by measuring their free energy of extension, $G_L(q)$, separately, and assuming that they are "in series" with the molecular construct without any further interaction. The combined free energy surface is then obtained as a convolution of the molecular surface $G_{\text{mol}}(q)$ and the linker surface,

$$e^{-\beta G_0(q)} = \int e^{-\beta G_{\text{mol}}(q')} e^{-\beta G_L(q-q')} dq'. \tag{31}$$

To obtain $G_{\text{mol}}(q)$ alone, one needs to deconvolve this relation. Subtracting linker contributions can be accomplished approximately by using parametrized models,[36] or by performing separate measurements for the linkers alone.

Illustrative application. To illustrate the different free energy reconstruction procedures, we use a simple model[24] to analyze the RNA unfolding experiments of Liphardt *et al.*[15] In these experiments, a small piece of folded RNA was tethered to beads by means of DNA/RNA hybrid linkers. The RNA was then unfolded by pulling the beads apart using laser tweezers. In our model, the free energy surface $G_0(q)$ is given by the convolution of contributions $G_{\text{linker}}(q)$ and $G_{\text{mol}}(q)$ from the molecular linkers and RNA, respectively, according to Eq. (31). If RNA unfolding is instantaneous on the time scale of the pulling experiment and simply leads to an increase of Δq in the RNA end-to-end distance, then the integral in Eq. (31) can be written as a sum of two terms for the folded and unfolded states, respectively: $\exp[-\beta G_0(q)] = \exp[-\beta G_f(q)] + \exp[-\beta G_u(q)]$. The experimental force-extension curves immediately before and immediately after the RNA unfolding transitions are essentially linear, corresponding to effectively harmonic linkers. As a consequence, we approximated the free energies of the folded (G_f) and unfolded (G_u) states of the system as[24]

$$G_f(q) = \frac{\kappa}{2}q^2 \tag{32}$$

$$G_u(q) = \frac{\kappa}{2}(q - \Delta q)^2 + \Delta G_u \tag{33}$$

where $\kappa = 0.22$ pN/nm is the effective force constant of the linkers, $\Delta q = 15$ nm is the apparent increase in length due to RNA unfolding, and $\Delta G_u = 34\, k_B T$ is the unfolding free energy. These parameters were estimated from the slopes of the experimental force-extension curves, and the force and measured increase in length at unfolding.[15]

We performed Brownian dynamics simulations on the combined free energy surface $G_0(q) + \kappa_S(q - vt)^2/2$ at the experimental pulling speeds,[15] and with a friction coefficient chosen to reproduce the experimental dissipation. As shown before,[24] the resulting simulation trajectories closely resemble those of the experiments. However, for the model parameters, the resulting free energy surface $G_0(q)$ is *not* bistable. Instead, RNA unfolding only leads to a small change in slope at $q \approx 50$ nm. The reason for this is that the overall potential of mean force is dominated by the linker molecules. Moreover, our estimate for ΔG_u is only about half of that extracted originally by Liphardt *et al.*[15] from their experiments. The reason for this difference has been discussed elsewhere.[24]

In Fig. 2, we compare the results of different free energy reconstruction procedures for the highest experimental pulling speed of 53 pN/s. Shown are the estimated free energy surfaces obtained from histogram reweighting, Eq. (24); the quasi-harmonic approximation, Eq. (21); and the stiff-spring approximation,[25] Eq. (17). Results are shown for two different spring constants: a soft spring, $\kappa_S = 0.1$ pN/nm, as used in the laser tweezer experiments[15] and in our original analysis;[24] and a stiff-spring, $\kappa_S = 10$ pN/nm, typical of AFMs.

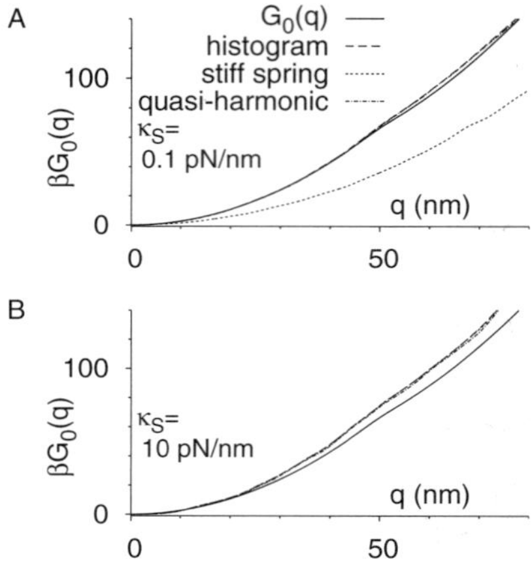

Fig. 2. Reconstruction of RNA plus RNA/DNA-linker free energy surface. Results from histogram reweighting (long-dashed line), the stiff-spring approximation (dotted line), and the quasi-harmonic approximation (short-dashed line) are compared to the exact surface $G_0(q)$ (solid line). In the reconstructions, 250 trajectories were used with a pulling speed of 53 pN/s and pulling spring constants of (A) $\kappa_S = 0.1$ pN/nm, and (B) 10 pN/nm. In (A), all reconstructed curves superimpose, except the one obtained from the stiff-spring approximation, which is substantially below the exact $G_0(q)$. In (B) all reconstructed curves are nearly identical and slightly above $G_0(q)$.

We find that for both pulling springs, histogram reweighting and the quasi-harmonic approximation produce free energy surfaces that agree well with the exact $G_0(q)$. The largest deviations occur for a stiff-spring for distances $q > 50$ nm beyond the RNA unfolding transition. The stiff-spring approximation, in contrast, works accurately only for a stiff spring, but fails for the soft spring typical of a laser tweezer setup. Moreover, with derivatives of $G(t)$ evaluated by finite differences, the stiff-spring approximation tends to produce substantially noisier curves.

Based on the results in Fig. 2, we conclude that the quasi-harmonic approximation, Eq. (21), is superior to the stiff-spring approximation because, on one hand, it converges to the latter for stiff springs, and on the other hand it is also applicable for relatively soft springs. Histogram re-weighting, Eq. (24), works well in all cases. The effect of the bias in the exponential work averages was found to be more pronounced with stiff springs. In summary, we recommend the quasi-harmonic approximation, Eq. (21), with and without the logarithmic term, if only relatively few trajectories are available. The time derivatives of the free energy should be calculated from the work-weighted moments of the force (see Eqs. (19) and (20)) rather than by numerical differentiation. If a large number of trajectories is available, histogram reweighting is to be preferred.

3. Kinetics from single-molecule pulling experiments

Exerting mechanical force on a molecular system can accelerate molecular transitions such as unfolding or dissociation. We will now develop a theoretical framework for the force-induced acceleration of molecular rupture required to extract kinetic information from constant-force or force-ramp pulling experiments.

Our central assumption is that the system can be described by two states (e.g., bound and dissociated, or folded and unfolded) that are connected by a quasi-irreversible transition (as, for example, when refolding transitions can be neglected for forces at which unfolding typically occurs). This assumption is not as restrictive as it may seem at first glance. If multiple states are populated and can be separated based on their molecular extensions, then one can simply use the formalism for each individual transition. The formalism is applicable both for the forward and backward transition (such as unfolding and folding), as long as in the experimental force regime these transitions are quasi-irreversible.

3.1. *Pulling with a constant force*

For a quasi-irreversible transition, we first consider the case of pulling with a constant force. Experimentally, this can be achieved through fast feedback loops[14,42] that quickly restore the target force upon fluctuations by moving the distance between the anchoring points of the molecule and the pulling spring. Alternatively, constant force can also be achieved by using nonlinear springs[43] that operate in a plateau regime where $dF/dx = 0$.

To describe the kinetics of rupture, we define the survival probability $S(t)$ as the probability that molecular rupture has not occurred at time t in a measurement that started at time $t = 0$. In general, the survival probability $S(t)$ will be nonexponential and dependent on how the force was applied. Such nonexponential rupture kinetics under constant force is indicative of other slow processes in the system, such as the crossing of multiple barriers before rupture occurs.

For forces that are not too high, rupture is in many systems dominated by a single rare event, so that the resulting distribution of life times is essentially exponential. In this case, the survival probability $S(t)$ satisfies a first-order rate equation:

$$\dot{S}(t) = -k(F)S(t) \tag{34}$$

where $S(0) = 1$ and $\dot{S} \equiv dS/dt$. $k(F)$ is the rate of molecular rupture in the presence of a constant mechanical force F. By integration, we obtain an exponentially decaying survival probability

$$S(t) = \exp[-k(F)t]. \tag{35}$$

The distribution of rupture times is then also exponential, $p(t) \equiv -\dot{S}(t) = k(F)\exp[-k(F)t]$.

Experimentally, $k(F)$ can be estimated by performing repeated life-time measurements under a constant force. Suppose that N such measurements were made. Rupture occurred at times τ_i in M of the measurements. The $N - M$ remaining measurements had to be stopped at times T_j before rupture occurred. Assuming uncorrelated measurements, we have the following maximum-likelihood estimate of the time to rupture under force[44]:

$$\tau(F) = \frac{1}{k(F)} = M^{-1}\left(\sum_{i=1}^{M}\tau_i + \sum_{j=1}^{N-M}T_j\right). \tag{36}$$

Thus one sums all times, but normalizes only with the number of actual rupture events. Alternatively, one can, for example, fit an exponential function to the survival probability but this requires more extensive data.

3.2. *Pulling with time-dependent forces: Force-ramp experiments*

In practice, it is often easier to perform experiments in which the force is ramped up so that it depends on time. This procedure avoids problems with force feedback loops and drift by reducing the overall measurement time. In the analysis of such force-ramp experiments, we further assume that the force-induced transitions are quasi-adiabatic. In the present context, "quasi-adiabatic" means that the probability for a molecular construct to rupture in a time interval $(t, t + dt)$ is only a function of the instantaneous force $F(t)$, and not of the preceding history of the measurement. This assumption of a memory-less (Markovian) dynamics is closely linked to the assumption of (nearly) exponential life times under constant force. If the force loading rate (i.e., the rate at which the force increases in a force-ramp experiment) is not too large, the barrier to rupture will remain high until most rupture events have occurred. Then, other degrees of freedom of the molecular system will effectively remain in an equilibrium that parametrically depends on the instantaneous molecular extension.

Under the quasi-adiabatic assumption, Eq. (34) for the survival probability becomes:

$$\dot{S}(t) = -k[F(t)]S(t), \tag{37}$$

where now the rate $k[F(t)]$ depends on the force $F(t)$ acting on the molecular system at time t. Note that $k(F)$ in Eqs. (34) and (37) is the same function of force.

We obtain $S(t)$ by integration, with Eq. (35) now becoming

$$S(t) = \exp\left[-\int_0^t k[F(t')]dt'\right]. \tag{38}$$

For a time-dependent and monotonically increasing force $F(t)$, the distribution of forces at rupture obtained from $p(F)dF = -\dot{S}\,dt$ is

$$p(F) = \frac{k(F)}{\dot{F}} \exp\left(-\int_0^F \frac{k(F')}{\dot{F}} dF'\right), \tag{39}$$

where $\dot{F}$ is a function of force, $\dot{F}(F) = \dot{F}[t(F)]$ with $t(F)$ being the time at which force F is reached (i.e., the inverse of $F(t)$, such that $F(t(F)) = F$). In an experiment in which the force increases linearly with time, $F(t) = \kappa v t$ and $t(F) = F/\kappa v$, $\dot{F}$ is the constant force-loading rate κv. Below, we will give an explicit formula for $\dot{F}$ in the presence of a worm-like chain linker.

In principle, the assumption of an explicitly time-dependent force $F(t)$ is approximate because the force is itself fluctuating as the system evolves on the combined potentials of molecule, linker, and spring.[45] However, in many practical cases a theory in which only the molecular coordinates fluctuate, but the external force is a prescribed function of time, is found to be adequate.

3.3. *Relating constant-force and force-ramp experiments*

We are now in a position to derive a relation that connects experiments performed under constant and time-dependent forces. For a force $F(t)$ that increases monotonically with time, the survival probability $S(t)$ is directly related to the cumulative distribution of rupture forces,

$$\int_0^F p(F')dF' = 1 - S[t(F)]. \tag{40}$$

As a consequence, the distribution (probability density) of rupture forces can also be written as

$$p(F) = -\frac{\dot{S}[t(F)]}{\dot{F}[t(F)]}. \tag{41}$$

Equations (37)–(41) can now be combined into an expression for the force-dependent rate of rupture in terms of $\dot{F}$ and the rupture-force distributions,

$$k(F) = \frac{\dot{F} p(F)}{1 - \int_0^F p(F')dF'}. \tag{42}$$

This expression[46] relates the rate at constant force on the left-hand side to the distribution of rupture forces on the right-hand side. Under the same assumptions, the product $v \ln S[t(F)]$ as a function of F is independent of v for $F(t) = \kappa v t$.[47]

Dudko *et al.*[48] recently illustrated the utility of Eq. (42) for unzipping of DNA hairpins[49] in a nanopore. In the experiments, a DNA hairpin with a single-stranded overhang is threaded into a membrane-bound α-hemolysin channel, thus blocking the ionic current through its pore. The voltage across the channel is ramped up at a constant rate. When the voltage reaches a critical value, the negatively charged DNA is pulled through the pore, an event marked by an increase in the measured ionic current. This critical voltage is analogous to the rupture force, while the voltage ramp rate is analogous to the pulling speed in mechanical force-spectroscopy experiments.

Figure 3 shows the collapse of experimental "rupture-force" distributions collected at "pulling speeds" covering more than two orders of magnitude. The histograms not only collapsed onto a single master curve, but also on the independently measured[49] life times at constant force (or voltage).[48] Thus for this system, the two different experiments can be related in an essentially model-free way.

3.4. *Effects of anharmonic molecular linkers*

We will now show that even for an anharmonic worm-like-chain linker, the resulting distribution of rupture forces $p(F)$ can be obtained by straightforward quadrature, without the need of simulations. This can be accomplished by using Eq. (39) for the rupture force distribution if we can express $\dot{F}$ as an explicit function of the force.

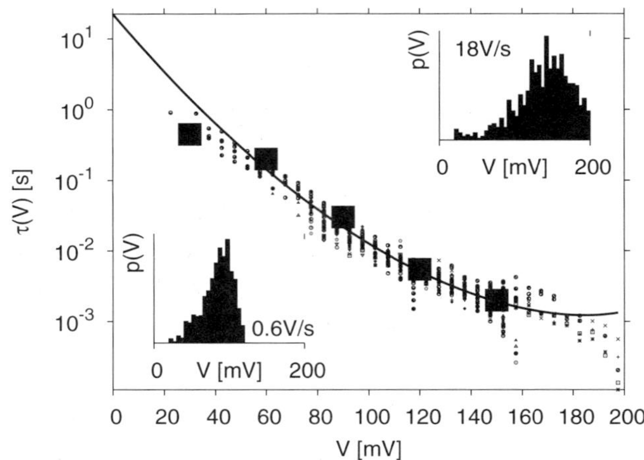

Fig. 3. Collapse of experimental force-ramp data for nucleic-acid unzipping in a nanopore.[48,49] In the experiments,[49] nucleic acid hairpins with a single stranded overhang are unfolded by threading them through a narrow nanopore under the influence of an electric potential. Voltage V thus corresponds directly to force F in the theory. Rupture-voltage histograms (small symbols) are collapsed using Eq. (42) and compared to independently measured life times of the hairpin under constant voltage (large squares).[48] The solid line is the theoretical life time, $\tau(F) = 1/k(F)$, obtained from a maximum-likelihood fit of rupture histograms for ramp speeds up to 12 V/s with the harmonic-cusp model, $v = 1/2$ in Eq. (67). The two insets show individual rupture-voltage histograms at low and high voltage-ramp speeds.

For simplicity, we assume that the pulling apparatus consists of a harmonic spring with spring constant κ_S. The anchoring point of the spring is moved at a constant velocity v. The molecular construct is tethered to the spring by a linker that has a force-extension curve $F_L(x)$. It can be shown[50] that the derivative of the applied force with respect to time can be written as

$$\dot{F} = v\left(\frac{1}{\kappa_S} + \frac{1}{\kappa_L(F)}\right)^{-1} \equiv v\,\kappa_{\text{eff}}(F). \tag{43}$$

The term in parentheses is the effective spring constant of the pulling spring in series with the molecular linker. The effective spring constant of the (anharmonic) linker at a force F is given by $\kappa_L(F) = F_L'[x(F)]$, where the linker extension $x(F)$ as a function of force can be obtained by solving $F = F_L(x)$ for x.

For a harmonic linker, $F_L(x) = \kappa_L x$, $\dot{F}$ simply becomes

$$\dot{F} = v({\kappa_S}^{-1} + {\kappa_L}^{-1})^{-1}. \tag{44}$$

For a worm-like chain with persistence length l_p and contour length L, the force-extension curve is approximately given by[51]

$$F_{WLC}(x) = \frac{1}{\beta l_p}\left[\frac{x}{L} - \frac{1}{4} + \frac{1}{4}\left(1 - \frac{x}{L}\right)^{-2}\right]. \tag{45}$$

For this system, $\kappa_{WLC}(F)$ can be expressed in terms of a root of a cubic equation. Alternatively, we can use an approximation[50]

$$\kappa_{WLC}(F) = \frac{3 + 5\beta l_p F + 8(\beta l_p F)^{5/2}}{2\beta l_p L(1 + \beta l_p F)} \tag{46}$$

that is accurate to within <3.5%. This relation, when combined with Eq. (43), results in an explicit expression of $\dot{F}$ in terms of the force F that can then be numerically integrated according to Eq. (39) to get the rupture-force distribution without the need of kinetic simulations. Of course, this formalism requires the specification of $k(F)$, for instance by using Bell's formula, Eq. (49), or more realistic expressions, Eq. (67).

It should be pointed out that to a good approximation one can use the formulas for *constant* force loading rates even in the presence of anharmonic linkers.[45] This requires using an apparent spring constant that depends on the pulling speed (i.e., $\dot{F} = v\kappa(v)$). For each pulling speed, $\kappa(v)$ can be estimated as the slope of the force extension curves near the point where most rupture events occur. In many practical situations, rupture occurs over a relatively narrow window in force so that the distributions of slopes will be strongly peaked.

We are now in a position to recast Eq. (42) connecting constant-force and force-ramp experiments in a way that accounts for the presence of anharmonic molecular linkers. By combining Eqs. (42) and (43), we find that the rate at constant force, $k(F)$, can be expressed in terms of the rupture-force distribution $p(F)$ and the corresponding cumulative distribution,

$$\frac{vp(F)}{1-\int_0^F p(F')dF'} = k(F)\left[\frac{1}{\kappa_S} + \frac{1}{\kappa_L(F)}\right]. \tag{47}$$

With this relation, rupture-force histograms collected at different pulling speeds v can be collapsed onto a single master curve. The resulting master curve is the product of the rate at constant force, $k(F)$, and the term in square brackets corresponding to the reciprocal of the effective force-dependent spring constant of the molecular linker and pulling spring in series. If the molecular linker is harmonic, $F_L(x) = \kappa_L x$, then that second term is simply $\kappa_S{}^{-1} + \kappa_L{}^{-1}$ (see Eq. (44)). If the molecular linker behaves like a worm-like chain, then Eq. (47) becomes

$$\frac{vp(F)}{1-\int_0^F p(F')dF'} = k(F)\left[\frac{1}{\kappa_S} + \frac{2\beta l_p L(1+\beta l_p F)}{3+5\beta l_p F + 8(\beta l_p F)^{5/2}}\right] \tag{48}$$

where we used Eq. (46), $\kappa_L(F) = \kappa_{WLC}(F)$.

To use these results in practice, one first collects rupture force histograms at different pulling speeds v, and then plots them according to the left-hand side of Eq. (47). In the resulting plot, the data should overlap. (Otherwise, the assumption of quasi-adiabatic rupture kinetics may be violated.) This plot is a graph of $k(F)$ times a force-dependent function accounting for linker effects. This formalism can be used to combine (and cross validate) data collected at constant force and in force-ramp experiments. In particular, after correction for the linker terms, $k(F)$ values estimated from life-time measurements at constant force should superimpose on the $k(F)$ curves from force-ramp experiments.

3.5. *Models for the rate of molecular rupture under force*

So far, we have not specified a functional form for the force-dependent rate $k(F)$. We will now introduce the widely used phenomenological expression for $k(F)$ of Bell.[52] We will then consider a simple microscopic description of force-induced molecular rupture based on the assumption of diffusive crossing of a barrier in one dimension. We first derive a general relation between $k(F)$ and the location of the transition state. Then, we will consider simple free energy surfaces for which one can obtain analytical expressions for the rupture-force distributions. We will show that the resulting expressions (1) encompass the phenomenological theory, with Bell's expression recovered as a special case, (2) aid in the extrapolation of $k(F)$ to zero force, and provide estimates (3) of the transition state location as well as (4) the height of the free energy barrier to molecular rupture.

3.5.1. *Bell–Evans model of molecular rupture under time-dependent force*

In the most widely used model,[52–55] the rate of rupture $k(F)$ as a function of force is given by the phenomenological Bell relation

$$k(F) = k_0 \exp(\beta F x^{\ddagger}) \tag{49}$$

where k_0 is the rate of rupture without force. In a microscopic interpretation, the parameter $x^{\ddagger}$ is identified with the distance between the stable state and the transition state in the direction of the pulling coordinate. If the force loading rate is constant, $\dot{F} = \kappa v$, by using Eq. (49) in Eq. (39) one finds[56]

$$p(F) = \frac{k_0}{\kappa v} \exp\left[\beta F x^{\ddagger} - \frac{k_0}{\beta \kappa v x^{\ddagger}}(e^{\beta F x^{\ddagger}} - 1)\right] \tag{50}$$

for the distribution of rupture forces. The corresponding mean force at rupture is

$$\langle F \rangle = \frac{1}{\beta x^{\ddagger}} \exp\left(\frac{k_0}{\beta \kappa v x^{\ddagger}}\right) E_1\left(\frac{k_0}{\beta \kappa v x^{\ddagger}}\right) \tag{51}$$

where $E_1(x) = \int_x^\infty e^{-t}t^{-1}dt$ is the exponential integral.[57] In the regime of high velocities relevant for most experiments, the average force at rupture is given by

$$\langle F \rangle \approx \frac{\ln(\kappa v x^\ddagger e^{-\gamma} {k_0}^{-1})}{\beta x^\ddagger} \tag{52}$$

where $\gamma = 0.5772\ldots$ is the Euler–Mascheroni constant. The mode of the rupture force distribution, corresponding to the most probable force with $dp(F)/dF = 0$, is given by this relation with γ set to zero.[45,56] In the same regime, the variance of the rupture force distribution is essentially independent of the pulling speed,

$$\sigma_F{}^2 = \langle F^2 \rangle - \langle F \rangle^2 \approx \frac{\pi^2}{6\beta^2 {x^\ddagger}^2}. \tag{53}$$

In summary, the phenomenological model predicts that the mean and the mode of the rupture force distribution are linear functions of the logarithm of the force loading rate κv, and that the variance is essentially independent of the ramp speed.

3.5.2. *General expression for the force-dependent rate of molecular rupture within Kramers theory*

At least qualitatively, the Bell–Evans model often captures the observed dependence of rupture forces on the force-ramp speed. However, if probed quantitatively, one frequently finds that the most probable rupture force is not perfectly linear in $\ln v$, or that the variance of rupture forces increases with $\ln v$ instead of remaining constant. Although such behavior has at times been interpreted as a change in mechanism (for instance, switching from one dominant transition state to another), simulations showed that simple microscopic models with a single free energy minimum could explain, for instance, curvature in $\langle F \rangle$ versus $\ln v$ plots.[45,58] Moreover, simulations of simple models also showed that even in cases where the Bell–Evans formalism fits the data well, the fitted parameters can be off substantially from the actual ones (e.g., by a factor 100 in the rate).[45]

In the following, we consider a microscopic theory of $k(F)$ that is based on Kramers theory of activated barrier crossing in the presence of a force. We will assume diffusive dynamics on a one-dimensional free energy surface $G(x) = G_0(x) - Fx$ with an effective diffusion coefficient D, where $G_0(x)$ is the profile in the absence of force. If a sufficiently high free energy barrier along the free-energy profile $G(x)$ separates the unruptured states from the ruptured states, we can use the Kramers high-barrier approximation for the rate of escape,

$$k(F) = \frac{D}{(\int_{\ddagger} e^{\beta G(x)} dx)(\int_{\text{well}} e^{-\beta G(x)} dx)}, \tag{54}$$

where the first integral is over the barrier region, and the second integral is over the free-energy minimum.

Before considering specific forms for $G_0(x)$, we derive some general relations between the force-dependent rate and microscopic properties.[50] By taking the logarithm of Eq. (54), the rate can be written as

$$\ln \frac{k(F)}{D} = -\ln \int_{\ddagger} e^{\beta[G_0(x)-Fx]} dx - \ln \int_{\text{well}} e^{-\beta[G_0(x)-Fx]} dx. \tag{55}$$

Differentiation with respect to F gives

$$\begin{aligned} \frac{\partial}{\partial F} \ln k(F) &= \beta \left[\frac{\int_{\ddagger} x e^{\beta[G_0(x)-Fx]} dx}{\int_{\ddagger} e^{\beta[G_0(x)-Fx]} dx} - \frac{\int_{\text{well}} x e^{-\beta[G_0(x)-Fx]} dx}{\int_{\text{well}} e^{-\beta[G_0(x)-Fx]} dx} \right] \\ &= \beta[\langle x \rangle_{\ddagger} - \langle x \rangle_{\text{well}}] \\ &\equiv \beta \left\langle x^{\ddagger}(F) \right\rangle \end{aligned} \tag{56}$$

where $\langle x \rangle_{\ddagger}$ and $\langle x \rangle_{\text{well}}$ are the force-dependent average values of x in the well and at the barrier, evaluated with Boltzmann and inverse-Boltzmann weights, respectively.

Equation (56) relates the force-dependent rate of rupture, $k(F)$, as obtained from constant-force or constant-velocity experiments, to the force-dependent average positions of the transition state and the free energy minimum:

$$\frac{\partial}{\partial F} \ln k(F) = \beta \langle x^{\ddagger}(F) \rangle. \tag{57}$$

By fitting the slope of the collapse-plot $k(F)$, one can thus estimate how the location of the transition state changes with force. Conversely, by integrating Eq. (57), we obtain an expression for the force-dependent rate as a function of the transition state location,

$$k(F) = k_0 \exp\left(\beta \int_0^F \langle x^{\ddagger}(F') \rangle dF'\right). \tag{58}$$

This expression is valid for forces that are sufficiently small so that the barrier is always high. Equation (58) is a generalization of Bell's formula, Eq. (49), in the framework of Kramers theory. It reduces to Bell's formula, Eq. (49), in the special case that the average distance between the well and transition state is independent of the applied force, with the trivial distinction that $x^{\ddagger}$ is the difference between averages rather than extrema. Clearly, a force-independent transition state location is unphysical: for any well of finite depth, the minimum will become unstable above a certain critical force F_c where $x^{\ddagger}(F_c) = 0$, as illustrated in Fig. 4. Thus, $x^{\ddagger}(F)$ must decrease as the force increases, and consequently $\ln k(F)$ must be a nonlinear function of F. A corollary is that Bell's formula can be accurate only at small forces, and that $k_{\mathrm{Bell}}(F) \geq k(F)$. To remove the assumption of a force-independent transition state location embodied in Bell's model, we now consider simple free energy surfaces $G_0(x)$.

3.5.3. *Microscopic models of force-induced molecular rupture rate $k(F)$*

Simple, analytically tractable microscopic models of molecular rupture in the presence of force can be constructed on the basis of

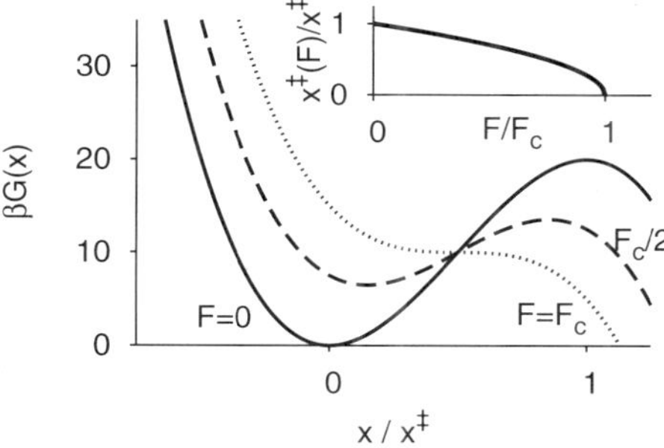

Fig. 4. Free energy surface as a function of force ($F = 0$, $F_c/2$, and F_c). The inset shows the force-dependent distance from the minimum to the transition state, $x^\ddagger(F)/x^\ddagger(F=0)$, up to the critical force F_c at which the barrier disappears.

Kramers theory for the rate of escape from a free energy minimum in one dimension. For a smooth free energy surface $G(x)$, near the minimum we have $G(x) \approx G(x_m) + G''(x_m)(x - x_m)^2/2$, and near the maximum $G(x) \approx G(x_M) - |G''(x_M)|(x - x_M)^2/2$. Substituting these into Eq. (54) and letting the upper and lower limits of both integrals go to $\pm\infty$, one obtains

$$k(F) = \frac{\beta D[G''(x_m)|G''(x_M)|]^{1/2}}{2\pi} \exp[-\beta\Delta G^\ddagger(F)] \tag{59}$$

where $\Delta G^\ddagger(F) = G(x_M) - G(x_m)$ is the force-dependent activation free energy. For a cusp-like barrier at x_M,[45] by expanding the free energy surface to linear order, $G(x) = G(x_M) + (x - x_M)G'(x_M)$, one can show that

$$k(F) = \frac{\beta^{3/2} D[G''(x_m)]^{1/2} G'(x_M)}{(2\pi)^{1/2}} \exp[-\beta\Delta G^\ddagger(F)]. \tag{60}$$

For a given $k(F)$ and force-ramp protocol given by $\dot{F}$, the distribution $p(F)$ of rupture forces can now be calculated from Eq. (39).

If the force is a linear function of time, $F(t) = \kappa v t$, accurate analytical expressions for $p(F)$ have been obtained for (1) the Bell–Evans phenomenological model described above, (2) a harmonic-cusp free energy surface,[45,46] and (3) a linear-cubic free energy surface.[46,58]

In the following, we will describe results for these models in more detail.

Harmonic-cusp model. In an effort to improve on the phenomenological model, we adopted a microscopic model for which the rupture-force distribution could still be found analytically. The harmonic-cusp model[45,46] uses what is arguably the simplest single-well potential, a quadratic free-energy surface with a sharp, cusp-like barrier,

$$\beta G_0(x) = \begin{cases} \Delta G^{\ddagger}\left(x/x^{\ddagger}\right)^2 & (x < x^{\ddagger}) \\ -\infty & (x \geq x^{\ddagger}) \end{cases}. \tag{61}$$

This potential has a minimum at $x = 0$, a cusp-like transition state at $x = x^{\ddagger}$, and an activation free energy of $\Delta G^{\ddagger}$. In the presence of an external force, the combined potential is $G(x) = G_0(x) - Fx$. The rate of rupture (or escape) is then obtained by using Kramers theory (Eq. (60)). In this way, $k(F)$ can be expressed in terms of k_0, $x^{\ddagger}$, and $\Delta G^{\ddagger}$. The harmonic-cusp model has one more parameter than Bell's formula (i.e., $\Delta G^{\ddagger}$) and reduces to it in the limit $\Delta G^{\ddagger} \to \infty$. This model of force-induced rupture happens to be precisely the continuum limit of the one used by Zwanzig *et al.*[59] in their analysis of Levinthal's paradox in protein folding. The external force corresponds to the energetic bias of the folded state which reduces the entropic barrier to folding.

We note that within the Kramers high-barrier approximation, an *identical* $k(F)$ curve is obtained for a cusp-like minimum with a quadratic barrier,

$$\beta G_0(x) = \begin{cases} \infty & (x < 0) \\ \Delta G^{\ddagger}[1 - (1 - x/x^{\ddagger})^2] & (x > 0) \end{cases}. \tag{62}$$

This equivalence shows that $k(F)$ is not particularly sensitive to all details of the underlying free energy surface.

Linear-cubic model. In an alternative approach, Dudko *et al.*[58] treated force-induced rupture starting with Garg's theory.[60] Garg argued that at the critical force, F_c, where the barrier vanishes, the combined free energy surface to lowest order is $G(x) = -\alpha x^3/3$. Then at forces somewhat less than the critical force, the free energy surface can be approximated by

$$\beta G(x) = -\alpha x^3/3 - a(F - F_c)x. \tag{63}$$

This potential has a minimum at $-[a(F_c - F)/\alpha]^{1/2}$ and a maximum at $[a(F_c - F)/\alpha]^{1/2}$ and the force-dependent activation free energy is

$$\beta \Delta G^{\ddagger}(F) = \frac{4a^{3/2}}{3\alpha^{1/2}}(F_c - F)^{3/2}. \tag{64}$$

By introducing a characteristic activation free energy $\Delta G_c = \Delta G^{\ddagger}(F = 0)$ the parameter α can be expressed in terms of ΔG_c and F_c. By applying Kramers theory (Eq. (59)) for diffusive crossing of this barrier, $k(F)$ can be expressed in terms of F_c and ΔG_c, and the constant diffusion coefficient D.

At first sight, it seems that this theory may have a limited range of applicability because Kramers theory is valid for forces at which the barrier is high, whereas the above model is constructed for forces at which the barrier nearly vanishes. Moreover, it is not clear how this theory is related or reduces to the phenomenological approach that involves k_0 and $x^{\ddagger}$. However, this theory can be reformulated from several different points of view.[46] First, consider the model potential

$$G_0(x) = \frac{3}{2}\Delta G^{\ddagger}\frac{x}{x^{\ddagger}} - 2\Delta G^{\ddagger}\left(\frac{x}{x^{\ddagger}}\right)^3 \tag{65}$$

which has a single well, minimum-to-maximum distance $x^{\ddagger}$, and activation free energy $\Delta G^{\ddagger}$. Then in the presence of an external force, the combined surface becomes $G(x) = G_0(x) - Fx$. If one applies Kramers theory to this potential one finds that $k(F)$ has the same functional form as found by Dudko *et al.*,[58] but now it can be expressed in terms of the "microscopic" parameters k_0, $x^{\ddagger}$, and $\Delta G^{\ddagger}$.

There is yet another point of view from which this "high-force" theory appears even more reasonable.[46] Any smooth potential will have an inflection point x^* (where $G''(x^*) = 0$) somewhere in between the single minimum and the ruptured state. As the external force F increases, the minima and maxima converge to this point. It seems natural to use a combined potential that is obtained from an intrinsic potential approximated by a Taylor expansion about this point x^*,

$$G(x) = G_0(x^*) + (x - x^*)G_0'(x^*) - (x - x^*)^3 G_0'''(x^*)/6 - Fx. \tag{66}$$

Now if one applies Kramers theory to this surface, one can write $k(F)$ in the same functional form as that for the above linear-cubic model potential, Eq. (65), by introducing *apparent* $\Delta G_c^\ddagger$ and $x_c^\ddagger$ defined as $x_c^\ddagger = 2[2G_0'(x^*)]^{1/2}/[-G_0'''(x^*)]^{1/2}$ and $\Delta G_c^\ddagger = (2/3)x_c^\ddagger G_0'(x^*)$.

3.5.4. *Unified theory of molecular rupture*

In collaboration with Olga Dudko, we showed that the final results of the three approaches (Bell–Evans, harmonic-cusp, and linear-cubic) can be unified.[46] Bell's formula Eq. (49) and the rates of rupture obtained using Eqs. (60) and (61), and using Eqs. (59) and (65), can be written as

$$k(F) = k_0\left(1 - \frac{\mu F x^\ddagger}{\Delta G^\ddagger}\right)^{1/\mu - 1} e^{\beta\Delta G^\ddagger[1-(1-\mu F x^\ddagger/\Delta G^\ddagger)^{1/\mu}]} \tag{67}$$

at constant force F, where $\mu = 1$ corresponds to the Bell formula Eq. (49), $\mu = 1/2$ to the harmonic-cusp model, and $\mu = 2/3$ to the linear-cubic model or, equivalently, the "high-force" theory. This relation is valid for forces that are small relative to the critical force $F_c = \Delta G^\ddagger/(\mu x^\ddagger)$.

When $F(t) = \kappa v t$, the distribution of rupture forces is obtained by using Eq. (67) in Eq. (39). By evaluating the integral analytically one finds

$$p(F) = (\kappa v)^{-1} k(F) e^{k_0/\beta x^\ddagger \kappa v} e^{-[k(F)/\beta x^\ddagger \kappa v][1-(\mu F x^\ddagger/\Delta G^\ddagger)]^{1-1/\mu}}. \tag{68}$$

The average of the rupture force is approximately

$$\langle F\rangle \approx \frac{\Delta G^{\ddagger}}{\mu x^{\ddagger}}\left\{1-\left[\frac{1}{\beta\Delta G^{\ddagger}}\ln\frac{k_0 e^{\beta\Delta G^{\ddagger}+\gamma}}{\beta x^{\ddagger}\kappa v}\right]^{\mu}\right\} \tag{69}$$

where $\gamma \approx 0.5772$ is the Euler–Mascheroni constant. The variance of the rupture-force distribution is

$$\sigma_{\mathrm{F}}^2 \approx \frac{\pi^2}{6\beta^2 x^{\ddagger 2}}\left[\frac{1}{\beta\Delta G^{\ddagger}}\ln\frac{k_0 e^{\beta\Delta G^{\ddagger}+\tilde{\gamma}}}{\beta x^{\ddagger}\kappa v}\right]^{2\mu-2} \tag{70}$$

where $\tilde{\gamma} = \gamma^2 - 3/\pi^2\psi''(1) \approx 1.064$. These expressions were obtained from Garg's asymptotic expansions[60] for $\mu = 2/3$ by assuming that the first two terms form a geometric series, summing this series, and then analytically continuing the result to all μ. When γ is set to zero, Eq. (69) closely approximates the maximum (mode) of the rupture force distribution. The average rupture force is a nonlinear function of the logarithm of the loading rate, $\ln \kappa v$, for all values of μ, except $\mu = 1$ corresponding to the phenomenological model. Even though Eq. (69) contains a term of the form $(\ln \kappa v)^{\mu}$, the average force does not actually scale as $\langle F\rangle \sim (\ln v)^{\mu}$ (see Fig. 5) for $\mu \neq 1$, although this is sometimes used as short-hand. Instead, the average force is almost linear in $\ln \kappa v$, with a slight upward curvature.

We note that when rupture occurs at forces close to the critical force F_{c}, Kramers high-barrier theory is not valid because the barrier is too low. In this case, one should use rate expressions determined from mean first passage times instead of the above analytical formulas.[48] However, then it is no longer possible to obtain the rupture force distribution analytically.

As discussed above, if the molecular system and pulling spring are connected by anharmonic linker molecules, one can still use these analytical expressions with an effective spring constant that depends on the pulling speed. This constant can be determined from the slope of the experimental force versus extension curve at rupture.[45] Alternatively, one can fit the force-extension curves to a worm-like-chain model and then use the corresponding force-dependent

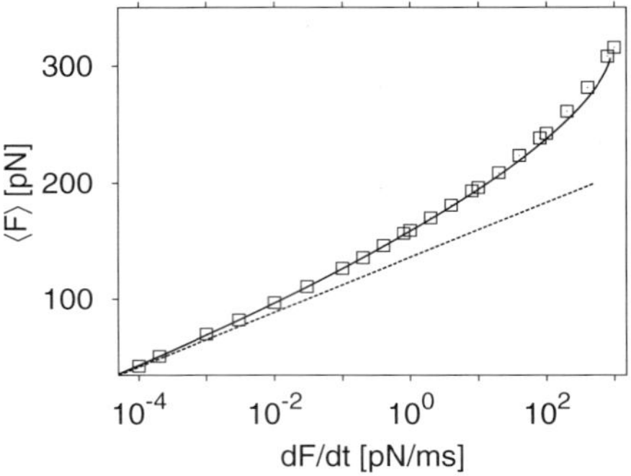

Fig. 5. Average rupture force as a function of the force loading rate. The Brownian dynamics simulation data (symbols) cover force-loading rates that range from 10^{-4} to 10^3 pN/ms. Also shown are the approximate mean forces from the linear-cubic theory, Eq. (69), (solid line) and the phenomenological model (dashed line) for the parameters of the simulation model. The latter approaches the simulation results only at the lowest pulling speeds. Linear fits at higher pulling speeds would produce incorrect estimates for k_0 and $x^\ddagger$, with rates that have been found to be off by more than two orders of magnitude.[45,46] In the Brownian dynamics simulations,[46] a linear-cubic free energy surface, Eq. (65), was used with a barrier height of $\Delta G^\ddagger = 20\ k_\mathrm{B}T$ and a transition state at $x^\ddagger = 0.4$ nm, with the friction coefficient chosen to give an intrinsic rate of escape $k_0 \approx 10^{-4}\ \mathrm{s}^{-1}$.

effective spring constant. The latter approach requires numerical quadrature, as discussed in Sec. 3.4.

Figure 6 shows simulated rupture-force histograms for the escape from a linear-cubic well.[46] Also shown is their collapse onto a master curve $\tau(F) = 1/k(F)$ by using Eq. (42). As for the experimental data in Fig. 3, the collapsed histograms superimpose nearly perfectly for pulling speeds that cover seven orders of magnitude. Moreover, the collapsed histograms agree very well with the theoretical $1/k(F)$ from the linear-cubic theory. This agreement suggests that from the collapsed histograms alone one can obtain accurate estimates of $k(F)$.

3.6. *Analysis of force-ramp experiments*

We advocate two complementary approaches to extracting microscopic information from rupture force distributions obtained at different pulling speeds. In the first approach,[48] a maximum-likelihood

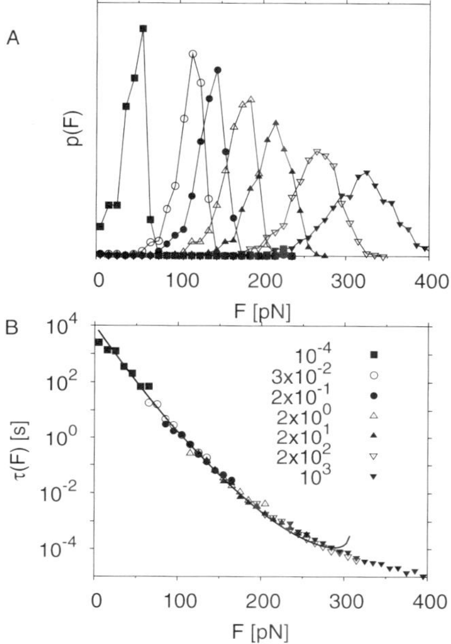

Fig. 6. Constant-force life time $\tau(F) = 1/k(F)$ from collapse of simulated force-ramp histograms $p(F)$ of the rupture force. (A) Histograms. (B) Collapse onto master curve by using Eq. (42). The solid line shows the predicted $\tau(F) = 1/k(F)$ from Eq. (67) for $\mu = 2/3$. At forces close to $F_c = 308.25$ pN, the analytical expression for $k(F)$, based on Kramers high-barrier approximation, breaks down. See Fig. 5 for simulation details.

(or Bayesian inference) formalism is used to fit all the available experimental data, namely the rupture forces F_i ($i = 1, \ldots, K$) at each of M loading rates v_j ($j = 1, \ldots, M$). The likelihood function L that needs to be maximized can be expressed in terms of the rupture-force distribution $p(F|v)$ at pulling speed v as

$$L = \prod_{i=1}^{K} \prod_{j=1}^{M} p(F_{ij}|v_j). \tag{71}$$

Clearly, to implement this approach it is convenient to have analytic expressions for $p(F|v)$. Our unified formalism[46] provides such an expression in Eq. (67) as a function of μ (which specifies different microscopic models), the intrinsic rate k_0, the distance to the transition state $x^{\ddagger}$, and the free energy of activation $\Delta G^{\ddagger}$. Given experimental measurements, the strategy is to find the optimum values of k_0, $x^{\ddagger}$, and $\Delta G^{\ddagger}$ for different μs (e.g., $\mu = 1/2$ for the harmonic-cusp model, and $\mu = 2/3$ for the linear-cubic model) by maximizing L or, equivalently, $\ln L$.[48] If the resulting parameters are relatively insensitive to the value of μ in the range of $1/2 \leq \mu \leq 2/3$, then the extracted parameters do not depend on the precise form of the underlying free energy surface, and hence can be considered meaningful. An example of this approach can be found in Ref. 48.

An alternate[50] approach is to "collapse" the rupture-force distributions measured at different loading rates by using Eq. (42). This approach is simpler but less rigorous, and it requires binning of the data. If the data do not collapse, then one has established that the mechanism of rupture cannot be described as an irreversible, Markovian escape over a single barrier. Such nonadiabatic behavior may also be evident in nonexponential distributions of the life times in constant-force experiments. If the data do collapse onto a single master curve, one immediately obtains the force-dependent rate of rupture, $k(F)$, or equivalently, the rupture life time under force, $\tau(F) = 1/k(F)$, after correcting for linker contributions by using Eq. (47). If rupture can be described by a one-dimensional reaction coordinate, then one can obtain the distance to the transition state as a function of force from Eq. (57) *independent* of the shape of the free energy surface. To obtain additional information, one must adopt a model (e.g., the harmonic-cusp and linear-cubic potentials) and hope that the model parameters are relatively insensitive to the assumed form of the free energy surface. In the framework of our unified approach, one can use Eq. (67) to least-square fit the collapsed histogram data to extract k_0, $x^{\ddagger}$, and $\Delta G^{\ddagger}$. As in the case of the maximum-likelihood analysis described above, if the three parameters are insensitive to μ, then they may be considered meaningful.

4. Concluding remarks

The major strength of single-molecule force spectroscopy is the ability to induce and monitor mechanical transitions in single molecules. However, the resulting nonequilibrium conditions complicate the quantitative analysis of the experiments. Nevertheless, as shown in the first part of the chapter, one can extract the underlying free energy profile rigorously from repeated nonequilibrium pulling experiments by using an extension[23,24] of Jarzynski's identity.[21,22]

In the second part of the chapter, we showed how one can extract kinetic information from single-molecule pulling experiments. Under the assumption of adiabatic rupture, we derived a quantitative relation between constant-force and force-ramp experiments.[46] Using this relation, Eq. (42), one can extract the rate of rupture $k(F)$ (or the force-dependent life time, $\tau(F) = 1/k(F)$) at constant force F from force-ramp experiments. If rupture can be described as escape from a deep free energy well along a one-dimensional reaction coordinate, we showed[50] that independent of the free energy surface the slope of $\ln k(F)$ with respect to F gives the force-dependent location of the transition state, $x^{\ddagger}(F)$. We also considered simple microscopic models of force-induced rupture for which the rupture-force distributions can be found analytically. These models encompass the Bell–Evans formalism [52–55] as a special case valid in the limit of infinitely high barriers. In contrast to Bell's formula, $\ln[k(F)]$ in general depends nonlinearly on F. This is simply due to the fact that in one dimension, the barrier and the well must move closer as the force increases, since the well eventually vanishes beyond a critical force F_c. The microscopic models can account for both linear and nonlinear dependences of the mean rupture force on the force-loading rate in force-ramp experiments, and for a loading-rate-dependent variance of the rupture force distribution. Finally, we outlined procedures to analyze experimental rupture statistics and extract intrinsic rates k_0, transition state locations $x^{\ddagger}$, and activation free energies $\Delta G^{\ddagger}$.

As the resolution and accuracy of force spectroscopy rapidly improves, the theories discussed in this chapter should prove useful in routine measurements of thermodynamic and kinetic properties of single molecules.

Acknowledgments

This research was supported by the Intramural Programs of the NIDDK, NIH. We want to thank Prof Olga Dudko, Prof Amit Meller, Prof Jérôme Mathé, and Dr Artur Adib for discussions and collaborations.

References

1. T. T. Perkins, D. E. Smith and S. Chu, Direct observation of tube-like motion of a single polymer chain, *Science* **264**(5160) (1994) 819–822.
2. E. L. Florin, V. T. Moy and H. E. Gaub, Adhesion forces between individual ligand-receptor pairs, *Science* **264**(5157) (1994) 415–417.
3. T. R. Strick, J. F. Allemand, D. Bensimon, A. Bensimon and V. Croquette, The elasticity of a single supercoiled DNA molecule, *Science* **271**(5257) (1996) 1835–1837.
4. S. B. Smith, Y. J. Cui and C. Bustamante, Overstretching B-DNA. The elastic response of individual double-stranded and single-stranded DNA molecules, *Science* **271**(5250) (1996) 795–799.
5. M. S. Z. Kellermayer, S. B. Smith, H. L. Granzier and C. Bustamante, Folding-unfolding transitions in single titin molecules characterized with laser tweezers, *Science* **276**(5315) (1997) 1112–1116.
6. L. Tskhovrebova, J. Trinick, J. A. Sleep and R. M. Simmons, Elasticity and unfolding of single molecules of the giant muscle protein titin, *Nature* **387** (6630) (1997) 308–312.
7. M. Rief, M. Gautel, F. Oesterhelt, J. M. Fernandez and H. E. Gaub, Reversible unfolding of individual titin immunoglobulin domains by AFM, *Science* **276** (5315) (1997) 1109–1112.
8. A. F. Oberhauser, P. E. Marszalek, H. P. Erickson and J. M. Fernandez, The molecular elasticity of the extracellular matrix protein tenascin, *Nature* **393** (6681) (1998) 181–185.
9. P. E. Marszalek, H. Lu, H. B. Li, M. Carrion-Vazquez, A. F. Oberhauser, K. Schulten and J. M. Fernandez, Mechanical unfolding intermediates in titin modules, *Nature* **402**(6757) (1999) 100–103.

10. R. Merkel, P. Nassoy, A. Leung, K. Ritchie and E. Evans, Energy landscapes of receptor-ligand bonds explored with dynamic force spectroscopy, *Nature* **397**(6714) (1999) 50–53.
11. M. Carrion-Vazquez, A. F. Oberhauser, S. B. Fowler, P. E. Marszalek, S. E. Broedel, J. Clarke and J. M. Fernandez, Mechanical and chemical unfolding of a single protein. A comparison, *Proceedings of the National Academy of Sciences of the United States of America* **96**(7) (1999) 3694–3699.
12. F. Oesterhelt, D. Oesterhelt, M. Pfeiffer, A. Engel, H. E. Gaub and D. J. Müller, Unfolding pathways of individual bacteriorhodopsins, *Science* **288** (5463) (2000) 143–146.
13. Y. Cui and C. Bustamante, Pulling a single chromatin fiber reveals the forces that maintain its higher-order structure, *Proceedings of the National Academy of Sciences of the United States of America* **97**(1) (2000) 127–132.
14. J. Liphardt, B. Onoa, S. B. Smith, I. Tinoco Jr. and C. Bustamante, Reversible unfolding of single RNA molecules by mechanical force, *Science* **292** (2001) 733–737.
15. J. Liphardt, S. Dumont, S. B. Smith, I. Tinoco and C. Bustamante, Equilibrium information from nonequilibrium measurements in an experimental test of Jarzynski's equality, *Science* **296**(5574) (2002) 1832–1835.
16. K. C. Neuman and S. M. Block, Optical trapping, *Review of Scientific Instruments* **75**(9) (2004) 2787–2809.
17. H. Grubmüller, B. Heymann and P. Tavan, Ligand binding molecular mechanics calculation of the streptavidin biotin rupture force, *Science* **271**(5251) (1996) 997–999.
18. B. Isralewitz, S. Izrailev and K. Schulten, Binding pathway of retinal to bacterio-opsin. A prediction by molecular dynamics simulations, *Biophysical Journal* **73**(6) (1997) 2972–2979.
19. E. Paci and M. Karplus, Forced unfolding of fibronectin type 3 modules. An analysis by biased molecular dynamics simulations, *Journal of Molecular Biology* **288**(3) (1999) 441–459.
20. M. O. Jensen, S. Park, E. Tajkhorshid and K. Schulten, Energetics of glycerol conduction through aquaglyceroporin Glpf, *Proceedings of the National Academy of Sciences of the United States of America* **99**(10) (2002) 6731–6736.
21. C. Jarzynski, Nonequilibrium equality for free energy differences, *Physical Review Letters* **78**(14) (1997) 2690–2693.
22. C. Jarzynski, Equilibrium free energy differences from nonequilibrium measurements. A master-equation approach, *Physical Review E* **56**(5/pt.A) (1997) 5018–5035.
23. G. Hummer and A. Szabo, Free energy reconstruction from nonequilibrium single-molecule pulling experiments, *Proceedings of the National Academy of Sciences of the United States of America* **98**(7) (2001) 3658–3661.

24. G. Hummer and A. Szabo, Free energy surfaces from single-molecule force spectroscopy, *Accounts of Chemical Research* **38** (2005) 504–513.
25. S. Park, F. Khalili-Araghi, E. Tajkhorshid and K. Schulten, Free energy calculation from steered molecular dynamics simulations using Jarzynski's equality, *Journal of Chemical Physics* **119**(6) (2003) 3559–3566.
26. G. Hummer, *Free Energy Calculations. Theory and Applications in Chemistry and Biology* (Springer, New York, 2007), Chap. 5, pp. 171–198.
27. A. M. Ferrenberg and R. H. Swendsen, Optimized Monte Carlo data analysis, *Physical Review Letters* **63** (1989) 1195–1198.
28. G. M. Torrie and J. P. Valleau, Monte Carlo free energy estimates using non-Boltzmann sampling: Application to the sub-critical Lennard–Jones fluid, *Chemical Physics Letters* **28** (1974) 578–581.
29. R. H. Wood, W. C. F. Mühlbauer and P. T. Thompson, Systematic errors in free energy perturbation calculations due to a finite sample of configuration space. Sample-size hysteresis, *Journal of Physical Chemistry* **95**(17) (1991) 6670–6675.
30. G. Hummer, Fast-growth thermodynamic integration error and efficiency analysis, *Journal of Chemical Physics* **114**(17) (2001) 7330–7337.
31. J. Gore, F. Ritort and C. Bustamante, Bias and error in estimates of equilibrium free-energy differences from nonequilibrium measurements, *Proceedings of the National Academy of Sciences of the United States of America* **100**(22) (2003) 12564–12569.
32. D. M. Zuckerman and T. B. Woolf, Theory of a systematic computational error in free energy differences, *Physical Review Letters* **89**(18) (2002) 180602.
33. D. Wu and D. A. Kofke, Asymmetric bias in free-energy perturbation measurements using two hamiltonian-based models, *Physical Review E* **70**(6) (2004) 066702.
34. G. E. Crooks, Nonequilibrium measurements of free energy differences for microscopically reversible Markovian systems, *Journal of Statistical Physics* **90**(5–6) (1998) 1481–1487.
35. G. E. Crooks, Entropy production fluctuation theorem and the nonequilibrium work relation for free energy differences, *Physical Review E* **60**(3) (1999) 2721–2726.
36. D. Collin, F. Ritort, C. Jarzynski, S. B. Smith, I. Tinoco and C. Bustamante, Verification of the Crooks fluctuation theorem and recovery of RNA folding free energies, *Nature* **437**(7056) (2005) 231–234.
37. C. H. Bennett, Efficient estimation of free energy differences from Monte Carlo data, *Journal of Computational Physics* **22** (1976) 245–268.
38. M. R. Shirts, E. Bair, G. Hooker and V. S. Pande, Equilibrium free energies from nonequilibrium measurements using maximum-likelihood methods, *Physical Review Letters* **91** (14) (2003) 140601.

39. A. B. Adib, Free energy surfaces from nonequilibrium processes without work measurement, *Journal of Chemical Physics* **124**(14) (2006) 144111.
40. G. Hummer and I. G. Kevrekidis, Coarse molecular dynamics of a peptide fragment free energy kinetics and long-time dynamics computations, *Journal of Chemical Physics* **118**(23) (2003) 10762–10773.
41. D. D. L. Minh, Free-energy reconstruction from experiments performed under different biasing programs, *Physical Review E* **74**(6) (2006) 061120.
42. A. F. Oberhauser, P. K. Hansma, M. Carrion-Vazquez and J. M. Fernandez, Stepwise unfolding of titin under force-clamp atomic force microscopy, *Proceedings of the National Academy of Sciences of the United States of America* **98**(2) 468–472 (2001).
43. W. J. Greenleaf, M. T. Woodside, E. A. Abbondanzieri and S. M. Block, Passive all-optical force clamp for high-resolution laser trapping, *Physical Review Letters* **95**(20) (2005) 208102.
44. I.-C. Yeh and G. Hummer, Nucleic acid transport through carbon nanotube membranes, *Proceedings of the National Academy of Sciences of the United States of America* **101** (2004) 12171–12182.
45. G. Hummer and A. Szabo, Kinetics from nonequilibrium single-molecule pulling experiments, *Biophysical Journal* **85**(1) (2003) 5–15.
46. O. K. Dudko, G. Hummer and A. Szabo, Intrinsic rates and activation free energies from single-molecule pulling experiments, *Physical Review Letters* **96**(10) (2006) 108101.
47. M. Raible, M. Evstigneev, P. Reimann, F. Bartels and P. Ros, Theoretical analysis of dynamic force spectroscopy experiments on ligand-receptor complexes, *Journal of Biotechnology* **112** (2004) 13–23.
48. O. K. Dudko, J. Mathé, A. Szabo, A. Meller and G. Hummer, Extracting kinetics from single-molecule force spectroscopy: Nanopore unzipping of DNA hairpins, *Biophysical Journal* **92** (2007) 4188–4195.
49. J. Mathé, H. Visram, V. Viasnoff, Y. Rabin and A. Meller, Nanopore unzipping of individual DNA hairpin molecules, *Biophysical Journal* **87** (2004) 3205–3212.
50. O. K. Dudko, A. Szabo and G. Hummer, in preparation.
51. J. F. Marko and E. D. Siggia, Stretching DNA, *Macromolecules* **28**(26) (1995) 8759–8770.
52. G. I. Bell, Models for the specific adhesion of cells to cells, *Science* **200** (1978) 618–627.
53. E. Evans, D. Berk and A. Leung, Detachment of agglutinin-bonded red blood cells. I. Forces to rupture molecular-point attachments, *Biophysical Journal* **59** (1991) 838–848.
54. E. Evans and K. Ritchie, Dynamic strength of molecular adhesion bonds, *Biophysical Journal* **72**(4) (1997) 1541–1555.

55. E. Evans, Probing the relation between force — lifetime — and chemistry in single molecular bonds, *Annual Review of Biophysics and Biomolecular Structure* **30** (2001) 105–128.
56. S. Izrailev, S. Stepaniants, M. Balsera, Y. Oono and K. Schulten, Molecular dynamics study of unbinding of the avidin-biotin complex, *Biophysical Journal* **72**(4) (1997) 1568–1581.
57. C. Gergely, J. C. Voegel, P. Schaaf, B. Senger, M. Maaloum, J. K. H. Horber, and J. Hemmerle, Unbinding process of adsorbed proteins under external stress studied by atomic force microscopy spectroscopy, *Proceedings of the National Academy of Sciences of the United States of America* **97**(20) (2000) 10802–10807.
58. O. K. Dudko, A. E. Filippov, J. Klafter and M. Urbakh, Beyond the conventional description of dynamic force spectroscopy of adhesion bonds, *Proceedings of the National Academy of Sciences of the United States of America* **100**(20) (2003) 11378–11381.
59. R. Zwanzig, A. Szabo and B. Bagchi, Levinthals paradox, *Proceedings of the National Academy of Sciences of the United States of America* **89**(1) (1992) 20–22.
60. A. Garg, Escape-field distribution for escape from a metastable potential well subject to a steadily increasing bias field, *Physical Review B* **51**(21) (1995) 15592–15595.

CHAPTER 6

Theory of Photon Counting in Single-Molecule Spectroscopy

Irina V. Gopich and Attila Szabo

1. Introduction

Single-molecule spectroscopy can monitor conformational changes of a macromolecule containing fluorophores whose photophysics is directly influenced by such changes, e.g., Förster resonance energy transfer (FRET) and fluorescence quenching.[1–22] The experimental output is a photon trajectory which contains information about the nature and time scale of the underlying conformational changes. Classically, the emission of a photon is associated with a kinetic transition between two states in a multistate kinetic scheme involving both photophysical and conformational microstates. Thus, the photon counting problem is reduced to the problem of determining the statistics of certain transitions in a kinetic scheme. To analyze experiments, one must be able to describe the statistics of such transitions within the framework of a microscopic model of the dynamics. Of interest are the probability distribution of the number of transitions in a time bin, the distribution of the time intervals between consecutive transitions and the transition number correlation function. This chapter will show how such quantities can be readily obtained when the dynamics is described by multistate rate (master) equations, or their continuum analogue, multidimensional reaction-diffusion equations.

Various theoretical aspects of the photon counting problem have recently attracted much attention.[23–45] This chapter presents a simple yet general approach that ignores quantum effects and is based on counting specific transitions in a multistate kinetic scheme.[46–51] We illustrate the general formalism by applying it to examples of increasing complexity. We thus provide self-contained derivations of many simple results that have been previously obtained using a variety of methods.

Various kinetic schemes that describe the experiments of interest are shown in Fig. 1. Figure 1(a) describes a system with a fluorophore (D) that is excited by a continuous laser beam with the rate k_{ex}, which is proportional to the laser intensity. The excited state D^* decays with the rate k_{D}. This decay can be radiative with the rate k_{R} or nonradiative with the rate k_{NR} (see Fig. 1(b)). The monitored events, i.e., emitted photons, are coincident with the radiative transition in scheme 1(b) (the dotted arrow). To account for the fact that not every emitted photon is detected, the radiative rate is split into parts corresponding to the detected and nondetected photons. In this way, the decay rate k_{D} is the sum of $k'_{\mathrm{R}} = \phi k_{\mathrm{R}}$ (ϕ is the detection efficiency), which corresponds to the detected photons, and $k'_{\mathrm{D}} = k_{\mathrm{D}} - k'_{\mathrm{R}}$. This is shown in Fig. 1(c) with the dotted arrow indicating the transition that leads to a photon that is actually detected.

Figure 1(d) describes triplet blinking where an excited fluorophore can go (with intersystem crossing rate k_{ISC}) into a long-lived "dark" triplet state. As in the previous case, the monitored transition corresponds to the detected photons (the dotted arrow). Figure 1(e) describes FRET between a donor and an acceptor. A donor is excited and the excitation can be transferred (with distance-dependent rate k_{TR}) to an acceptor. There can be two kinds of monitored events here, i.e., donor and acceptor photons, which are associated with the transitions form the excited donor and excited acceptor states. For simplicity, in Fig. 1(e), only the transition corresponding to the donor photons is designated as a monitored transition (the dotted arrow). Figure 1(f) shows how the previous scheme is modified to take into account the re-excitation of the donor when the acceptor is excited.

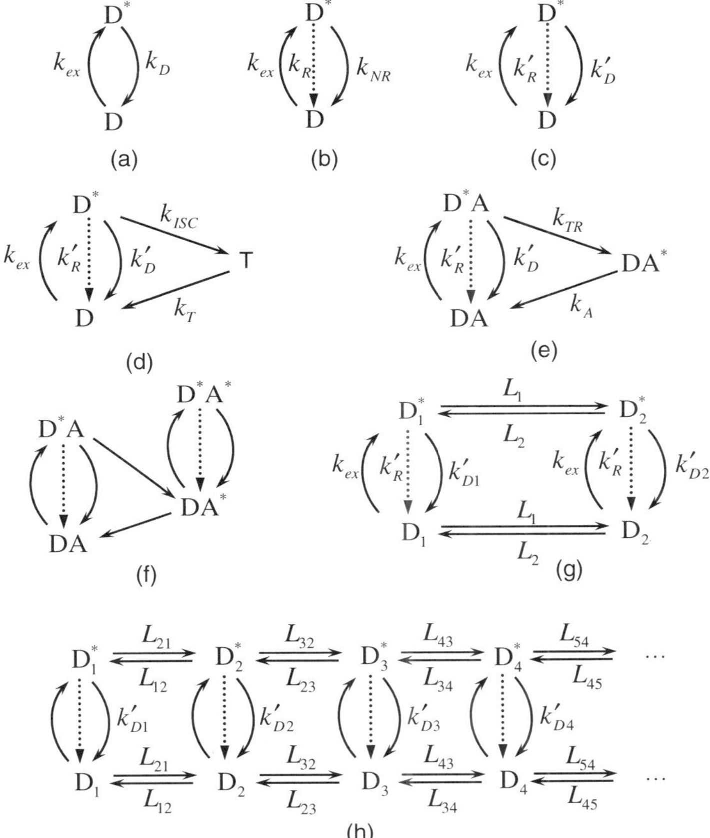

Fig. 1. Kinetic schemes that describe electronic and conformational transitions in fluorescent systems. *Fluorescence*: (a) k_{ex} and k_{D} are the excitation and decay rates. (b) The radiative transition is monitored (dotted arrow) k_{R} and k_{NR} are the radiative and nonradiative rate constants, $k_{\mathrm{R}} + k_{\mathrm{NR}} = k_{\mathrm{D}}$. (c) The transition corresponding to the detected photons is monitored (dotted arrow), $k'_{\mathrm{R}} = \phi k_{\mathrm{R}}$, ϕ is the detection efficiency, $k'_{\mathrm{D}} = k_{\mathrm{D}} - k'_{\mathrm{R}} = (1-\phi)k_{\mathrm{R}} + k_{\mathrm{NR}}$. *Triplet blinking*: (d) the excited fluorophore goes into the "dark" triplet state with intersystem crossing rate k_{ISC}, k_{T} is the decay rate of the triplet state. *Förster resonance energy transfer*: (e) excitation is transferred from the donor to the acceptor with rate k_{TR}; donor photons (dotted arrow) are monitored. (f) The donor is re-excited when the acceptor is excited. *Fluorescence quenching*: (g) the fluorophore interconverts (with rates L_1 and L_2) between two conformations with different decay rates k'_{D1} and k'_{D2}. (h) Many conformations with different nonradiative decay rates.

The nonradiative decay rate of a fluorophore can depend on the distance r between the fluorophore and a quencher such as tryptophan. This distance and hence the nonradiative rate can fluctuate because of conformational changes. The kinetic schemes that describe fluorophore quenching and conformational dynamics are shown in Figs. 1(g) and 1(h). These kinetic schemes contain several interconverting copies of the simple scheme in Fig. 1(c). The monitored events are now associated with several transitions (one for each conformation). Figure 1(g) shows how Fig. 1(c) is modified when the fluorophore can exist in two conformations. We have assumed for the sake of simplicity that the rates of conformational changes are the same in the ground and excited states. The transitions $D_1^* \to D_1$ and $D_2^* \to D_2$ (the dotted arrows in Fig. 1(g)) are experimentally indistinguishable and constitute the same event. Figure 1(h) shows the scheme with many discrete conformations.

This chapter considers the analysis of a photon (event) trajectory (see Fig. 2). Each event marks the time when a photon is detected or, equivalently, a transition in a multistate kinetic scheme (the dotted arrows in Fig. 1) occurs. The data can be processed in a number of ways. The simplest procedure (which is commonly used in photon counting and FRET studies) is to divide the trajectory into equal bins of duration T (see Fig. 2(a)) and then calculate the probability distribution of detecting N photons in a bin, $P(N \mid T)$. This is often called

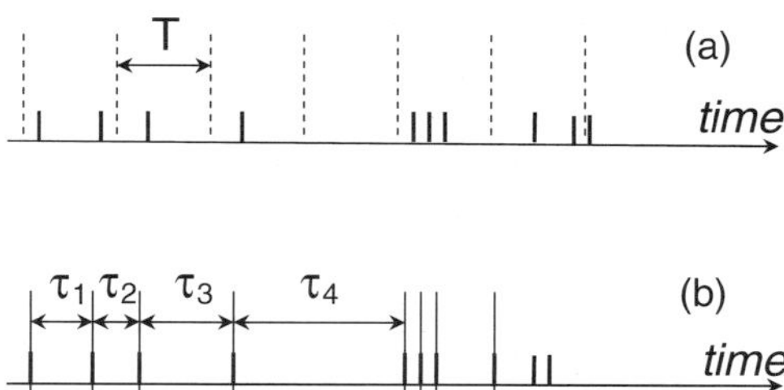

Fig. 2. Processing of photon trajectories. (a) By dividing the time trajectory into bins of equal size T, the distribution of the number of events in a bin, $P(N \mid T)$, can be found. (b) The photon trajectory is characterized by the time intervals between consecutive photons. The interphoton time distribution $\varphi(\tau)$ can be determined.

a photon counting histogram.[52–54] If the time intervals are chosen small enough so that there is at most one photon in each bin, one can obtain the time dependence of the correlation function of the number of photons, or the intensity correlation function.[55–57] Alternatively, one can focus on the time intervals between consecutive photons (see Fig. 2(b)) and find, for example, the interphoton time distribution $\varphi(\tau)$, i.e., the distribution of the time intervals between consecutive photons. Finally, the whole trajectory can be analyzed using the maximum likelihood method (see for example Refs. 41–43).

Since the formalism developed here is based on counting state-to-state transitions in a kinetic scheme, it is applicable to any problem that can be formulated in this way. Figure 3 shows two examples where monitored transitions are not related to the counting of single photons. In the first example (see Fig. 3(a)), an enzyme E binds a substrate with rate k_1. The enzyme–substrate complex (ES) can dissociate with rate k_{-1} or form a product with catalytic rate k_{CAT}. The enzyme–product complex can dissociate with rate k_{DIS} to yield free product. Recent single-enzyme experiments employed systems where only the product fluoresces.[58–60] In this case, the catalytic transition (i.e., product formation) is followed by a burst of photons. The statistics of these transitions can be determined. Figure 3(b) shows the simplest kinetic scheme that describes a molecular motor. A motor protein M binds ATP with the rate k_1. This complex can dissociate

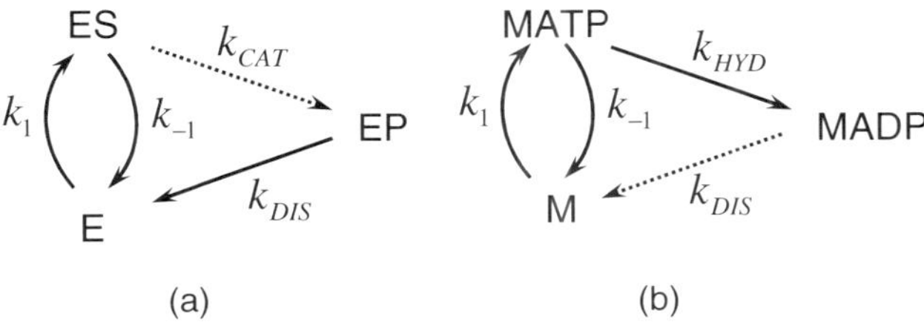

Fig. 3. (a) *Enzymes*: when the product P is fluorescent, the formation of the product coincides with a burst of photons and thus the catalytic transition can be monitored. (b) *Motors*: the dissociation of ADP from a motor protein M is associated with movement (i.e., a "step"). The statistics of steps is then the same as the statistics of the MADP→M transition.

with rate k_{-1} or the ATP can be hydrolyzed to ADP with the rate k_{HYD}. The ADP then dissociates with rate k_{DIS} and the cycle begins anew. If it is assumed that the motion of the motor is associated with the dissociation of ADP, then the problem of counting the steps the motor makes is equivalent to the problem of determining the statistics of the "dotted" transitions in Fig. 3(b). Finally, it is interesting to note that the kinetic schemes that describe triplet blinking (Fig. 1(d)), FRET (Fig. 1(e)), enzyme catalysis (Fig. 3(a)), and motors (Fig. 3(b)) are formally identical but the monitored transitions are different.

Section 2 of this chapter presents a formalism for calculating the statistics of photon counts given an arbitrary kinetic scheme. Our focus is on the distribution of the number of the photons in a bin, the intensity correlation function, and the interphoton time distribution. The theory is applied to fluorophore quenching in the presence of conformational dynamics in Sec. 3. When there is a separation of time scales, general formalism can be simplified. In this case, the dimensionality of the system can be reduced, and one needs to consider only slow conformational changes (see Secs. 3.2 and 3.4). Finally, as a nontrivial example of slow dynamics, we consider the diffusion of fluorescent molecules through the laser spot in Sec. 4.

2. General formalism

Suppose that the dynamics of the system of interest can be described by a multistate kinetic scheme. The detected events (i.e., photons) are associated with one or more transitions in this scheme (e.g., the transitions denoted by the dotted arrows in Fig. 1). We are interested in the statistics of these *monitored* transitions, namely, the probability of the number of transitions during a time interval, the distribution of the time between transitions, and the transition number correlation function. In this section, general formalism is illustrated by application to the simplest scheme with a single excited state (Figs. 1(a)–1(c)). At the end of this section, we show how the general formalism of photon counting is related to renewal theory and examine the conditions under which the statistics becomes Poissonian.

2.1. *Rate equations*

The populations of the states in a kinetic scheme satisfy a set of rate or master equations. Let K_{ij} be the rate constant for the transition from state j to state i. Then, the probability that the system is in state i at time t, $p_i(t)$, satisfies a set of rate equations, which in matrix notation is

$$\frac{d}{dt}\mathbf{p} = \mathbf{K}\mathbf{p}, \tag{1}$$

where $\mathbf{p}$ is a vector with components $p_i(t)$. The off-diagonal elements of $\mathbf{K}$ are K_{ij} and its diagonal elements are $K_{ii} = -\sum_{j\neq i} K_{ji}$ (or, in matrix notation, $\mathbf{1}^\top\mathbf{K} = \sum_j K_{ji} = 0$, where $\mathbf{1}$ is the unit vector and $\top$ denotes transpose). The solution of Eq. (1) can be formally written as

$$\mathbf{p}(t) = \exp(\mathbf{K}t)\mathbf{p}(0). \tag{2}$$

The matrix exponential can be expressed in terms of the eigenvalues and eigenvectors of $\mathbf{K}$ as $\exp(\mathbf{K}t) = \mathbf{T}\exp(\mathbf{k}t)\mathbf{T}^{-1}$ where $\mathbf{KT} = \mathbf{Tk}$ and $\mathbf{k}$ is a diagonal matrix with the eigenvalues on the diagonal.

At long times, $\mathbf{p}(t)$ approaches its steady-state value, $\mathbf{p}_{\mathrm{ss}}$. The normalized vector of steady-state probabilities $\mathbf{p}_{\mathrm{ss}}$ (with elements $p_{\mathrm{ss}}(i)$) satisfies

$$\mathbf{K}\mathbf{p}_{\mathrm{ss}} = 0, \quad \mathbf{1}^\top\mathbf{p}_{\mathrm{ss}} = 1. \tag{3}$$

The solution of Eq. (2) with the initial condition that at $t = 0$ the system is in the state j (i.e., $p_i(0) = \delta_{ij}$) is called the Green's function, or propagator, and is denoted by $G^0_{ij}(t)$. It is the probability that the system is in state i at time t, provided it was in state j initially. The probabilities $G^0_{ij}(t)$ are the elements of the matrix $\mathbf{G}^0(t)$, which can be formally expressed as

$$\mathbf{G}^0(t) = \exp(\mathbf{K}t), \tag{4}$$

or, in Laplace space ($\hat{f}(s) \equiv \int_0^\infty f(t)\exp(-st)dt$),

$$\hat{\mathbf{G}}^0(s) = (s\mathbf{I} - \mathbf{K})^{-1}, \tag{5}$$

where $\mathbf{I}$ is the identity matrix. At long times, the probability $G^0_{ij}(t)$ approaches the probability of finding the system in state i at steady-state, $p_{ss}(i)$.

In the case of the two-state system in Fig. 1(a), the rate matrix $\mathbf{K}$ in the basis (D, D^*) is

$$\mathbf{K} = \begin{pmatrix} -k_{ex} & k_D \\ k_{ex} & -k_D \end{pmatrix} \tag{6}$$

and the steady-state populations of the ground and excited states are

$$p_{ss}(D) = \frac{k_D}{k_{ex} + k_D}, \tag{7a}$$

$$p_{ss}(D^*) = \frac{k_{ex}}{k_{ex} + k_D}. \tag{7b}$$

2.2. *Dynamics in the absence of detected photons*

The first step in obtaining the statistics of photons is to find the probability that no photons are detected in a time interval. To do this one has to eliminate the possibility that transitions resulting in photons occur. This can be done by making the transitions resulting in detected photons (the monitored transitions) irreversible.[31,34,47] Let $G_{ij}(t)$ be the probability of going from state j to state i in time t without making a monitored transition. The matrix of these probabilities,

$\mathbf{G}(t)$, satisfies[51]

$$\frac{d}{dt}\mathbf{G} = (\mathbf{K} - \mathbf{V})\mathbf{G}, \tag{8}$$

with $\mathbf{G}(0) = \mathbf{I}$. Here, $\mathbf{V}$ is the matrix of the monitored transition rates, i.e., the rates corresponding to the detected photons. It is constructed by setting all the elements of $\mathbf{K}$ equal to zero except those off-diagonal ones that correspond to detected photons. For example, if we are monitoring only the $m \to m'$ transition, then $\mathbf{V}$ has the only nonzero element $V_{m'm} = K_{m'm}$.

In the case of the two-state system in Fig. 1(b) where every emitted photon is detected, the monitored transition is the radiative transition. In this case, the only nonzero element of $\mathbf{V}$ is the off-diagonal element of $\mathbf{K}$ that corresponds to the radiative transition k_{R}:

$$\mathbf{V} = \begin{pmatrix} 0 & k_{\mathrm{R}} \\ 0 & 0 \end{pmatrix}. \tag{9}$$

When the photons are detected with an efficiency $\phi < 1$, the radiative rate constant can be split into parts corresponding to detected and nondetected photons. This is analogous to splitting the decay rate into the sum of the radiative and nonradiative rates. This simple trick allows one to account for the detection, which is a stochastic process, in the same framework as other processes. Thus for the two-state system when not every emitted photon is detected, Fig. 1(c), the monitored transition is associated with the dotted arrow and $\mathbf{V}$ is

$$\mathbf{V} = \begin{pmatrix} 0 & k'_{\mathrm{R}} \\ 0 & 0 \end{pmatrix}, \tag{10}$$

where $k'_{\mathrm{R}} = \phi k_{\mathrm{R}}$ is the rate of emitting those photons that are eventually detected. The rate matrix

$$\mathbf{K} - \mathbf{V} = \begin{pmatrix} -k_{\mathrm{ex}} & k_{\mathrm{D}} - k'_{\mathrm{R}} \\ k_{\mathrm{ex}} & -k_{\mathrm{D}} \end{pmatrix} \tag{11}$$

describes the irreversible kinetic scheme shown in Fig. 4.

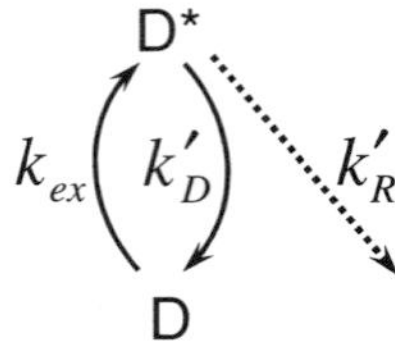

Fig. 4. Two-state kinetic scheme which describes the kinetics of fluorophore populations when no photons are detected. Figure 1(c) describes the kinetics without this restriction.

The formal solutions of Eq. (8) for $\mathbf{G}$ in time and Laplace domains are

$$\mathbf{G}(t) = \exp((\mathbf{K} - \mathbf{V})t), \tag{12a}$$

$$\hat{\mathbf{G}}(s) = (s\mathbf{I} - \mathbf{K} + \mathbf{V})^{-1}. \tag{12b}$$

As shown below, all quantities of interest can be expressed in terms of the matrices $\mathbf{G}^0$, $\mathbf{G}$, and $\mathbf{V}$.

2.3. *Distribution of the number of photons in a time interval*

Consider the probability $P_{ij}(N \mid T)$ that N photons are detected in a time interval (bin) T, given that the system was in state j in the beginning of the interval and in state i at the end. These probabilities can be considered as elements of the matrix $\mathbf{P}(N \mid T)$. To obtain the probabilities, we use the "perturbation" expansion of $\mathbf{G}^0$. Using Eqs. (5) and (12b) in Laplace domain, it is easy to show that $\hat{\mathbf{G}}^0(s) = (\hat{\mathbf{G}}^{-1}(s) - \mathbf{V})^{-1} = \hat{\mathbf{G}}(s)(\mathbf{I} - \mathbf{V}\hat{\mathbf{G}}(s))^{-1}$. Expanding this in powers of $\mathbf{V}$, we get

$$\hat{\mathbf{G}}^0(s) = \hat{\mathbf{G}}(s) + \hat{\mathbf{G}}(s)\mathbf{V}\hat{\mathbf{G}}(s) + \hat{\mathbf{G}}(s)\mathbf{V}\hat{\mathbf{G}}(s)\mathbf{V}\hat{\mathbf{G}}(s) + \cdots \tag{13}$$

In quantum mechanics one usually expands the full propagator ($\mathbf{G}$) in terms of the unperturbed propagator ($\mathbf{G}^0$) and the perturbation ($\mathbf{V}$). Here, we have done exactly the opposite. The reason is that the above expansion has a simple interpretation. The first term in the sum on the right-hand side ($\hat{\mathbf{G}}(s)$) is the Laplace transform of the probability that the system goes from one state to another without making monitored transitions (i.e., without photons being detected). The second term corresponds to the probability that there is one detected photon, i.e., the system propagates without emitting photons that are detected ($\hat{\mathbf{G}}(s)$), then a photon is detected ($\mathbf{V}$), after which the system again evolves without emitting detected photons ($\hat{\mathbf{G}}(s)$). Thus, the total probability that the system goes from one state to another ($\hat{\mathbf{G}}^0(s)$) is the sum of the probabilities that it does so by emitting no (first term), one (second term), two, etc., detected photons. The Nth term on the right-hand side of Eq. (13) is the probability of N photons during time T in Laplace space:

$$\hat{\mathbf{P}}(N \mid s) = \hat{\mathbf{G}}(s)(\mathbf{V}\hat{\mathbf{G}}(s))^N. \tag{14}$$

In the time domain, this can be expressed as successive convolutions. A more rigorous derivation of this is given in Ref. 51.

To calculate the distribution of photons $P(N \mid T)$ for a steady-state trajectory, we sum $\mathbf{P}(N \mid T)$ over all final states (i.e., multiply by the column vector $\mathbf{1}^\top$ on the left) and a steady-state distribution of initial states:

$$P(N \mid T) = \mathbf{1}^\top \mathbf{P}(N \mid T)\mathbf{p}_{\text{ss}}. \tag{15}$$

When $N = 0$, this gives the probability that no photons were detected during time T. It is interesting that this is equal to the survival probability of a system with irreversible monitored transitions (i.e., Fig. 4).

2.4. *Generating function*

An important tool for obtaining and analyzing the properties of $P(N \mid T)$ is its generating function.[47] This was independently introduced into single-molecule spectroscopy by Brown.[37,39] The generating function is defined as

$$F(\lambda, T) = \sum_{N=0}^{\infty} \lambda^N P(N \mid T). \tag{16}$$

To find an explicit expression for the generating function, we multiply the Laplace transform $\hat{\mathbf{P}}(N \mid s)$, Eq. (14), by λ^N and sum the series.

As a result we find that in Laplace space

$$\begin{aligned}\hat{F}(\lambda, s) &= \mathbf{1}^{\top}\hat{\mathbf{G}}(s)(\mathbf{I} - \lambda\mathbf{V}\hat{\mathbf{G}}(s))^{-1}\mathbf{p}_{ss} \\ &= \mathbf{1}^{\top}([\hat{\mathbf{G}}(s)]^{-1} - \lambda\mathbf{V})^{-1}\mathbf{p}_{ss} \\ &= \mathbf{1}^{\top}(s\mathbf{I} - \mathbf{K} + (1-\lambda)\mathbf{V})^{-1}\mathbf{p}_{ss},\end{aligned} \tag{17}$$

where we used Eq. (12b) for $\hat{\mathbf{G}}(s)$ to get the last equality. Inverting the Laplace transform, we find

$$F(\lambda, T) = \mathbf{1}^{\top}e^{[\mathbf{K}-(1-\lambda)\mathbf{V}]T}\mathbf{p}_{ss}. \tag{18}$$

This generating function can be expressed as

$$F(\lambda, T) = \mathbf{1}^{\top}\mathbf{f}(T \mid \lambda), \tag{19}$$

where the vector of the generating functions $\mathbf{f}(t \mid \lambda)$ is the solution of the differential equation:

$$\frac{d}{dt}\mathbf{f} = (\mathbf{K} - (1-\lambda)\mathbf{V})\mathbf{f}, \tag{20}$$

with the initial condition $\mathbf{f}(0 \mid \lambda) = \mathbf{p}_{ss}$. This equation has the same structure as the usual rate equation (1), but involves a rate matrix that depends on the counting parameter λ. When $\lambda = 1$, the two equations coincide. In this case $\mathbf{f}(T \mid 1) = \mathbf{p}_{ss}$ and $F(1, T) = 1$. This means that the probabilities $P(N \mid T)$ are normalized. When $\lambda = 0$, the generating function results in the probability that no photons were detected in the time interval T, $F(0, T) = P(0 \mid T)$.

Let us expand $\mathbf{f}$ in Eqs. (19) and (20) in powers of λ as

$$\mathbf{f} = \sum_{N=0}^{\infty} \lambda^{N}\mathbf{f}_{N} \tag{21}$$

so that $P(N \mid T) = \mathbf{1}^{\top}\mathbf{f}_{N}$. Substituting this into Eq. (20) and equating the coefficients of λ^{N}, we find

$$\frac{d}{dt}\mathbf{f}_{N} = (\mathbf{K} - \mathbf{V})\mathbf{f}_{N} + \mathbf{V}\mathbf{f}_{N-1}, \quad N = 0, 1, \ldots \tag{22}$$

with $\mathbf{f}_{-1} \equiv 0$ and $\mathbf{f}_{N}(t = 0) = \delta_{N0}\mathbf{p}_{ss}$. The components of the vector $\mathbf{f}_{N}$ are the populations of the various states when precisely N

photons have been detected. These states are produced from states with one fewer detected photon ($\mathbf{V}\mathbf{f}_{N-1}$) and evolve without emitting another photon ($(\mathbf{K}-\mathbf{V})\mathbf{f}_N$). Since this equation has such a physically appealing interpretation, one could have used it as a starting point to derive Eq. (20) by multiplying by λ^N and summing over N (i.e., by simply reversing our steps).

For the two-state system in Fig. 1(c), the kinetic scheme that is described by Eq. (22) is shown in Fig. 5. This was constructed by stringing together the irreversible scheme shown in Fig. 4. In this way, we can follow how many photons (labelled by index N) have already been detected. When the excited state with $N-1$ detected photons undergoes a transition that leads to a detected photon, one ends up in the ground state of the system with N detected photons.

As an example, consider the generating function for the two-state system in Fig. 1(c). Using Eq. (6) for $\mathbf{K}$ and Eq. (10) for $\mathbf{V}$ in Eq. (20), we have $\hat{F}(\lambda, s) = \hat{f}_D + \hat{f}_{D^*}$, where

$$\frac{d}{dt} f_D = -k_{\mathrm{ex}} f_D + k_{\mathrm{D}} f_{D^*} - (1-\lambda) k'_{\mathrm{R}} f_{D^*}, \tag{23a}$$

$$\frac{d}{dt} f_{D^*} = k_{\mathrm{ex}} f_D - k_{\mathrm{D}} f_{D^*}, \tag{23b}$$

with the initial conditions $f_D(0\,|\,\lambda) = p_{\mathrm{ss}}(D)$ and $f_{D^*}(0\,|\,\lambda) = p_{\mathrm{ss}}(D^*)$, where the steady-state populations are given by Eq. (7).

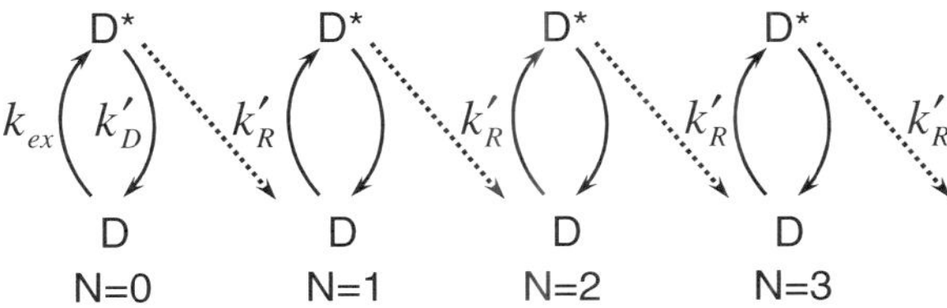

Fig. 5. The kinetic scheme corresponding to Eq. (22) which can be used to find the probability of finding the system in various states on condition that precisely N photons have been detected.

Solving this pair of equations in Laplace space, expanding the generating function in powers of λ and inverting the Laplace transform, we find that[47,48]:

$$P(N \mid T) = \frac{(1-\gamma^2)^N t^N e^{-t}}{(2\gamma)^N N! \sqrt{8\gamma t/\pi}} \times \{2\gamma\,(N+t) I_{N-1/2}(\gamma t) + (1+\gamma^2) t I_{N+1/2}(\gamma t)\}, \quad (24)$$

where $k \equiv k_{\text{ex}} + k_{\text{D}}$, $t = kT/2$, $\gamma^2 = 1 - 4k_{\text{ex}}k'_{\text{R}}/k^2$, $I_n(z)$ are the modified Bessel functions of the first kind. Note that even in this simple case the distribution of photon counts is quite complex. However, as we will see in Sec. 2.8, it simplifies when the detection efficiency is small.

2.5. *Moments and the intensity correlation function*

Given the generating function, one can readily find the moments of the distribution. The mean number of photons in a bin of size T is equal to the derivative of the generating function at $\lambda = 1$:

$$\langle N \rangle = \sum_{N=0}^{\infty} N P(N \mid T) = \frac{\partial}{\partial \lambda} F(\lambda, T)|_{\lambda=1}. \quad (25)$$

To evaluate this and higher derivatives, we first represent the Laplace transform of the generating function as $\hat{F}(\lambda, s) = \mathbf{1}^{\top}([\hat{\mathbf{G}}^0(s)]^{-1} - (\lambda-1)\mathbf{V})^{-1}\mathbf{p}_{\text{ss}}$ (see Eqs. (5) and (17)) and then expand it in powers of $(\lambda - 1)\mathbf{V}$:

$$\hat{F}(\lambda, s) = \frac{1}{s^2}(s + (\lambda-1)\mathbf{1}^{\top}\mathbf{V}\mathbf{p}_{\text{ss}} + (\lambda-1)^2\mathbf{1}^{\top}\mathbf{V}\hat{\mathbf{G}}^0(s)\mathbf{V}\mathbf{p}_{\text{ss}} + \cdots), \quad (26)$$

where we have used the facts that $\mathbf{1}^{\top}\hat{\mathbf{G}}^0(s) = \mathbf{1}^{\top}/s$ and $\hat{\mathbf{G}}^0(s)\mathbf{p}_{\text{ss}} = \mathbf{p}_{\text{ss}}/s$. Differentiating this with respect to λ, setting $\lambda = 1$ and inverting the Laplace transform, we find that the mean number of photons in a bin of size T is

$$\langle N \rangle = \mathbf{1}^\top \mathbf{V}\mathbf{p}_{ss}\, T = \langle n \rangle T, \tag{27}$$

where $\langle n \rangle$ is the mean number of detected photons per unit time (the mean count rate)

$$\langle n \rangle = \mathbf{1}^\top \mathbf{V}\mathbf{p}_{ss}. \tag{28}$$

The mean number of photons per unit time is simply related to the steady-state populations of the excited states. For a single monitored transition, say, $m \to m'$, $\mathbf{V}$ has a single off-diagonal element, $K_{m'm}$, and it follows from Eq. (28) that the mean number of $m \to m'$ transitions per unit time is $K_{m'm}\, p_{ss}(m)$.

The mean square number of photons in a bin is calculated analogously by taking the second derivative of the generating function in Eq. (26), $\langle N(N-1) \rangle = (\partial^2/\partial\lambda^2) F(\lambda, T)|_{\lambda=1}$:

$$\langle N(N-1) \rangle = 2 \int_0^T (T-t)\mathbf{1}^\top \mathbf{V}\mathbf{G}^0(t)\mathbf{V}\mathbf{p}_{ss} dt. \tag{29}$$

This can be expressed in terms of the intensity correlation function

$$\langle N(N-1) \rangle = 2 \int_0^T (T-t)\langle n(t)n(0) \rangle dt. \tag{30}$$

where the intensity correlation function $\langle n(t)n(0) \rangle$ is defined as

$$\langle n(t)n(0) \rangle = \mathbf{1}^\top \mathbf{V}\mathbf{G}^0(t)\mathbf{V}\mathbf{p}_{ss} \tag{31a}$$

$$= \mathbf{1}^\top \mathbf{V} e^{\mathbf{K}t} \mathbf{V}\mathbf{p}_{ss}. \tag{31b}$$

This expression, when read from right to left has transparent physical interpretation. A photon is selected from a steady-state trajectory at

$t = 0$ ($\mathbf{V}\mathbf{p}_{ss}$), the system propagates until time t emitting an arbitrary number of photons ($\mathbf{G}^0(t)$), and then a photon is detected ($\mathbf{V}$). At long times, $G^0_{ij}(t) \to p_{ss}(i)$ and $\langle n(t)n(0)\rangle \to \langle n\rangle^2$, as is expected, because the photons become uncorrelated.

The intensity correlation function is obtained from a photon trajectory by dividing the trajectory in equal size bins in such a way that each bin contains at most one photon. This converts the trajectory into a sequence of 0's (no photons) and 1's (one photon). Then $\langle n(t)n(0)\rangle$ for times longer than the bin size is just the autocorrelation function of the 1's. The nonzero contributions to this correlation function occur only when two 1's are separated by a time t. This suggests that the correlation function is related to the conditional probability density of detecting a photon at time t given that a photon was detected at $t = 0$. In fact this probability density is equal to $\langle n(0)n(t)\rangle/\langle n\rangle$ (for example, see Ref. 12).

Using Eq. (30), the variance of the photon count distribution can be written as

$$\langle N^2\rangle - \langle N\rangle^2 = \langle N\rangle + 2\int_0^T (T-t)\langle \delta n(t)\delta n(0)\rangle dt, \qquad (32)$$

where we have defined $\delta n \equiv n - \langle n\rangle$. The variance is related to the Mandel "Q" parameter, which is a measure for the deviation from Poisson statistics.[61] The Mandel parameter is defined as

$$Q(T) = \frac{\langle N^2\rangle - \langle N\rangle^2}{\langle N\rangle} - 1. \qquad (33)$$

For Poisson statistics (considered in Sec. 2.8), the Mandel parameter is zero. Using Eqs. (27) and (32), we find that the Mandel parameter is related to the intensity correlation function by[62]

$$Q(T) = \frac{2}{\langle n\rangle T}\int_0^T (T-t)\langle \delta n(t)\delta n(0)\rangle dt. \qquad (34)$$

In the case of a two-state fluorophore (Fig. 1(c)), the mean number of photons per unit time is

$$\langle n \rangle = k'_{\rm R} p_{\rm ss}(D^*) = \frac{k'_{\rm R} k_{\rm ex}}{k_{\rm ex} + k_{\rm D}}. \tag{35}$$

It can also be readily shown that the intensity correlation function is

$$\langle n(t)n(0) \rangle = \langle n \rangle^2 (1 - e^{-kt}), \tag{36}$$

where $k = k_{\rm ex} + k_{\rm D}$ is the relaxation time of the two-state system. For very short times $kt \ll 1$, the correlation function approaches zero. This is due to antibunching, i.e., after emitting a photon, it takes time to emit the next photon. For long times $kt \gg 1$, the normalized correlation function $\langle n(t)n(0) \rangle / \langle n \rangle^2$ approaches unity.

The Mandel parameter is always negative for the two-state system shown in Fig. 1(c):

$$Q(T) = -\frac{2\langle n \rangle}{k^2 T}(kT - 1 + \exp(-kT)). \tag{37}$$

This is a signature of antibunching. Its value at long bin times ($kT \gg 1$) depends on $\langle n \rangle / k$, which is the ratio of the relaxation time of the system, k^{-1}, to the mean time between photons, $\langle n \rangle^{-1}$. When this ratio is small (e.g., due to small detection efficiency) and $kT \gg 1$, then Q approaches zero and the statistics of photons becomes Poissonian.

Finally, we note that the higher-order terms of the expansion in Eq. (26) are related to the factorial cumulants (and therefore to the higher moments) of photon counts.[50] Thus, the power series expansion of the generating function around $\lambda = 1$ provides the moments of photon counts. The power series expansion of the generating function around $\lambda = 0$ provides the probability of photon counts in a bin.

2.6. *Interphoton time distribution*

As has been shown above, $\langle n(0)n(t) \rangle / \langle n \rangle$ is the conditional probability density of finding two photons separated by time t irrespective of how many photons there are in between. Mathematically this is reflected in the fact that Eq. (31a) involves the propagator $\mathbf{G}^0$ which does not care how many photons were or were not detected in the interval t. This suggests that to get the distribution of times between consecutively detected photons (i.e., the interphoton time distribution, denoted here by $\varphi(\tau)$) we can simply replace $\mathbf{G}^0$ by $\mathbf{G}$ which is the propagator when there are no detected photons in the time interval τ. A rigorous argument[51] shows that indeed

$$\varphi(\tau) = \mathbf{1}^\top \mathbf{V}\mathbf{G}(\tau)\mathbf{V}\mathbf{p}_{ss}/\langle n\rangle \tag{38a}$$

$$= \mathbf{1}^\top \mathbf{V} e^{(\mathbf{K}-\mathbf{V})\tau}\mathbf{V}\mathbf{p}_{ss}/\langle n\rangle . \tag{38b}$$

The structure of this expression is similar to that of the intensity correlation function in Eq. (31). A photon is selected from a steady-state trajectory at $\tau = 0$ ($\mathbf{V}\mathbf{p}_{ss}$), the system propagates until time τ without emitting detected photons ($\mathbf{G}(\tau)$), and then a photon is detected ($\mathbf{V}$). The factor $\langle n\rangle$ is a normalization constant that guarantees that $\int_0^\infty \varphi(\tau)d\tau = 1$.

It can be shown[51] that the mean time between photons $\langle\tau\rangle = \int_0^\infty \tau\varphi(\tau)d\tau$ is equal to $\langle n\rangle^{-1}$, as it should be. Note that the interphoton time distribution $\varphi(\tau)$ is proportional to the "correlation" function of $\mathbf{V}$ when the dynamics is irreversible (i.e., described by the rate matrix $\mathbf{K} - \mathbf{V}$, see Eq. (38)). On the other hand, Eq. (31) shows that $\langle n(t)n(0)\rangle$ is the correlation function of $\mathbf{V}$ for the system described by the rate matrix $\mathbf{K}$.

In the case of a two-state fluorophore, Fig. 1(c), $\varphi(\tau)$ is biexponential

$$\varphi(\tau) = \frac{2\langle n\rangle}{\sqrt{1-4\langle n\rangle/k}}\, e^{-k\tau/2} \sinh\left(\frac{1}{2}k\tau\sqrt{1-4\langle n\rangle/k}\right) \tag{39}$$

where $\langle n\rangle$ is given by Eq. (35) and $k = k_{ex} + k_D$. When $\tau \to 0$, $\varphi(\tau) \to 0$ due to antibunching. When the relaxation time k^{-1} is much shorter than the mean time between photons $\langle n\rangle^{-1}$, the interphoton time distribution becomes single exponential at $k\tau \gg 1$, $\phi(\tau) = \langle n\rangle \exp(-\langle n\rangle\tau)$.

Now consider how the interphoton time distribution is related to the intensity correlation function and to the generating function. When all dynamics are on the time scale much shorter than the mean time between consecutive photons, the interphoton time distribution can be approximately related to the intensity correlation function

$$\varphi(\tau) \approx \frac{\langle n(\tau)n(0)\rangle}{\langle n\rangle} e^{-\langle n\rangle \tau}. \tag{40}$$

This relation allows one to determine the intensity correlation function at short times (compared to the mean time between photons) from the interphoton time distribution.[20] The latter is measured using the start–stop detection technique and then corrected by the exponential decay factor (the pile-up correction).[1,5]

The interphoton time distribution is rigorously related to the generating function. By differentiating Eq. (18) for the generating function with respect to time twice, it can be shown[51] that

$$\varphi(\tau) = \langle n\rangle^{-1} \left. \frac{d^2 F(\lambda, \tau)}{d\tau^2} \right|_{\lambda=0}. \tag{41}$$

This relation will be used later to obtain the interphoton time distribution of diffusing fluorophores.

Finally, consider how the above formalism can be used to find the likelihood of the entire photon trajectory. When one tries to interpret single-molecule data in the framework of some model, one can vary the model parameters so as to fit, for example, $\langle n(t)n(0)\rangle$, $\varphi(\tau)$, and $P(N \mid T)$. These are reduced descriptions of the photon trajectory. Alternately, one can fit *all* the data by maximizing the likelihood function with respect to the model parameters.[28,41–43] By generalizing Eq. (38), one can express the probability, or likelihood, of the entire photon trajectory shown in Fig. 2(b) as:

$$\begin{aligned} \text{Likelihood} &= \mathbf{1}^{\top}\mathbf{V} \cdots e^{(\mathbf{K}-\mathbf{V})\tau_3}\mathbf{V} e^{(\mathbf{K}-\mathbf{V})\tau_2}\mathbf{V} e^{(\mathbf{K}-\mathbf{V})\tau_1}\mathbf{V}\mathbf{p}_{\text{ss}}/\langle n\rangle \\ &= \mathbf{1}^{\top}\mathbf{V}\left(\prod_i e^{(\mathbf{K}-\mathbf{V})\tau_i}\mathbf{V}\right)\mathbf{p}_{\text{ss}}/\langle n\rangle \end{aligned} \tag{42}$$

where τ_i is the time interval between the ith and $(i+1)$th photon.

For the simple scheme in Fig. 1(c), using Eqs. (10), (11), (35), and (38), we find

$$\text{Likelihood} = \prod_i \varphi(\tau_i), \tag{43}$$

where the interphoton time distribution $\varphi(\tau)$ is given in Eq. (39).

2.7. *Relation to renewal theory*

Here, we consider how the above general formalism is related to renewal theory. Renewal theory considers the statistics of events when the time intervals between successive events are independently and identically distributed.[63,64] It describes, for example, successive replacements of light bulbs: when a bulb fails it is immediately replaced or renewed. Renewal theory relates the properties of random variables to a single function, the distribution of the time between consecutive events. Renewal theory can describe the photon statistics of the two-state system shown in Fig. 1(c). In this case the time intervals between consecutive photons are uncorrelated, as indicated by the factorization of the likelihood function in Eq. (43). In the single-molecule context, approaches based on renewal theory have been used in Refs. 31 and 34.

Our formalism reduces to renewal theory if the matrix of the monitored transitions (i.e., detected photons), $\mathbf{V}$, is separable, namely, when it can be represented as

$$\mathbf{V} = \mathbf{u}\mathbf{v}^{\top}, \tag{44}$$

where $\mathbf{u}$ and $\mathbf{v}$ are column vectors. This is possible only when all transitions that result in detected photons originate from or go to a single state. In this case successive interphoton times are uncorrelated.[36] For example, the statistics of the photons in the schemes in Figs. 1(b)–1(e) (in the absence of conformational changes) can be described in terms of renewal theory. However, renewal theory does not work for the scheme in Fig. 1(f) that describes donor reexcitation, and, more significantly, those in Figs. 1(g) and 1(h), i.e., the schemes that include conformational dynamics.

When matrix **V** is separable, matrix multiplications are simplified and all distributions and the correlation function can be expressed in terms of the interphoton time distribution, $\varphi(\tau)$. This quantity is the input of renewal theory. In this case, we have shown elsewhere[51] that our formalism for the probability of the number of photons in a bin reduces to:

$$\begin{aligned} \hat{P}(0 \mid s) &= \frac{1}{s} - \frac{1-\hat{\varphi}(s)}{s^2\langle\tau\rangle}, \\ \hat{P}(N \mid s) &= \frac{\hat{\varphi}(s)^{N-1}(1-\hat{\varphi}(s))^2}{s^2\langle\tau\rangle}, \qquad N = 1,\ 2, \ldots \end{aligned} \tag{45}$$

where $\langle\tau\rangle = \langle n\rangle^{-1}$ is the mean time between photons. In addition, the Laplace transform of the intensity correlation function is related to $\hat{\varphi}(s)$ as[12]

$$\int_0^\infty \frac{\langle n(t)n(0)\rangle}{\langle n\rangle} e^{-st}\, dt = \frac{\hat{\varphi}(s)}{1-\hat{\varphi}(s)}. \tag{46}$$

These are standard results of renewal theory.[63] However, it should be emphasized that in the presence of conformational changes renewal theory does not work.

2.8. *Poisson statistics*

Even for the two-state system (Fig. 1(c)), the distribution of the number of photons is not Poissonian (see Eq. (24)) because after emitting a photon, the system must be reexcited before the next photon can be emitted. Since this takes time, successive photons are correlated. However, when the relaxation to steady-state is fast compared to the time between two consecutively detected photons (e.g., due to small detection efficiency), the distribution of photons can be well approximated by the Poisson distribution at all but the shortest times.

Let us now show how the Poisson distribution of photon counts arises for a simple system described by the kinetic scheme shown in Fig. 1(c). Consider Eq. (23) which determines the generating function of the photon count probabilities. When the detection efficiency is small, the "escape" rate $(1-\lambda)k'_R$ is small compared to other terms. This means that the system makes many cycles between the ground and excited states before a photon is detected. Consequently, f_D and f_{D^*} can be approximated by the steady-state populations (in the absence of decay) corrected by the decay term, which is the same for both states. Mathematically, we set $f_D \approx p_{ss}(D)f(t\,|\,\lambda)$ and $f_{D^*} \approx p_{ss}(D^*)f(t\,|\,\lambda)$, where the steady-state populations p_{ss} are given in Eq. (7). Summing up the equations for f_D and f_{D^*}, Eq. (23), and then using the above approximation, we find that

$$\frac{d}{dt}f(t\,|\,\lambda) = -(1-\lambda)\langle n\rangle f(t\,|\,\lambda), \tag{47}$$

where $\langle n\rangle$ is given by Eq. (35) and $f(0\,|\,\lambda) = 1$. The solution of this equation is a single exponential and thus

$$\sum_{N=0}^{\infty}\lambda^N P(N\,|\,T) = f(T\,|\,\lambda) = \exp(-(1-\lambda)\langle n\rangle T). \tag{48}$$

Expanding this generating function in powers of λ, we get the Poisson distribution

$$P(N\,|\,T) = \frac{(\langle n\rangle T)^N}{N!}e^{-\langle n\rangle T}. \tag{49}$$

For future reference note that the generating function for a Poisson process in Eq. (48) is formally identical to the decay of the concentration of A in a simple irreversible reaction $A \to 0$ with rate constant given by $(1-\lambda)\langle n\rangle$.

The above derivation can be extended to arbitrary kinetic schemes. The distribution of photon counts is Poissonian when the relaxation to steady state is faster than the mean time between photons, $\langle n\rangle^{-1}$, which is defined in Eq. (28).

The mean of the Poisson distribution is given by Eq. (27). The variance is equal to the mean,

$$\langle N^2 \rangle - \langle N \rangle^2 = \langle N \rangle, \tag{50}$$

and hence the Mandel "Q" parameter is zero (see Eq. (33)).

Analogously, in the fast relaxation limit, the interphoton time distribution becomes exponential:

$$\varphi(\tau) = \langle n \rangle \exp(-\langle n \rangle \tau). \tag{51}$$

This is valid at times longer than the time it takes the system to come to steady state. At these times, the photons are no longer correlated, i.e., $\langle n(t) n(0) \rangle = \langle n \rangle^2$.

3. Fluorescence quenching and conformational dynamics

A simple example of how photon statistics can be influenced by conformational dynamics is fluorescence quenching. A quencher changes the nonradiative decay rate of a fluorophore by an amount that depends on the fluorophore-quencher distance. In the simplest model there are only two such distances and the corresponding kinetic scheme is given in Fig. 1(g). This scheme will be considered in detail in Secs. 3.1 and 3.2. The extension to a continuum description of many conformational states is given in Secs. 3.3 and 3.4. In both cases we start with the exact description of photon statistics obtained by directly applying the general formalism in Sec. 2 to fluorescence quenching (Secs. 3.1 and 3.3). In Secs. 3.2, and 3.4, we consider conformational dynamics that is much slower than photophysical relaxation. In this case, by using a generalization of the procedure used above to derive Poisson statistics, we can eliminate photophysical states and thus simplify the formalism. Finally, in Sec. 3.5 we show that the distribution of photon counts in a bin in the presence of slow conformational dynamics can be written in the form of Mandel's formula.

3.1. *Two conformational states*

Consider statistics of photon counts when the fluorophore can be in two conformational states that have different lifetimes. The corresponding kinetic scheme is shown in Fig. 1(g). The transition rates between the two conformational states are L_1 and L_2. For this model the rate matrix $\mathbf{K}$ is a 4×4 matrix. In the basis (D_1, D_2, D_1^*, D_2^*) it is

$$\mathbf{K} = \begin{pmatrix} -(L_1 + k_{\text{ex}}) & L_2 & k_{\text{D1}} & 0 \\ L_1 & -(L_2 + k_{\text{ex}}) & 0 & k_{\text{D2}} \\ k_{\text{ex}} & 0 & -(L_1 + k_{\text{D1}}) & L_2 \\ 0 & k_{\text{ex}} & L_1 & -(L_2 + k_{\text{D2}}) \end{pmatrix}. \tag{52}$$

The matrix $\mathbf{V}$ is obtained by deleting all the elements of $\mathbf{K}$, except those *off-diagonal* ones that correspond to the detected photons, and hence has two nonzero elements:

$$\mathbf{V} = \begin{pmatrix} 0 & 0 & k'_{\text{R}} & 0 \\ 0 & 0 & 0 & k'_{\text{R}} \\ 0 & 0 & 0 & 0 \\ 0 & 0 & 0 & 0 \end{pmatrix}. \tag{53}$$

All quantities of interest can be obtained by using these matrices in the general expressions presented in Sec. 2. For instance, the interphoton time distribution, $\varphi(\tau)$, is the sum of four exponentials with the exponents equal to the eigenvalues of $\mathbf{K} - \mathbf{V}$ (see Eq. (38b)). The intensity correlation function is the sum of three exponentials with the exponents equal to the nonzero eigenvalues of $\mathbf{K}$ (see Eq. (31b)). The distribution of photon counts can be found, for instance, by inverting the Laplace transform in Eq. (14) numerically.

Figure 6 shows the intensity correlation function $\langle n(t)n(0)\rangle / \langle n \rangle^2$ for different rates of conformational changes $L_0 = L_1 + L_2$. The dashed line shows the interphoton time distribution $\varphi(\tau)/\langle n \rangle$. The curves are obtained by substituting $\mathbf{K}$ and $\mathbf{V}$, Eqs. (52) and (53), into Eqs. (31) and (38) and calculating the matrix products and exponentials using Mathematica (Wolfram Research, Champaign, IL).

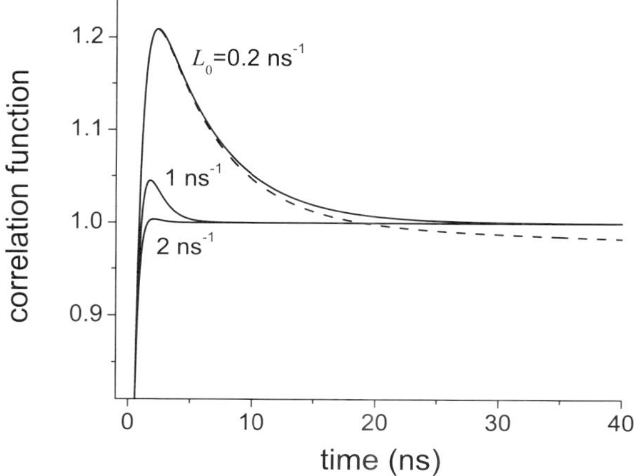

Fig. 6. Intensity correlation function $\langle n(t)n(0)\rangle/\langle n\rangle^2$ (full lines) for the fluorophore with two interconverting conformations, Fig. 1(g). $k_{ex} = 1\,\text{ns}^{-1}$, $k_{D1} = 0.5\,\text{ns}^{-1}$, $k_{D2} = 5\,\text{ns}^{-1}$, $k'_R = 1\,\mu\text{s}^{-1}$, $L_1 = L_2$, the rate of conformational relaxation $L_0 = L_1 + L_2$ is $0.2\,\text{ns}^{-1}$, $1\,\text{ns}^{-1}$, and $2\,\text{ns}^{-1}$. Dashed line shows the interphoton time distribution $\varphi(\tau)/\langle n\rangle$ for $L_0 = 0.2\,\text{ns}^{-1}$.

The vector of the steady-state probabilities $\mathbf{p}_{ss}$ is the eigenvector of $\mathbf{K}$ that corresponds to the zero eigenvalue. At short times, both the interphoton time distribution and the correlation function show a dip due to antibunching. At long times, the normalized correlation function approaches unity and the interphoton time distribution decays to zero. At intermediate times, the correlation function has a maximum due to conformational dynamics.[20,40] The maximum becomes smaller and eventually disappears as the rate of the conformational changes approaches the photophysical rates. The mean time between photons (about $3\,\mu$s for the parameters used here) is much longer than any relaxation time in the system. Consequently, the interphoton time distribution is related to the correlation function by Eq. (40). This relation is very accurate for the parameters used in Fig. 6.

3.2. *Slow conformational dynamics: Two states*

Although the exact formalism is easy to implement numerically in the case of two conformations, we now show how it can be simplified when the photophysical relaxation in each conformational state is fast compared to the rate of conformational changes. This is the case when excitation and decay are on the nanosecond time scale and the conformational relaxation time is on the order of tens of nanoseconds and longer. We show that in this limit the four-state kinetic scheme in Fig. 1(g) can be reduced to a two-state one. This simplification will allow us to get analytic expressions for the quantities of interest. In Sec. 3.4, we will use the generalization of the approach, which we have illustrated here in the simplest possible context, to simplify the formalism when the number of conformations is arbitrary.

When the transitions between the ground and excited states are much faster than conformational changes, the steady state of the excited and ground states is established in each conformation before conformational changes occur. This is the key idea that we shall exploit. The steady-state probabilities of the ground and excited states given the fluorophore is in conformation i ($i = 1, 2$) are

$$
\begin{aligned}
p^0_{\text{ss}}(D, i) &= \frac{k_{Di}}{k_{\text{ex}} + k_{Di}}, \\
p^0_{\text{ss}}(D^*, i) &= \frac{k_{\text{ex}}}{k_{\text{ex}} + k_{Di}},
\end{aligned}
\tag{54}
$$

where k_{Di} is the conformation-dependent decay rate. The steady-state probabilities in the presence of slow conformational changes are approximated by $p_{\text{ss}}(D, i) = p^0_{\text{ss}}(D, i)p_{\text{eq}}(i)$ and $p_{\text{ss}}(D^*, i) = p^0_{\text{ss}}(D^*, i)p_{\text{eq}}(i)$, where $p_{\text{eq}}(i)$ is the equilibrium probability to be in the conformational state i:

$$
p_{\text{eq}}(1) = \frac{L_2}{L_1 + L_2}, \quad p_{\text{eq}}(2) = \frac{L_1}{L_1 + L_2}. \tag{55}
$$

When the fluorophore is in conformational state i ($i = 1, 2$), the photophysical relaxation rate is $k_i = k_{\text{ex}} + k_{Di}$, and the photons are detected with the rate (number of photons per unit time) n_i:

$$
n_i = k'_{\text{R}} p^0_{\text{ss}}(D^*, i) = \frac{k'_{\text{R}} k_{\text{ex}}}{k_{\text{ex}} + k_{Di}}. \tag{56}
$$

The average number of photons is the equilibrium average of n_1 and n_2:

$$\langle n \rangle = \langle n \rangle_{\mathrm{c}} \equiv n_1 p_{\mathrm{eq}}(1) + n_2 p_{\mathrm{eq}}(2) \tag{57}$$

where the subscript "c" indicates averaging over conformations. Thus all information about the photophysical transitions is packed into the photon count rates n_1 and n_2. The problem is reduced to a simpler one, where light intensity switches between the values n_1 and n_2 with the transition rates L_1 and L_2. Below, we show how the general formalism for the generating function, correlation function, and the interphoton time distribution is simplified by exploiting the separation of time scales.

3.2.1. *Distribution of the number of photons in a bin*

We start with the four coupled equations that determine the generating function. These are obtained by using **K** and **V** given by Eqs. (52) and (53) in Eq. (20). Then, we add the equations for $f_{D_i}(t \mid \lambda)$ and $f_{D_i^*}(t \mid \lambda)$ first for $i = 1$ and then for $i = 2$. Finally, we use the local steady-state approximation, $f_{D_i}(t \mid \lambda) \approx p_{\mathrm{ss}}^0(D, i) f_i(t \mid \lambda)$ and $f_{D_i^*}(t \mid \lambda) \approx p_{\mathrm{ss}}^0(D^*, i) f_i(t \mid \lambda)$ where $p_{\mathrm{ss}}^0(D, i)$ and $p_{\mathrm{ss}}^0(D^*, i)$ are the steady-state probabilities of the ground and excited states in frozen conformation i, $i = 1, 2$ (see Eq. (54)). In this way, we find that the generating function is given by

$$\sum_{N=0}^{\infty} \lambda^N P(N \mid T) = f_1(T \mid \lambda) + f_2(T \mid \lambda), \tag{58}$$

where $f_1(t \mid \lambda)$ and $f_2(t \mid \lambda)$ satisfy

$$\frac{df_1}{dt} = -L_1 f_1 + L_2 f_2 - (1 - \lambda) n_1 f_1 \tag{59a}$$

$$\frac{df_2}{dt} = L_1 f_1 - L_2 f_2 - (1 - \lambda) n_2 f_2 \tag{59b}$$

where the count rate in the ith conformational state, n_i, is given in Eq. (56). Equations (59a) and (59b) must be solved subject to the initial conditions $f_i(0 \mid \lambda) = p_{\text{eq}}(i)$, $i = 1, 2$. These equations describe two-state conformational dynamics with the irreversible decay terms proportional to the photon count rates n_1 and n_2. Thus we have reduced the four-state system in Fig. 1(g) to an effective two-state system. This reduction is valid when photophysical relaxation is faster than conformational changes, the mean count rate, and the reciprocal of the bin size.

The solution of the above equations for the generating function can be presented in matrix form as in Eq. (18), when $\mathbf{K}$ is identified with the rate matrix describing conformational dynamics $\mathcal{L}$

$$\mathbf{K} \to \mathcal{L} = \begin{pmatrix} -L_1 & L_2 \\ L_1 & -L_2 \end{pmatrix}, \tag{60}$$

$\mathbf{V}$ with the diagonal matrix $\mathcal{N}$

$$\mathbf{V} \to \mathcal{N} = \begin{pmatrix} n_1 & 0 \\ 0 & n_2 \end{pmatrix}, \tag{61}$$

and $\mathbf{p}_{\text{ss}}$ with $\mathbf{p}_{\text{eq}}$, where $\mathbf{p}_{\text{eq}}$ is the vector of equilibrium probabilities which has the elements given in Eq. (55). Note that $\mathcal{L}\mathbf{p}_{\text{eq}} = 0$, in complete analogy with $\mathbf{K}\mathbf{p}_{\text{ss}} = 0$.

Solving Eq. (59) and expanding the solution in powers of λ, we can find the distribution of the number of photons in a bin. This can be expressed analytically as $(n_2 > n_1)$[51]:

$$\begin{aligned} P(N \mid T) &= p_{\text{eq}}(1)e^{-L_1 T}\frac{(n_1 T)^N}{N!}e^{-n_1 T} + p_{\text{eq}}(2)e^{-L_2 T}\frac{(n_2 T)^N}{N!}e^{-n_2 T} \\ &\quad + \frac{2L_0 T p_{\text{eq}}(1) p_{\text{eq}}(2)}{n_2 - n_1}\int_{n_1}^{n_2}\frac{(\eta T)^N}{N!}e^{-(\eta + z L_0)T}(I_0(y) \\ &\quad + L_0 T(1-z) I_1(y)/y) d\eta \end{aligned} \tag{62}$$

where $L_0 = L_1 + L_2$, $y = 2L_0T\sqrt{p_{\text{eq}}(1)p_{\text{eq}}(2)x(1-x)}$, $x = (\eta - n_1)/(n_2 - n_1)$, $z = p_{\text{eq}}(2)(1-x) + p_{\text{eq}}(1)x$, and $I_n(y)$ are the modified Bessel function of the first kind of order n.

The first two terms in Eq. (62) are Poisson distributions weighted by the equilibrium probabilities of the conformational states. These describe events that occur when the system remains in the state 1 or 2 during the bin time. The last term in Eq. (62) is due to transitions between the conformations. The first two terms are dominant when the bin size is very short compared to conformational relaxation time ($L_0T \ll 1$). Therefore, when the bin size T is short, the distribution of photon counts contains information only about the equilibrium populations. The conformational dynamics is reflected in $P(N \mid T)$ when the bin size T is comparable to the conformation relaxation time, L_0^{-1}. When the bin size is long compared to the conformational relaxation ($L_0T \gg 1$), the distribution of photons, $P(N \mid T)$, is a Gaussian centered on average number of photons, $\langle N \rangle = \langle n \rangle_c T$, where $\langle n \rangle_c$ is given in Eq. (57). The variance is equal to

$$\langle N^2 \rangle - \langle N \rangle^2 = \langle N \rangle + 2\frac{(n_1 - n_2)^2 p_{\text{eq}}(1)p_{\text{eq}}(2)}{L_0^2} \times (L_0T + e^{-L_0T} - 1). \tag{63}$$

When conformational dynamics is faster than the mean time between photons, $L_0 \gg \langle n \rangle$, the second term in Eq. (63) is negligible; therefore, the variance is equal to the mean and the distribution of photons is approximately Poissonian.

Figure 7 shows the distribution of photon counts in a bin, $P(N \mid T)$, calculated using Eq. (62). The distribution is plotted as a function of $N/\langle N \rangle = N/(\langle n \rangle T)$ so that it does not shift when the bin size is increased. The conformational relaxation time, $L_0^{-1} = (L_1 + L_2)^{-1} = 2$ ms, is the same for all distributions. When the bin size is shorter than the conformational relaxation time, $L_0T \ll 1$, the distribution is the superposition of Poisson distributions (the first two terms in Eq. (62)). As bin size increases, the Poissonian peaks narrow.

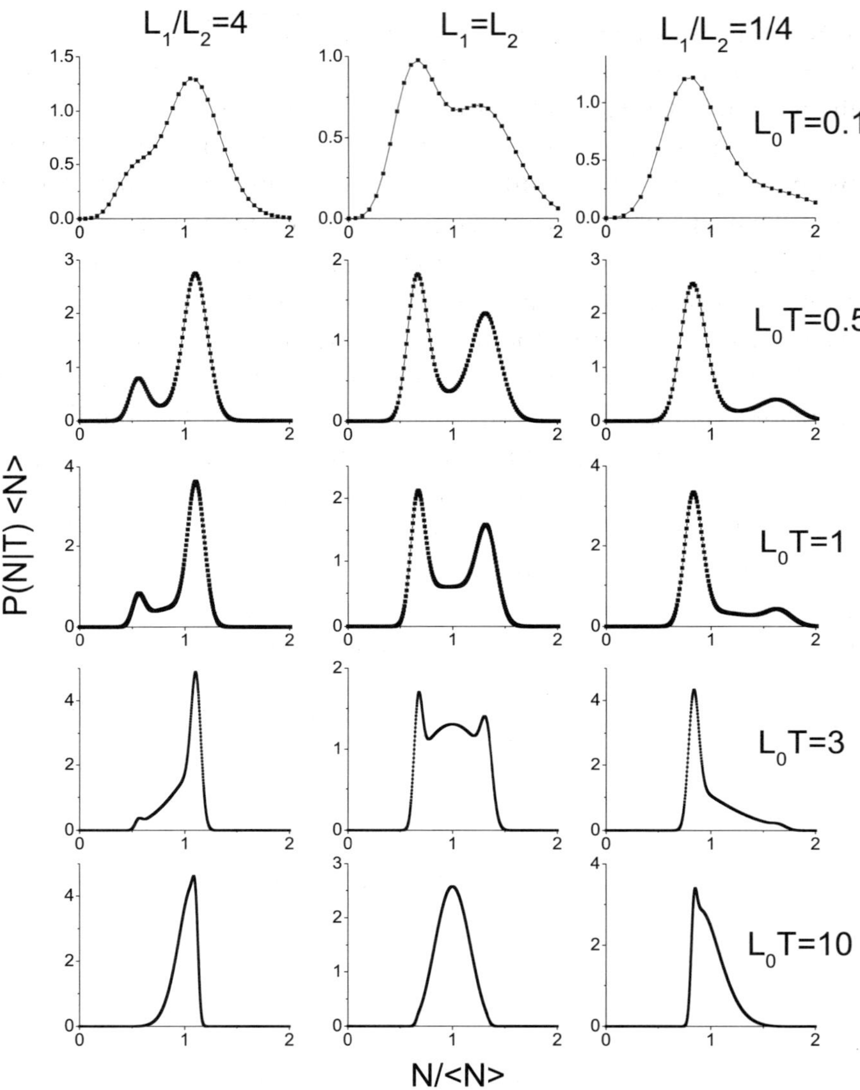

Fig. 7. Distribution of the number of photons, $P(N \mid T)\langle N\rangle$, as a function of $N/\langle N\rangle = N/\langle n\rangle T$. The photons are emitted by a fluorophore with two slowly interconverting conformations with two different lifetimes. $n_1 = 50\,\mathrm{ms}^{-1}$, $n_2 = 100\,\mathrm{ms}^{-1}$, $L_0 = L_1 + L_2 = 0.5\,\mathrm{ms}^{-1}$. Left ($L_1/L_2 = 4$), center ($L_1 = L_2$), and right ($L_1/L_2 = 1/4$) columns correspond to different equilibrium populations. Bin size is $T = 0.2, 1, 2, 6, 20$ ms.

When the bin size becomes comparable to the conformational relaxation time, a plateau appears between the peaks due to transitions between the conformations. As the bin size increases further, the distribution eventually becomes a Gaussian centered at $N/\langle N\rangle_T = 1$. The left and right columns show how the distribution behaves when the conformational states are unequally populated.

In summary, when all dynamics are much faster than the time between photons, the distribution of photon counts is Poissoinian, as discussed in Sec. 2.8. Conformational dynamics that are slow compared to the time between photons alters this distribution. The shape of the distribution depends on the bin size. When the bin size is small, the conformational state does not change during the bin and the distribution is a superposition of Poissonian ones. When the bin size is large, a molecule samples different conformational states during the bin time and the distribution is a Gaussian.

3.2.2. *Intensity correlation function*

Now consider the intensity correlation function when two-state conformational dynamics is slower than photophysical relaxation.

At times t short compared to the conformational relaxation time, $L_0 t \ll 1$, the conformational coordinate does not change during t. Therefore, the correlation function is the equilibrium average of the correlation functions of the two states, Eq. (36):

$$\langle n(t)n(0)\rangle = \sum_{i=1}^{2} n_i^2(1 - e^{-k_i t})p_{\text{eq}}(i), \tag{64}$$

where $k_i = k_{\text{ex}} + k_{Di}$ is the photophysical relaxation rate of conformation i.

At times longer than the photophysical relaxation time, the problem is reduced to finding the correlation function of the light intensity that switches between the values n_1 and n_2 with the transition rates

L_1 and L_2:

$$\langle n(t)n(0)\rangle_c = \langle n\rangle_c^2 + (\langle n^2\rangle_c - \langle n\rangle_c^2)e^{-L_0 t}$$
$$= \langle n\rangle_c^2 \left(1 + \frac{(n_1 - n_2)^2 L_1 L_2}{(n_1 L_2 + n_2 L_1)^2} e^{-L_0 t}\right). \tag{65}$$

Here the subscript "c" indicates the averaging over conformational fluctuations, $L_0 = L_1 + L_2$ is the rate of conformational relaxation, $\langle n\rangle_c$ is given in Eq. (57), and $\langle n^2\rangle_c = n_1^2 p_{eq}(1) + n_2^2 p_{eq}(2)$. This can be formally obtained from Eq. (31b) by replacing **K** and **V** by $\mathcal{L}$ (Eq. (60)) and $\mathcal{N}$ (Eq. (61)), respectively.

If one defines δn as the deviation of n from its average value, $\delta n = n - \langle n\rangle_c$, then Eq. (65) can be written as

$$\frac{\langle \delta n(t)\delta n(0)\rangle_c}{\langle \delta n^2\rangle_c} = e^{-L_0 t}. \tag{66}$$

Thus in this case, the decay of the fluctuations of the intensity about the mean is a direct measure of the conformational relaxation time.

Finally, we can combine Eqs. (64) and (65) to get an expression for the intensity correlation function which is approximately valid at all times:

$$\langle n(t)n(0)\rangle \approx \langle n\rangle_c^2 + [\langle n^2\rangle_c - \langle n\rangle_c^2]e^{-L_0 t} - \sum_{i=1}^{2} n_i^2 e^{-k_i t} p_{eq}(i). \tag{67}$$

At times short compared to the conformational relaxation time, $L_0 t \ll 1$, Eq. (64) is recovered. At times long compared to the photophysical relaxation time $k_i t \gg 1$, we get Eq. (65). Because of time scale separation, $k_i \gg L_0$, these ranges of validity overlap and Eq. (67) is valid at all times.

The above correlation function is essentially equivalent to

$$\frac{\langle n(t)n(0)\rangle}{\langle n\rangle^2} \approx \left(1 - \sum_{i=1}^{2} \frac{n_i^2}{\langle n^2\rangle_c} e^{-k_i t} p_{eq}(i)\right)$$
$$\times \left(1 + \frac{\langle n^2\rangle_c - \langle n\rangle_c^2}{\langle n\rangle_c^2} e^{-L_0 t}\right). \tag{68}$$

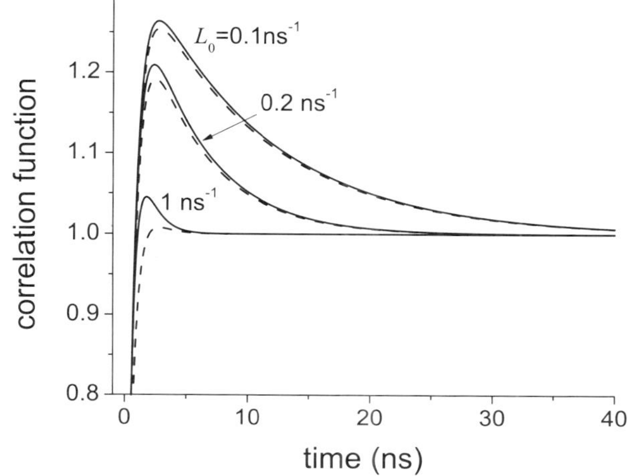

Fig. 8. Intensity correlation function $\langle n(t)n(0)\rangle/\langle n\rangle^2$ for the fluorophore with two conformations (see Fig. 1(g)). The exact results (full lines) are compared with the approximation in Eq. (68) (dashed lines). Parameters are the same as in Fig. 6, except the conformational relaxation rate $L_0 = 0.1\ \text{ns}^{-1}$, $0.2\ \text{ns}^{-1}$, and $1\ \text{ns}^{-1}$.

If we expand this out and use $k_i \gg L_0$, we recover Eq. (67). The first factor in Eq. (68) describes fast photophysical dynamics, i.e., antibunching, and the second factor describes bunching due to slow conformational dynamics.

Figure 8 compares the exact intensity correlation function (full line, obtained using Eqs. (52) and (53) in Eq. (31b)) and the approximate one (dashed lines, calculated using Eq. (68)), for various values of conformational relaxation rate. The photophysical relaxation rates in two conformations are $k_1 = 1.5\ \text{ns}^{-1}$ and $k_2 = 6\ \text{ns}^{-1}$. The approximation works well when the conformational relaxation rate L_0 is much smaller than k_1 and k_2, i.e., when the time scales are well separated. In this limit, the approximations in Eqs. (67) and (68) are indistinguishable.

The approximations for the correlation function discussed above can be used to describe photon counting in the presence of triplet blinking, Fig. 1(d). Because the relaxation of the ground and excited states is fast compared to intersystem crossing rate, one can reduce the three-state system in Fig. 1(d) to a two-state system with a bright and a dark (triplet) state. The photon count rate in the dark state is zero and in the bright state is $n_1 = k'_R k_{ex}/(k_{ex} + k_D)$.

Triplet blinking is a special case of the above formalism for quenching with slow two-state conformational changes where $k_{D2} \to \infty$ (i.e., state 2 is completely quenched). The interconverting rates for triplet blinking are

$$L_1 = k_{ISC} p_{ss}^0(D^*) = \frac{k_{ISC} k_{ex}}{k_{ex} + k_D}, \qquad L_2 = k_T, \tag{69}$$

so that the relaxation rate is $L_0 = k_{ISC} p_{ss}^0(D^*) + k_T$. Using the above parameters in Eq. (68), we find that

$$\frac{\langle n(t)n(0)\rangle}{\langle n\rangle^2} = (1 - e^{-(k_{ex}+k_D)t}) \left(1 + \frac{p_{ss}(T)}{1 - p_{ss}(T)} e^{-L_0 t}\right), \tag{70}$$

where $p_{ss}(T) = L_1/L_0 = (1 + k_T(k_{ex} + k_D)/k_{ISC} k_{ex})^{-1}$ is the steady-state population of the triplet state.

3.2.3. *Interphoton time distribution*

When conformational dynamics is slow compared to excitation and decay, but still fast compared to the mean time between photons, the interphoton time distribution and the intensity correlation function are related by Eq. (40) for all times. In this case, the interphoton time distribution can be found using the approximations for the correlation function in Eqs. (67) or (68).

This is not so when the time scale of conformational dynamics is comparable to or slower than the mean time between photons. In this case, the interphoton time distribution distribution can be found using the local steady-state approximation analogous to that in Eq. (59). It is given by Eq. (38) with **K** and **V** replaced by $\mathcal{L}$, Eq. (60), and $\mathcal{N}$, Eq. (61), respectively. The distribution is biexponential with the exponents equal to the eigenvalues of $\mathcal{L} - \mathcal{N}$:

$$\varphi(\tau) = \frac{a_- e^{-\lambda_- \tau} - a_+ e^{-\lambda_+ \tau}}{L_0 \langle n \rangle_c (\lambda_+ - \lambda_-)}$$

$$\lambda_\pm = \frac{1}{2}[L_0 + n_1 + n_2 \pm \sqrt{(L_1 - L_2 + n_1 - n_2)^2 + 4L_1 L_2}] \quad (71)$$

$$a_\pm = (L_0 \langle n \rangle_c + (n_1 - \lambda_\pm) n_2)(L_0 \langle n \rangle_c + (n_2 - \lambda_\pm) n_1).$$

This expression for the interphoton time distribution is valid when both τ and conformational relaxation time L_0^{-1} are longer than the photophysical relaxation time.

When conformational dynamics is slow compared to the mean time between photons ($L_0 \ll \langle n \rangle_c$), the two exponents in the above expression approach the count rates in each conformational state, n_1 and n_2:

$$\varphi(\tau) = \langle n \rangle^{-1} \left(p_{\text{eq}}(1) n_1^2 e^{-n_1 \tau} + p_{\text{eq}}(2) n_2^2 e^{-n_2 \tau} \right). \quad (72)$$

Note that in this limit φ does not depend on the time scale of conformational dynamics unlike the intensity correlation function (see Eq. (65)). In the opposite limit, when conformational dynamics is fast ($L_0 \gg \langle n \rangle_c$), the distribution becomes single exponential, Eq. (51).

3.3. *Many conformational states*

It is straightforward to generalize the above formalism to many discrete conformational states (see Fig. 1(h)). However, it is actually simpler to consider the continuum limit because the structure of the resulting theory is more transparent. In the continuum limit, the discrete label i for the conformational state is replaced by a conformational coordinate r. We describe conformational dynamics in D and D^* states as diffusion on one-dimensional free-energy surface $U(r)$, which is assumed here to be the same for the ground and excited states. The formalism can be further generalized to handle "non-Markovian" conformational dynamics by using multidimensional conformational coordinates. Although this is formally straightforward, for the sake of simplicity we restrict ourselves to the one-dimensional case.

The generalization of Eq. (52) to a continuum of conformational states is

$$\mathbf{K} = \begin{pmatrix} \mathcal{L}_r - k_{\mathrm{ex}} & k_{\mathrm{D}}(r) \\ k_{\mathrm{ex}} & \mathcal{L}_r - k_{\mathrm{D}}(r) \end{pmatrix}. \tag{73}$$

Here, the lifetime $k_{\mathrm{D}}(r)^{-1}$ depends on the fluorophore-quencher distance r that fluctuates because of conformational changes. $\mathcal{L}_r$ is the diffusion operator describing conformational dynamics:

$$\mathcal{L}_r \equiv \frac{\partial}{\partial r} De^{-\beta U(r)} \frac{\partial}{\partial r} e^{\beta U(r)} \tag{74}$$

where D is the diffusion coefficient and $\beta = (k_{\mathrm{B}}T)^{-1}$. The normalized conformational equilibrium distribution $p_{\mathrm{eq}}(r)$ satisfies $\mathcal{L}_r p_{\mathrm{eq}}(r) = 0$ and is given by

$$p_{\mathrm{eq}}(r) = \frac{\exp(-\beta U)}{\int \exp(-\beta U) dr}. \tag{75}$$

Matrix $\mathbf{V}$ is the same as that in Eq. (10):

$$\mathbf{V} = \begin{pmatrix} 0 & k'_{\mathrm{R}} \\ 0 & 0 \end{pmatrix}, \tag{76}$$

The radiative rate is assumed not to depend on conformation. All quantities of interest can be found by solving Eq. (20) with the above $\mathbf{K}$ and $\mathbf{V}$.

The steady-state populations $p_{\mathrm{ss}}(D, r)$ and $p_{\mathrm{ss}}(D^*, r)$ are solutions of $\mathbf{K}\mathbf{p}_{\mathrm{ss}} = 0$, which in this case is

$$\begin{aligned} &(\mathcal{L}_r - k_{\mathrm{ex}}) p_{\mathrm{ss}}(D, r) + k_{\mathrm{D}}(r) p_{\mathrm{ss}}(D^*, r) = 0 \\ &(\mathcal{L}_r - k_{\mathrm{D}}(r)) p_{\mathrm{ss}}(D^*, r) + k_{\mathrm{ex}} p_{\mathrm{ss}}(D, r) = 0. \end{aligned} \tag{77}$$

The steady-state populations are normalized so that $\int [p_{\mathrm{ss}}(D, r) + p_{\mathrm{ss}}(D^*, r)] dr = 1$. The mean number of detected photons per unit time is given by

$$\langle n \rangle = k'_{\mathrm{R}} \int p_{\mathrm{ss}}(D^*, r)dr \tag{78}$$

which is the generalization of Eq. (35). All other quantities of interest can be obtained by solving Eq. (20) for appropriate initial conditions and values of λ. Using Eq. (73) for **K** and Eq. (76) for **V** in Eq. (20), we have

$$\begin{aligned} \frac{\partial}{\partial t} f_D &= \mathcal{L}_r f_D - k_{\mathrm{ex}} f_D + k_{\mathrm{D}}(r) f_{D^*} - (1-\lambda) k'_{\mathrm{R}} f_{D^*}, \\ \frac{\partial}{\partial t} f_{D^*} &= \mathcal{L}_r f_{D^*} + k_{\mathrm{ex}} f_D - k_{\mathrm{D}}(r) f_{D^*}. \end{aligned} \tag{79}$$

When this is solved with steady-state initial conditions $f_D(r, 0 \mid \lambda) = p_{\mathrm{ss}}(D, r)$ and $f_{D^*}(r, 0 \mid \lambda) = p_{\mathrm{ss}}(D^*, r)$, then the generating function of the probability distribution of photon counts is given by

$$\sum_{N=0}^{\infty} \lambda^N P(N \mid T) = \int [f_D(r, T \mid \lambda) + f_{D^*}(r, T \mid \lambda)]dr. \tag{80}$$

When Eq. (79) is solved for $\lambda = 0$ with initial conditions $f_D(r, 0 \mid 0) = p_{\mathrm{ss}}(D^*, r)$ and $f_{D^*}(r, 0 \mid 0) = 0$, then the interphoton time distribution is given by

$$\varphi(\tau) = \frac{k'^{\,2}_{\mathrm{R}}}{\langle n \rangle} \int f_{D^*}(r, \tau \mid 0)dr. \tag{81}$$

Finally, when Eq. (79) is solved for $\lambda = 1$ with the same initial conditions, $f_D(r, 0 \mid 1) = p_{\mathrm{ss}}(D^*, r)$ and $f_{D^*}(r, 0 \mid 1) = 0$, then the intensity correlation function is given by

$$\langle n(t)n(0) \rangle = k'^{\,2}_{\mathrm{R}} \int f_{D^*}(r, t \mid 1)dr. \tag{82}$$

In general, Eq. (79) must be solved numerically. The most straightforward procedure is based on using a finite-difference approximation of the diffusion operator. This results in the kinetic scheme in

Fig. 1(h) in which the rate constants between different conformations are expressed in terms of D and $U(r)$. Then, we can simply use the matrix formalism developed in Sec. 2 to calculate quantities of interest.

3.4. *Slow conformational dynamics: Many states*

We can simplify the above formalism when conformational dynamics is much slower than photophysical relaxation by generalizing the approach used in Sec. 3.2. The basic idea is the same. The populations of the ground and excited states come to steady state before the conformation changes. For fixed conformational coordinate r, the photophysical relaxation rate is $k(r) = k_{\mathrm{ex}} + k_{\mathrm{D}}(r)$ and the mean number of photons detected per unit time $n(r)$ is

$$n(r) = \frac{k'_{\mathrm{R}} k_{\mathrm{ex}}}{k_{\mathrm{ex}} + k_{\mathrm{D}}(r)}, \tag{83}$$

which is the generalization of Eq. (56). In the limit of slow conformational changes, the mean number of photons per unit time is equal to $\langle n \rangle_{\mathrm{c}}$, which is the equilibrium average of $n(r)$:

$$\langle n \rangle = \langle n \rangle_{\mathrm{c}} \equiv \int n(r) p_{\mathrm{eq}}(r) dr \tag{84}$$

where $p_{\mathrm{eq}}(r)$ is the equilibrium distribution of the conformational coordinate. This is is the generalization of Eq. (57).

The photon count rate fluctuates because of conformational dynamics. Below, we show how the distribution of photon counts, intensity correlation function, and interphoton time distribution can be obtained given $n(r)$ and the operator describing conformational dynamics.

3.4.1. *Distribution of the number of photons in a bin*

When a conformational coordinate r is fixed and photophysical relaxation is fast compared to the mean time between photons, the distribution of photon counts is Poissonian (see Sec. 2.8). The corresponding generating function can be found by solving (see Eq. (47))

$$\frac{\partial}{\partial t} f = -(1-\lambda) n(r) f \tag{85}$$

to be

$$\sum_{N=0}^{\infty} \lambda^N P(N \mid T) = e^{-(1-\lambda) n(r) T} \tag{86}$$

In the presence of conformational dynamics described by an operator $\mathcal{L}_r$, $n(r)$ fluctuates because r fluctuates. In this case, the generating function is given by

$$\sum_{N=0}^{\infty} \lambda^N P(N \mid T) = \int f(r, T \mid \lambda) dr, \tag{87}$$

where $f(r, t \mid \lambda)$ is found by solving

$$\frac{\partial}{\partial t} f = \mathcal{L}_r f - (1-\lambda) n(r) f \tag{88}$$

with initial condition $f(r, 0 \mid \lambda) = p_{\text{eq}}(r)$. This is a straightforward generalization of the two-state conformation results in Eqs. (58) and (59) to a continuum of conformations. Formally, this equation can be derived from the exact equation (79) by using the local steady-state approximation. This formalism works when the photophysical relaxation time is short compared to the bin size, mean time between photons, and the conformational relaxation time.

Thus, by exploiting the separation of time scales we have reduced the solution of Eq. (79), which couples excited and ground states, to the solution of Eq. (88), in which all information about fast photophysical processes is contained in the conformation-dependent count rate $n(r)$. Solving Eq. (88), one can find the photon count distribution by expanding the generating function in powers of the λ.

3.4.2. *Intensity correlation function*

The intensity correlation function for many states is the generalization of the correlation function for two conformational states considered in Sec. 3.2.2. For times short compared to the conformational relaxation time, we have

$$\langle n(t)n(0)\rangle = \int n(r)^2(1 - e^{-k(r)t})p_{\mathrm{eq}}(r)dr \tag{89}$$

which is the generalization of Eq. (64). At times longer than $k(r)^{-1}$, we can express the intensity correlation function as

$$\langle n(t)n(0)\rangle_{\mathrm{c}} = \iint n(r)G^0(r, t\,|r_0)n(r_0)p_{\mathrm{eq}}(r_0)drdr_0 \tag{90}$$

where $G^0(r, t \mid r_0)$ is the Green's function (or propagator) describing conformational dynamics. It is the solution of

$$\frac{\partial}{\partial t}G^0 = \mathcal{L}_r G^0 \tag{91}$$

with initial condition $G^0(r, 0 \mid r_0) = \delta(r - r_0)$. This is the generalization of the result for two conformational states in Eq. (65). Note that the intensity correlation function for times longer than the photophysical relaxation time is a direct measure of conformational fluctuations.

By combining Eqs. (89) and (90), we can find an expression for the intensity correlation function that is approximately valid at all times:

$$\langle n(t)n(0)\rangle \approx \langle n(t)n(0)\rangle_{\mathrm{c}} - \int n^2(r)e^{-k(r)t}p_{\mathrm{eq}}(r)dr. \tag{92}$$

At times short compared to the conformational relaxation time, the first term is just the equilibrium average of $n(r)^2$ and Eq. (89) is recovered. At times long compared to the photophysical relaxation time $k(r)^{-1}$, the second term is zero and we obtain Eq. (90).

Alternatively, this approximate correlation function is essentially the same as

$$\frac{\langle n(t)n(0)\rangle}{\langle n\rangle^2} \approx \left(1 - \frac{\int n^2(r)e^{-k(r)t}p_{\text{eq}}(r)dr}{\int n^2(r)p_{\text{eq}}(r)dr}\right)\frac{\langle n(t)n(0)\rangle_{\text{c}}}{\langle n\rangle_{\text{c}}^2}. \tag{93}$$

This approximation is the generalization of Eq. (68) for two conformational states. The first factor on the right-hand side of Eq. (93) describes antibunching due to fast photophysical relaxation and the second factor describes bunching due to slow conformational dynamics.

3.4.3. *Interphoton time distribution*

Just as in the case of the two conformational states, Sec. 3.2.3, when conformational dynamics is slow compared to photophysical relaxation but still fast compared to the mean time between photons, the interphoton time distribution for all times can be found using the approximate correlation function in Eqs. (92) or (93), in Eq. (40).

When the time scale of conformational dynamics is comparable to or slower than the mean time between photons, one can use the local steady-state approximation, which is valid only for times longer than the photophysical relaxation time, to find

$$\varphi(\tau) = \langle n\rangle_{\text{c}}^{-1}\int n(r)G(r,\tau\mid r_0)n(r_0)p_{\text{eq}}(r_0)drdr_0 \tag{94}$$

where $G(r,t\mid r_0)$ is the solution of

$$\frac{\partial}{\partial t}G = \mathcal{L}_r G - n(r)G \tag{95}$$

with initial condition $G(r,0\mid r_0) = \delta(r - r_0)$.

When conformational dynamics is slower than the mean time between photons, $\varphi(\tau)$, approaches

$$\varphi(\tau) = \langle n\rangle_{\text{c}}^{-1}\int n^2(r)e^{-n(r)\tau}p_{\text{eq}}(r)dr \tag{96}$$

which is the generalization of the result in Eq. (72) for two conformations. This distribution can be a highly nonexponential function of time. However, since it is determined by the conformational equilibrium distribution, it does not contain information about the time scale of slow conformational dynamics, unlike the intensity correlation function.

3.5. *Mandel's formula in the presence of slow conformational fluctuations*

When the conformational dynamics is slow, the distribution of the number of photons can be represented in the form of Mandel's formula. This formula is commonly used to describe photon statistics when the incident light intensity fluctuates.[61] First, we note that the generating function in Eqs. (87) and (88) can formally be written as a path integral,

$$\sum_{N=0}^{\infty} \lambda^N P(N \mid T) = \langle e^{-(1-\lambda)\int_0^T n(r(t))dt} \rangle, \tag{97}$$

where the angular brackets denote averaging over all conformational trajectories starting from equilibrium. This is the generalization of the generating function of the Poisson distribution given in Eq. (86), when the photon count rate $n(r)$ fluctuates because of conformational dynamics. Remarkably, this generating function is formally identical to the survival probability of a unimolecular irreversible reaction with a fluctuating decay rate equal to $(1-\lambda)n(r)$.

Expanding the right-hand side of Eq. (97) in powers of λ and equating coefficients, we find that $P(N \mid T)$ can be written in the form of Mandel's formula:

$$\begin{aligned} P(N \mid T) &= \left\langle \frac{[\int_0^T n(r(t))dt]^N}{N!} e^{-\int_0^T n(r(t))dt} \right\rangle \\ &= \int_0^\infty \frac{(\eta T)^N}{N!} e^{-\eta T} P_c(\eta \mid T) d\eta, \end{aligned} \tag{98}$$

where $P_c(\eta \mid T)$ is the distribution of the average number of photons in a time bin T that fluctuates due to slow conformational changes:

$$P_c(\eta \mid T) = \left\langle \delta\left(\eta - \frac{1}{T}\int_0^T n(r(t))dt\right)\right\rangle. \tag{99}$$

Mandel's formula shows that, in the presence of slow conformational changes, the probability distribution of detecting N photons in time T is a superposition of Poisson distributions with the weights determined by the fluctuations of the count rate due solely to conformational dynamics.

Using the Fourier representation of the Dirac δ-function,

$$\delta(x) = (2\pi)^{-1}\int_{-\infty}^{\infty} \exp(iwx)dw,$$

$P_c(\eta \mid T)$ can be represented as

$$P_c(\eta \mid T) = \frac{T}{2\pi}\int_{-\infty}^{\infty}\int e^{iw\eta T} f'(r, T \mid w)drdw, \tag{100}$$

where f' satisfies

$$\frac{\partial}{\partial t} f' = \mathcal{L}_r f' - iwn(r) f'. \tag{101}$$

This is the same as Eq. (88) with $(1 - \lambda)$ being replaced by iw. We have described a numerical procedure for obtaining $P_c(\eta \mid T)$ in this way in Ref. 46.

Consider the properties of the distribution $P_c(\eta \mid T)$. The mean of the distribution is $\langle n\rangle_c$, Eq. (84), and the variance is

$$\sigma_c^2(T) = \langle\eta^2\rangle - \langle\eta\rangle^2 = \frac{2}{T^2}\int_0^T (T - t)\langle\delta n(t)\delta n(0)\rangle_c dt \tag{102}$$

where $\delta n = n - \langle n\rangle_c$ and the correlation function $\langle\delta n(t)\delta n(0)\rangle_c = \langle n(t)n(0)\rangle_c - \langle n\rangle_c^2$ can be found using Eq. (90). As $T \to \infty$,

$$\sigma_c^2(T) = \frac{2\langle\delta n^2\rangle_c \tau_c}{T} \tag{103}$$

where we have defined the conformational relaxation time as

$$\tau_c = \int_0^\infty \frac{\langle \delta n(t)\delta n(0)\rangle_c}{\langle \delta n^2\rangle_c} dt. \tag{104}$$

When $T \ll \tau_c$, the conformational coordinate does not change during the bin and thus $P_c(\eta \mid T) = \int \delta(\eta - n(r))p_{eq}(r)dr$. When $T \gg \tau_c$, the distribution is approximately a Gaussian with mean $\langle n\rangle_c$, and variance given in Eq. (102). For very long bins, the distribution approaches a delta-function centered on the mean number of photons, $P_c(\eta \mid T) = \delta(\eta - \langle n\rangle_c)$.

For a two-state conformational dynamics, an analytical expression for $P_c(\eta \mid T)$ can be obtained by rescaling the results obtained by Berezhkovskii *et al.*[44] (see Eq. (2.18) in Ref. 46). In this way we find $(n_2 \geq \eta \geq n_1)$ that

$$\begin{aligned} P_c(\eta \mid T) &= p_{eq}(1)e^{-L_1 T}\delta(\eta - n_1) + p_{eq}(2)e^{-L_2 T}\delta(\eta - n_2) \\ &\quad + \frac{2L_0 T p_{eq}(1)p_{eq}(2)}{n_2 - n_1} e^{-zL_0 T} \\ &\quad \times (I_0(y) + L_0 T(1 - z)I_1(y)/y). \end{aligned} \tag{105}$$

Here, $y = 2L_0 T\sqrt{p_{eq}(1)p_{eq}(2)x(1-x)}$, $x = (\eta - n_1)/(n_2 - n_1)$, $z = p_{eq}(2)(1 - x) + p_{eq}(1)x$, and $I_n(x)$ is the modified Bessel function of the first kind. The first two terms in Eq. (105) describe events that occur when the system remains in state 1 or 2 during the bin time. The last term is due to transitions between the conformations. For large bins this term is dominant and eventually becomes a delta function centered on the average number of photons $(P_c(\eta \mid T) \to \delta(\eta - \langle n\rangle_c)$ as $T \to \infty)$. Note that using this $P_c(\eta \mid T)$ in Mandel's formula, Eq. (98), we obtain the two-state result for the distribution of photon counts given in Eq. (62).

4. Influence of translational diffusion

Consider how the diffusion of molecules through the laser spot influences photon statistics (see Fig. 9). A photon trajectory generated by diffusing molecules at low concentrations consists of bursts of photons with long gaps between them. One burst corresponds to one molecule that diffuses through the laser spot. The molecule may enter the spot several times before it leaves the spot forever. The next burst of photons corresponds to a different molecule.

The simplest (but oversimplified) description of the influence of diffusion on photon statistics is based on a two-state kinetic model. The system containing many molecules at concentration c diffusing in and out of the laser spot is mapped onto a two-state system which can be in a "bright" (a fluorophore is inside the laser spot, state 1) or "dark" state (the spot is empty, state 2). The two states interconvert with the

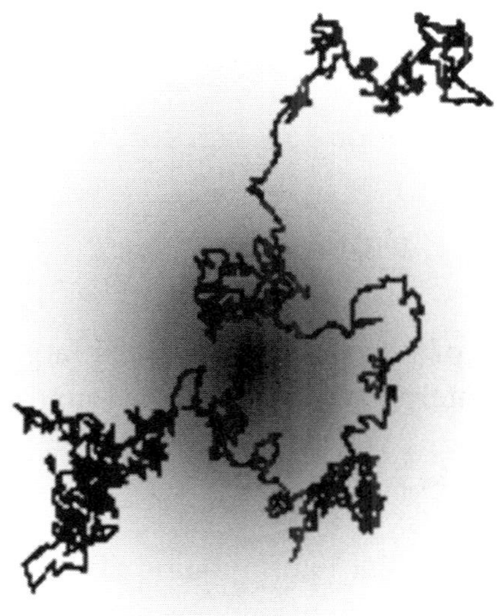

Fig. 9. Trajectory of a molecule diffusing through the laser spot.

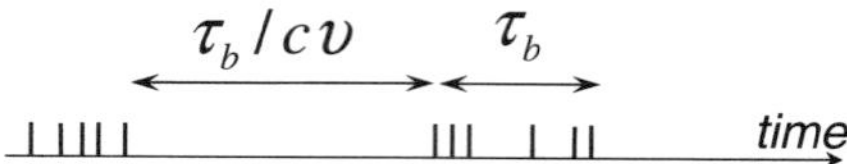

Fig. 10. Schematic representation of photons emitted by diffusing molecules of concentration c, v is the effective volume of the laser spot and τ_b is the burst duration time.

rates L_1 and L_2. Let n_1 be the average number of photons detected in state 1 and $n_2 = 0$. The equilibrium constant L_2/L_1 is equal to cv, where v is the effective volume of the spot. Because, we assume that c is small, $L_2 \ll L_1$. This means that the lifetime $(1/L_1)$ of the "bright" state is much shorter than that of the dark state. We can define the duration of a burst, τ_b, as the mean lifetime of the "bright" state. The time between bursts (i.e., the lifetime of the "dark state") will then be $1/L_2 = \tau_b/cv$. The relaxation rate is $L_0 = L_1 + L_2 \approx 1/\tau_b$. This model is the same as that for triplet blinking when the system is mostly in the triplet state, which corresponds to the molecule being outside of the laser spot. The resulting photon trajectory is shown schematically in Fig. 10.

The intensity correlation function for this model can be found from Eq. (66) to be

$$\frac{\langle \delta n(t) \delta n(0) \rangle}{\langle \delta n^2 \rangle} \approx e^{-t/\tau_b}. \tag{106}$$

When the number of photons in a burst is large ($n_1 \tau_b \gg 1$), the distribution of the interphoton times can be found using Eq. (71) to be

$$\varphi(\tau) \approx n_1 e^{-n_1 \tau} + \frac{1}{n_1 \tau_b} \cdot \frac{cv}{\tau_b} e^{-cv\tau/\tau_b}. \tag{107}$$

This has a simple physical interpretation. The first term is the normalized distribution of the time between photons within a bursts. The second term is the normalized distribution of times between bursts. The second term has a relative weight equal to the reciprocal mean number of photons in a burst, $(n_1 \tau_b)^{-1}$, which is the probability that two consecutive photons belong to different bursts.

This model, while qualitatively correct, is oversimplified for a variety of reasons. First, the laser spot is not homogeneous and is better approximated by a Gaussian shape rather than a sphere, so its "volume" cannot be rigorously defined. Second, diffusion is a stochastic process (see Fig. 9) in which a molecule can go in and out of a "fuzzy" laser spot. Thus, there is no rigorous definition of the duration of a burst. This is why, the intensity correlation function is not exponential as in Eq. (106) but rather decays as a power law ($t^{-3/2}$) for long times t, as is well known from fluorescence correlation spectroscopy.

In spite of these apparent difficulties, we will show below that it is possible to develop a rigorous theory of photon counting of diffusing molecules. To do so, one must abandon the simple two-state description and explicitly take into account not only diffusion but also the many-particle nature of the problem. In other words, we must consider a system that contains many molecules (at concentration c) diffusing essentially independently in and out of a laser spot with inhomogeneous intensity.

Diffusion gives rise to fluctuations of the count rate, $n(\mathbf{R})$, because the excitation rate $k_{\text{ex}}(\mathbf{R})$ and the detection efficiency $\phi(\mathbf{R})$ depend on the translational coordinate $\mathbf{R}$. Specifically, for quenching in the absence of conformational changes,

$$n(\mathbf{R}) = \frac{\phi(\mathbf{R}) k_{\text{R}} k_{\text{ex}}(\mathbf{R})}{k_{\text{ex}}(\mathbf{R}) + k_{\text{D}}}. \tag{108}$$

Since fluctuations due to translational diffusion are on the millisecond time scale, translational diffusion is a slow process which can be treated using the appropriate generalization of the formalism in Sec. 3.4.

4.1. *Distribution of the number of photons in a time bin*

We begin by considering the general case when the concentration of diffusing molecules is arbitrary and show how to get the equation for the generating function. Then, we take the small concentration limit to obtain the distribution of photon counts (or, the photon counting histogram) in the single-molecule limit.

Consider M fluorescent molecules diffusing in a large volume V at concentration $c = M/V$. The generating function for $P(N \mid T)$ can be found using Eqs. (87) and (88) by identifying r with $(\mathbf{R}_1, \mathbf{R}_2, \ldots, \mathbf{R}_M)$, where $\mathbf{R}_i$ is the position vector of the ith particle, $\mathcal{L}$ with $D\sum_{i=1}^{M} \nabla_i^2$, where D is the translational diffusion coefficient, and $n(r)$ with $\sum_{i=1}^{M} n(\mathbf{R}_i)$ where $n(\mathbf{R}_i)$ is the laser-profile-dependent average number of detected photons emitted by molecule i at position $\mathbf{R}_i$. Thus, Eqs. (87) and (88) become

$$F(\lambda, T) = \sum_{N=0}^{\infty} \lambda^N P(N \mid T) = \int f \, d\mathbf{R}_1 d\mathbf{R}_2 \cdots d\mathbf{R}_M, \qquad (109)$$

where $f(\mathbf{R}_1, \ldots, \mathbf{R}_M, t \mid \lambda)$ satisfies

$$\frac{\partial}{\partial t} f = \sum_{i=1}^{M} (D\nabla_i^2 - (1-\lambda) n(\mathbf{R}_i)) f, \qquad (110)$$

with initial condition $f(t = 0) = 1/V^M$. Since the molecules are independent, f factors and we can write the generating function in terms of a one-particle function $g(\mathbf{R}, T \mid \lambda)$ as

$$F(\lambda, T) = \left[\frac{\int g(\mathbf{R}, T \mid \lambda) d\mathbf{R}}{V} \right]^M, \qquad (111)$$

where

$$\frac{\partial}{\partial t} g = D\nabla^2 g - (1-\lambda) n(\mathbf{R}) g, \qquad (112)$$

with $g(\mathbf{R}, 0 \mid \lambda) = 1$. To take the thermodynamic limit, we first differentiate $F(\lambda, T)$ with respect to T,

$$\frac{\partial}{\partial T} F(\lambda, T) = \frac{M}{V} \left[\int \frac{\partial g}{\partial T} d\mathbf{R} \right] \left[\frac{\int g d\mathbf{R}}{V} \right]^{M-1}. \qquad (113)$$

Then, we let both M and $V \to \infty$ in such a way that $M/V = c$. In this limit there is no difference between M and $M - 1$, so that

$$\frac{\partial}{\partial T}F(\lambda, T) = -ck(T \mid \lambda)F(\lambda, T), \tag{114}$$

where we have defined

$$k(T \mid \lambda) \equiv -\int \frac{\partial g}{\partial T} d\mathbf{R}. \tag{115}$$

By integrating both sides of Eq. (112) with respect to $\mathbf{R}$, it can be shown that

$$k(T \mid \lambda) = (1 - \lambda)\int n(\mathbf{R})g(\mathbf{R}, T \mid \lambda)d\mathbf{R}. \tag{116}$$

Finally, solving Eq. (114) with the initial condition $F(\lambda, 0) = 1$, we find

$$\begin{aligned} F(\lambda, T) = \sum_{N=0}^{\infty} \lambda^N P(N \mid T) &= \exp\left(-c\int_0^T k(t \mid \lambda)dt\right) \\ &= \exp\left(c\int (g(\mathbf{R}, T \mid \lambda) - 1)d\mathbf{R}\right). \end{aligned} \tag{117}$$

This equation provides a rigorous solution to the problem of how translational diffusion influences the statistics of the photon counts. Such a rigorous theory can be developed only when the trajectory is divided into bins of equal size and does not require an *ad hoc* definition of what constitutes a burst.

Remarkably, this generating function is identical to the survival probability of A in the irreversible diffusion-influenced pseudo-first-order *bimolecular* reaction $A + B \to 0$ obtained within the framework of the Smoluchowski approach.[65–67] Specifically, $F(\lambda, t) = [A](t)/[A](0)$ when c is identified with the concentration of B and the reaction between A and B is described by an effective position-dependent reactivity $(1 - \lambda)n(\mathbf{R})$. This is the generalization of the result that the generating function of the Poisson distribution, Eq. (48), is the survival probability of A in the irreversible *unimolecular* reaction $A \to 0$ that occurs with the rate $(1-\lambda)\langle n\rangle$.

In the single-molecule limit, $c \to 0$,

$$F(\lambda, T) = 1 + c\int (g(\mathbf{R}, T \mid \lambda) - 1)d\mathbf{R} + \cdots. \tag{118}$$

The first term is unity because all bins are empty when the concentration is zero. Expansion of the second term in powers of λ provides the single-molecule distribution of photon counts.

As an illustration, let us consider the distribution of photon counts when the observation volume profile is described by the 3D Gaussian:

$$n(\mathbf{R}) = n(0)\exp\left(-\frac{2(x^2+y^2)}{a_{xy}^2}\right)\exp\left(-\frac{2z^2}{a_z^2}\right) \tag{119}$$

where a_{xy} and a_z are the lateral and axial dimensions of the observation spot. The effective volume of the spot is defined as

$$v = \frac{[\int n(\mathbf{R})d\mathbf{R}]^2}{\int n(\mathbf{R})^2 d\mathbf{R}}. \tag{120}$$

For the Gaussian spot, the observation volume is $v = \pi^{3/2}a_{xy}^2 a_z$. If $n(\mathbf{R}) = n(0)$ for $R \le a$ and 0 at $R > a$, then $v = 4\pi a^3/3$ is the physical volume of the spot.

The photon counting histograms are found by numerically solving the reaction-diffusion equation, Eq. (112), as we described elsewhere.[50] The solution is then used in Eq. (118) to obtain single-molecule histograms, or in Eq. (117) for multiple-molecule histograms. Figure 11 shows results obtained in this way for an isotropic Gaussian spot $a_{xy} = a_z = a$. The single-molecule (dashed lines) and multiple-molecule (full lines) histograms coincide at small photon counts. However, even at very small concentrations, they deviate significantly for large photon counts. The reason for this difference is that it is more likely to have several molecules in the laser spot each emitting few photons than to have one molecule emitting many photons.

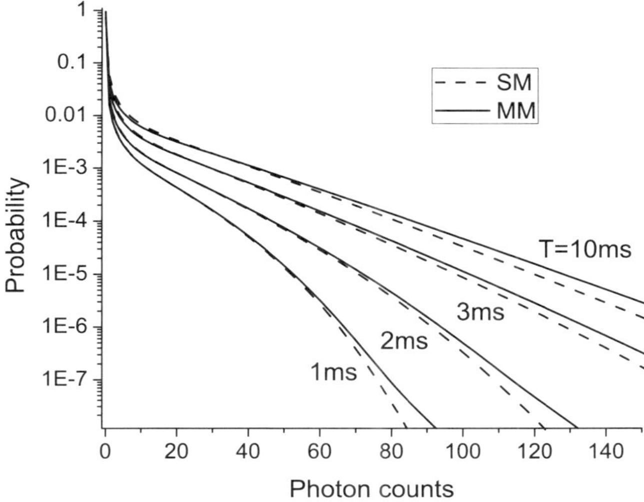

Fig. 11. Photon counting histograms of a diffusing fluorophore for different bin sizes $T = 1\,\text{ms}$, $2\,\text{ms}$, $5\,\text{ms}$, and $10\,\text{ms}$. Full lines take multiple-molecule (MM) effects into account, dashed lines correspond to the histograms in the single-molecule (SM) approximation. Concentration is 0.01 molecules per observation volume. Burst duration is $a^2/2D = 1.3\,\text{ms}$. The histograms are calculated by solving numerically the reaction-diffusion equation, Eq. (112), with the Gaussian observation volume profile $a = 0.5\,\mu\text{m}$, $D = 0.2\,\mu\text{m}^2/\text{ms}$, $n(0) = 100\,\text{ms}^{-1}$ using the procedure described in Ref. 50.

Consequently, in single-molecule measurements, if one chooses the threshold value for the number of photons too large, one actually makes measurements on several molecules, even when the concentration is very small.

4.2. *Moments and correlation function*

The average number of photons in a bin of size T can be obtained from the generating function in Eq. (117) via $\langle N\rangle = (\partial/\partial\lambda)F(\lambda, T)|_{\lambda=1}$ (see Eq. (25)). The result is

$$\begin{aligned} \langle N\rangle &= \langle n\rangle T \\ \langle n\rangle &= c\int n(\mathbf{R})d\mathbf{R} \equiv c\bar{n} \end{aligned} \tag{121}$$

where we have introduced the notation $\bar{n}$ for the integral of $n(\mathbf{R})$ over volume. $\langle N\rangle$ is linear in bin size and concentration, as expected.

Similarly, using $\langle N(N-1)\rangle = (\partial^2/\partial\lambda^2)F(\lambda,T)|_{\lambda=1}$ it can be shown that the variance is

$$\langle N^2\rangle - \langle N\rangle^2 = \langle N\rangle + 2c\int_0^T (T-t) \\ \times \iint n(\mathbf{R})G_d^0(\mathbf{R},t\,|\,\mathbf{R}_0)n(\mathbf{R}_0)d\mathbf{R}d\mathbf{R}_0 \qquad (122)$$

where $G_d^0(\mathbf{R},t\,|\,\mathbf{R}_0) = \exp(-(\mathbf{R}-\mathbf{R}_0)^2)/(4\pi Dt)^{3/2}$ is the Green's function for free diffusion. The first term in the above equation is the variance due to shot noise. The second term is due to diffusion through the laser spot. Comparing Eq. (122) with Eq. (32), it can be seen that

$$\frac{\langle\delta n(t)\delta n(0)\rangle_d}{\langle\delta n^2\rangle_d} = \frac{\int n(\mathbf{R})G_d^0(\mathbf{R},t\,|\,\mathbf{R}_0)n(\mathbf{R}_0)d\mathbf{R}d\mathbf{R}_0}{\int n(\mathbf{R})^2 d\mathbf{R}}, \qquad (123)$$

This is a standard result for the diffusion correlation function in fluorescence correlation spectroscopy[68] that describes fluctuations of the intensity about its mean due to translational diffusion (the subscript "d" indicates diffusion). For a Gaussian laser spot Eq. (119) this correlation function can be calculated analytically:

$$\frac{\langle\delta n(t)\delta n(0)\rangle_d}{\langle\delta n^2\rangle_d} = \left(1+\frac{4Dt}{a_{xy}^2}\right)^{-1}\left(1+\frac{4Dt}{a_z^2}\right)^{-1/2}. \qquad (124)$$

Note that this correlation function is not exponential as in the oversimplified model discussed in the beginning of this section (see Eq. (106)). A consequence of this is that it is not possible to uniquely define the burst duration τ_b. This is because a molecule can reenter the laser spot many times. However, even when a correlation function is not exponential, it is customary to define a correlation time as the area under the correlation function. Thus, one can define a burst duration time via

$$\tau_{\mathrm{b}} = \int_0^\infty \frac{\langle \delta n(t)\delta n(0)\rangle_d}{\langle \delta n^2\rangle_d} dt. \tag{125}$$

This is analogous to Eq. (104), which defines the conformational relaxation time. For the isotropic Gaussian laser spot (Eq. (119) with $a_{xy} = a_z = a$), $\tau_{\mathrm{b}} = a^2/(2D)$.

4.3. *Interphoton time distribution*

Let us consider the distribution of the times between photons when molecules diffuse through the laser spot. The interphoton time distribution is found from the concentration-dependent generating function, Eq. (117), by using Eq. (41). Differentiating the generating function with respect to τ twice and setting $\lambda = 0$ we find[51]

$$\varphi(\tau) = \frac{1}{\bar{n}}\left(-\frac{dk(\tau)}{d\tau} + ck(\tau)^2\right)\exp\left(-c\int_0^\tau k(t)dt\right), \tag{126}$$

where $\bar{n} \equiv \int n(\mathbf{R})d\mathbf{R}$ and $k(\tau) \equiv k(\tau\,|\,0)$ (see Eq. (116)), which is identical to the time-dependent rate coefficient of an irreversible diffusion-influenced bimolecular reaction with a distance-dependent reactivity $n(\mathbf{R})$. At long times, $k(\tau)$ approaches a steady-state value $k_\infty = \lim_{\tau\to\infty} k(\tau)$.

Consider this distribution in the small concentration limit when a photon trajectory consists of bursts of photons. Each burst is produced by a single molecule. The bursts are separated by a time which is much longer than the burst size. In this case, the term $dk(\tau)/d\tau$ in Eq. (126) is dominant at short times, while $ck(\tau)^2$ is dominant at long times. Replacing $k(\tau)$ by k_∞, Eq. (126) can be approximated by

$$\varphi(\tau) \approx \frac{1}{\bar{n}}\left|\frac{dk(\tau)}{d\tau}\right| + p_{\mathrm{b}}ck_\infty \exp(-ck_\infty\tau), \tag{127}$$

where we have defined $p_{\mathrm{b}} \equiv k_\infty/\bar{n}$.

This distribution has a simple structure. The second term describes the distribution of the time between photons in different bursts. The mean time between the bursts is equal to $1/ck_\infty$. The first term in Eq. (127) describes the distribution of the time between photons in the same burst. This distribution should be compared with the approximate two-state result in Eq. (107). The second term is the same if we identify k_∞ with v/τ_{b}. The first term is however quite different. It behaves as a power law at times comparable to the burst duration, $\varphi(\tau) \approx k_\infty^2(4\pi D\tau)^{-3/2}/\bar{n}$. Thus, the true distribution of interphoton times in a burst is actually highly nonexponential in contrast to Eq. (107).

In general, k_∞ is found numerically by solving the reaction-diffusion equation, Eqs. (112) and (116). A good estimate for k_∞ can be found using the Wilemski–Fixman approximation,[69] which in the present context gives

$$\begin{aligned}\frac{1}{k_\infty} &= \frac{1}{\bar{n}} + \int_0^\infty \frac{\langle \delta n(t)\delta n(0)\rangle_d}{v\langle \delta n^2\rangle_d}dt \\ &= \frac{1}{\bar{n}} + \frac{\tau_{\mathrm{b}}}{v}\end{aligned} \tag{128}$$

where we have used our definition of the burst size given in Eq. (125). When the number of photons in a burst is large, k_∞ is well approximated by v/τ_{b}. Then, the time between bursts is $(ck_\infty)^{-1} \approx \tau_{\mathrm{b}}/cv$. In this way, one can obtain the qualitative picture shown in Fig. 10 from the rigorous expression for the interphoton time distribution given in Eq. (126).

Figure 12 shows the exact interphoton time distribution calculated using Eq. (126) for the Gaussian laser spot with the same parameters as in Fig. 11. The distribution is a power law $t^{-3/2}$ for times comparable to the burst duration $\tau_{\mathrm{b}} = 1.3\,\mathrm{ms}$. For times comparable to the time between bursts, $\tau_{\mathrm{b}}/cv = 130\,\mathrm{ms}$, the distribution becomes exponential.

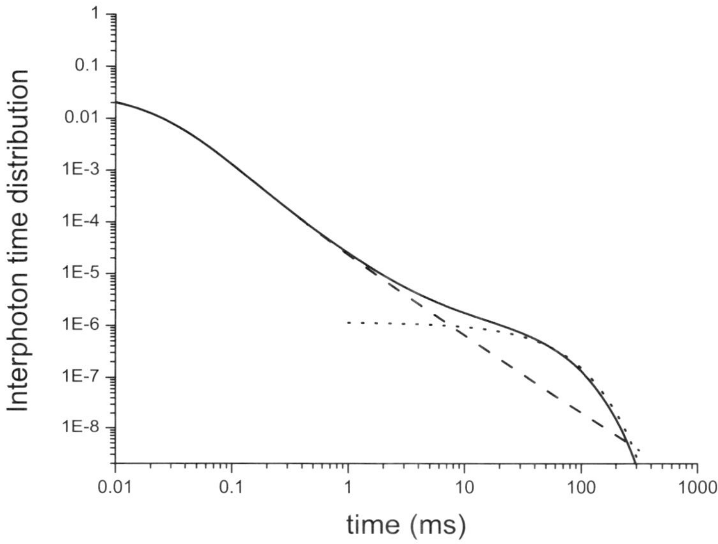

Fig. 12. Interphoton time distribution of diffusing fluorophores (full line) with the same parameters as in Fig. 11. The distribution is numerically calculated using Eq. (126). The dashed and the dotted lines show the first and the second terms in Eq. (127), respectively.

4.4. *Diffusing molecules undergoing conformational dynamics*

Now consider the case when the photon count rate is influenced by both translational diffusion and conformational dynamics:

$$n(\mathbf{R}, r) = \frac{\phi(\mathbf{R}) k_{\mathrm{R}} k_{\mathrm{ex}}(\mathbf{R})}{k_{\mathrm{ex}}(\mathbf{R}) + k_{\mathrm{D}}(r)}. \tag{129}$$

It can be shown[48] that the generating function now can be expressed as

$$\begin{aligned} F(\lambda, T) &= \sum_{N=0}^{\infty} \lambda^N P(N \mid T) \\ &= \exp\left(c \int (g(\mathbf{R}, r, T \mid \lambda) - p_{\mathrm{eq}}(r)) d\mathbf{R} dr \right) \end{aligned} \tag{130}$$

where $g(\mathbf{R}, r, t \mid \lambda)$ is the solution of the reaction-diffusion equation

$$\frac{\partial}{\partial t} g = (D\nabla^2 + \mathcal{L}_r)g - (1 - \lambda)n(\mathbf{R}, r)g \tag{131}$$

with the initial condition $g(t = 0) = p_{\text{eq}}(r)$. That is the generalization of Eq. (112). Here, the Laplacian describes diffusion through the laser spot and the $\mathcal{L}_r$ describes conformational dynamics. For diffusion in a potential of mean force $U(r)$, $\mathcal{L}_r$ is the Smoluchowski operator, Eq. (74).

For two-state conformational changes, $\mathcal{L}_r$ is a 2×2 matrix given in Eq. (60), and the above equations for the generating function become

$$F(\lambda, T) = \exp\left(c \int [g_1(\mathbf{R}, t \mid \lambda) - p_{\text{eq}}(1) + g_2(\mathbf{R}, t \mid \lambda) - p_{\text{eq}}(2)] d\mathbf{R}\right), \tag{132}$$

where $g_1(\mathbf{R}, t \mid \lambda)$ and $g_2(\mathbf{R}, t \mid \lambda)$ are solutions of

$$\begin{aligned} \frac{\partial}{\partial t} g_1 &= D\nabla^2 g_1 - L_1 g_1 + L_2 g_2 - (1 - \lambda)n_1(\mathbf{R})g_1 \\ \frac{\partial}{\partial t} g_2 &= D\nabla^2 g_2 + L_1 g_1 - L_2 g_2 - (1 - \lambda)n_2(\mathbf{R})g_2 \end{aligned} \tag{133}$$

with initial conditions $g_i(\mathbf{R}, 0 \mid \lambda) = p_{\text{eq}}(i), i = 1, 2$. When $n_2(\mathbf{R}) = 0$ and L_1 and L_2 are given by Eq. (69), these equations describe the influence of triplet blinking on the statistics of photons emitted by diffusing molecules. However, now L_1 depends on $\mathbf{R}$ through $k_{ex}(\mathbf{R})$.

In the presence of translational diffusion and conformational dynamics, the generating function is formally identical to the survival probability in an irreversible *stochastically-gated* diffusion-influenced bimolecular reaction.[70] Specifically, $F(\lambda, T) = [A(t)]/[A(0)]$ is the survival probability of A in the bimolecular reaction $A + B \to 0$ when $[B] = c$ and intrinsic reactivity is $(1 - \lambda)n(\mathbf{R}, r)$. This rate fluctuates not only because the distance between A and B changes, but also because the Bs can assume conformations with different reactivity. For example, B can have a reactive conformation (i.e., a "gate" is open) or an unreactive one (e.g., the "gate" is closed). This is analogous to triplet blinking if one identifies "open" with "bright" and "closed" with "dark."

When the concentration of the molecules c is small, the generating function is linear in concentration

$$F(\lambda, T) = 1 + c \int (g(\mathbf{R}, r, T \mid \lambda) - p_{\text{eq}}(r)) d\mathbf{R} dr + \cdots . \quad (134)$$

The above equations in principle determine how the distribution of photon counts is influenced by conformational dynamics and diffusion through the laser spot. However, in practice, the numerical implementation of this formalism to calculate photon counting histograms is rather involved.

The calculation of the intensity correlation function does simplify when the photon count rate can be factored as $n(\mathbf{R}, r) = n_{\text{d}}(\mathbf{R}) n_{\text{c}}(r)$. In the context of quenching, this happens when the laser intensity is sufficiently low ($k_{\text{ex}}(\mathbf{R}) \ll k_{\text{D}}(r)$ in Eq. (129)). In this case, the intensity correlation function generalizes to

$$\frac{\langle \delta n(t) \delta n(0) \rangle}{\langle \delta n^2 \rangle} = \frac{\langle \delta n(t) \delta n(0) \rangle_{\text{d}}}{\langle \delta n^2 \rangle_{\text{d}}} \cdot \frac{\langle n(t) n(0) \rangle_{\text{c}}}{\langle n \rangle_{\text{c}}^2} . \quad (135)$$

Here, the first factor is the intensity correlation function resulting from translational diffusion alone (see Eqs. (123) and (124))). The second factor is the intensity correlation function due to conformational dynamics (see Eq. (90)).

In fact, Eq. (135) can be shown to hold more generally. If it is used in conjunction with Eq. (93) (or (68) for the two-state conformational dynamics), it describes the influence of diffusion, antibunching, and conformational dynamics.

5. Concluding remarks

We presented a unified formalism to characterize photon statistics in the classical limit. The emission of photons is associated with state-to-state transitions in a kinetic scheme. Dynamics is described by a multistate master equation, or, in continuum limit, by a multidimensional reaction-diffusion equation. Given a kinetic scheme containing both photophysical and conformational states and specifying the transitions that result in detected photons, various statistical properties can be found using standard matrix algebra. Because a low-dimensional non-Markovian system can be described by a multidimensional Markovian one, the formalism is not restricted by Markovian systems.

In this chapter, we explicitly considered monitored transitions of a single type, i.e., photons of the same color. The formalism can be almost immediately generalized to describe FRET between a donor and an acceptor separates by a distance r. A donor is excited by a laser beam and can emit a photon or transfer the excitation to an acceptor which then can emit a photon of a different color. The rate of transfer depends on r^{-6} and this is why there is information about conformational dynamics. The output of such an experiment is a sequence of donor and acceptor photons. We are now interested in the distribution of both the donor and acceptor photons which are associated with transitions of different types.[48] The formalism can be generalized by introducing two counting parameters, λ_A and λ_D, and replacing $\mathbf{K} - (1 - \lambda)\mathbf{V}$ by $\mathbf{K} - (1 - \lambda_A)\mathbf{V}_A - (1 - \lambda_D)\mathbf{V}_D$ where $\mathbf{V}_A$ and $\mathbf{V}_D$ are matrices containing the rate coefficients associated with the acceptor and donor photons. Specifically, the generating function of the probability of having N_A acceptor photons and N_D donor photons during the time interval T is given by

$$\sum_{N_A, N_D=0}^{\infty} \lambda_A^{N_A} \lambda_D^{N_D} P(N_A, N_D \mid T) = \mathbf{1}^\top e^{[\mathbf{K}-(1-\lambda_A)\mathbf{V}_A-(1-\lambda_D)\mathbf{V}_D]T} \mathbf{p}_{ss} \tag{136}$$

which is the generalization of Eq. (18).

When photophysical relaxation is fast compared to conformational dynamics, the photophysical states can be eliminated and the generating function can be found from

$$\sum_{N_A,N_D=0}^{\infty} \lambda_A^{N_A}\lambda_D^{N_D} P(N_A, N_D \mid T) = \int f(r, T \mid \lambda_A, \lambda_D)dr, \qquad (137)$$

where $f(r, t \mid \lambda_A, \lambda_D)$ satisfies

$$\frac{\partial}{\partial t} f = \mathcal{L}_r f - (1 - \lambda_A)n_A(r)f - (1 - \lambda_D)n_D(r)f \qquad (138)$$

with initial condition $f(t = 0) = p_{eq}(r)$. This is generalization of Eq. (88) and involves only conformational states. The photophysical rate constants are packed into the count rates of acceptor, $n_A(r)$, and donor, $n_D(r)$, photons. We have used this formalism elsewhere[48] to analyze FRET efficiency distribution.

Acknowledgments

We thank the NIH Fellows Editorial Board for editorial assistance. This work was supported by the Intramural Research Program of the National Institutes of Health, NIDDK.

References

1. T. Basché, W. E. Moerner, M. Orrit and H. Talon, Photon antibunching in the fluorescence of a single dye molecule trapped in a solid, *Physical Review Letters* **69**(10) (1992) 1516–1519.
2. H. P. Lu, L. Xun and X. S. Xie, Single-molecule enzymatic dynamics, *Science* **282**(5395) (1998) 1877–1882.
3. C. Eggeling, J. R. Fries, L. Brand, R. Günther and C. A. M. Seidel, Monitoring conformational dynamics of a single molecule by selective fluorescence spectroscopy, *Proceedings of the National Academy of Science of the United States of America* **95**(4) (1998) 1556–1561.

4. A. A. Deniz, M. Dahan, J. R. Grunwell, T. Ha, A. E. Faulhaber, D. S. Chemla, S. Weiss and P. G. Schultz, Single-pair fluorescence resonance energy transfer on freely diffusing molecules: Observation of Forster distance dependence and subpopulations, *Proceedings of the National Academy of Science of the United States of America* **96**(7) (1999) 3670–3675.
5. L. Fleury, J. M. Segura, G. Zumofen, B. Hecht and U. P. Wild, Nonclassical photon statistics in single-molecule fluorescence at room temperature, *Physical Review Letters* **84**(6) (2000) 1148–1151.
6. D. S. Talaga, W. L. Lau, H. Roder, J. Tang, Y. Jia, W. F. DeGrado and R. M. Hochstrasser, Dynamics and folding of single two-stranded coiled-coil peptides studied by fluorescent energy transfer confocal microscopy, *Proceedings of the National Academy of Science of the United States of America* **97**(24) (2000) 13021–13026.
7. R. Rigler, S. Wennmalm and L. Edman, *Fluorescence Correlation Spectroscopy*, Chapter FCS in Single Molecule Analysis (Springer, Berlin, 2001), pp. 459–476.
8. B. Schuler, E. A. Lipman and W. A. Eaton, Probing the free-energy surface for protein folding with single-molecule fluorescence spectroscopy, *Nature* **419**(6908) (2002) 743–747.
9. H. Yang, G. Luo, P. Karnchanaphanurach, T. M. Louie, I. Rech, S. Cova, L. Xun and X. S. Xie, Protein conformational dynamics probed by single-molecule electron transfer, *Science* **302**(5643) (2003) 262.
10. M. Margittai, J. Widengren, E. Schweinberger, G. F. Schroder, S. Felekyan, E. Haustein, M. Konig, D. Fasshauer, H. Grubmuller, R. Jahn and C. A. M. Seidel, Single-molecule fluorescence resonance energy transfer reveals a dynamic equilibrium between closed and open conformations of syntaxin 1, *Proceedings of the National Academy of Science of the United States of America* **100**(26) (2003) 15516–15521.
11. P. J. Rothwell, S. Berger, O. Kensch, S. Felekyan, M. Antonik, B. M. Wöhrl, T. Restle, R. S. Goody and C. A. M. Seidel, Multiparameter single-molecule fluorescence spectroscopy reveals heterogeneity of HIV-1 reverse transcriptase: prime template complexes, *Proceedings of the National Academy of Science of the United States of America* **100**(4) (2003) 1655–1660.
12. M. Lippitz, F. Kulzer and M. Orrit, Statistical evaluation of single nano-object fluorescence, *ChemPhysChem* **6**(5) (2005) 770–789.
13. M. Orrit, Photon statistics in single molecule experiments, *Single Molecules* **3**(5–6) (2002) 255–265.
14. A. N. Kapanidis, N. K. Lee, T. A. Laurence, S. Doose, E. Margeat and S. Weiss, Fluorescence-aided molecule sorting: Analysis of structure and interactions by alternating-laser excitation of single molecules, *Proceedings of the National Academy of Science of the United States of America* **101**(24) (2004) 8936–8941.

15. E. Rhoades, M. Cohen, B. Schuler and G. Haran, Two-state folding observed in individual protein molecules, *Journal of American Chemical Society* **126**(45) (2004) 14686–14687.
16. W. Min, G. Luo, B. J. Cherayil, S. C. Kou and X. S. Xie, Observation of a power-law memory kernel for fluctuations within a single protein molecule, *Physical Review Letters* **94**(19) (2005) 198302.
17. T. A. Laurence, X. Kong, M. Jäger and S. Weiss, Probing structural heterogeneities and fluctuations of nucleic acids and denatured proteins, *Proceedings of the National Academy of Science of the United States of America* **102**(48) (2005) 17348–17353.
18. X. Michalet, S. Weiss and M. Jäger, Single-molecule fluorescence studies of protein folding and conformational dynamics, *Chemical Reviews* **106**(5) (2006) 1785–1813.
19. K. A. Merchant, R. B. Best, J. M. Louis, I. V. Gopich and W. A. Eaton, Characterizing the unfolded states of proteins using single-molecule FRET spectroscopy and molecular simulations, *Proceedings of the National Academy of Science of the United States of America* **104**(5) (2007) 1528–1533.
20. D. Nettels, I. V. Gopich, A. Hoffmann and B. Schuler, Ultrafast dynamics of protein collapse from single-molecule photon statistics, *Proceedings of the National Academy of Science of the United States of America* **104**(8) (2007) 2655–2660.
21. A. Hoffmann, A. Kane, D. Nettels, D. E. Hertzog, P. Baumgartel, J. Lengefeld, G. Reichardt, D. A. Horsley, R. Seckler, O. Bakajin and B. Schuler, Mapping protein collapse with single-molecule fluorescence and kinetic synchrotron radiation circular dichroism spectroscopy, *Proceedings of the National Academy of Science of the United States of America* **104**(1) (2007) 105–110.
22. S. Mukhopadhyay, R. Krishnan, E. A. Lemke, S. Lindquist and A. A. Deniz, A natively unfolded yeast prion monomer adopts an ensemble of collapsed and rapidly fluctuating structures, *Proceedings of the National Academy of Science of the United States of America* **104**(8) (2007) 2649–2654.
23. E. L. Elson and D. Magde, Fluorescence correlation spectroscopy. I. Conceptual basis and theory, *Biopolymers* **13**(1) (1974) 1–27.
24. E. Geva and J. L. Skinner, Two-state dynamics of single biomolecules in solution, *Chemical Physics Letters* **288**(2) (1998) 225–229.
25. J. Cao, Event-averaged measurements of single-molecule kinetics, *Chemical Physics Letters*, **327** (2000) 38–44.
26. S. Yang and J. Cao, Direct measurements of memory effects in single-molecule kinetics, *Journal of Chemical Physics* **117**(24) (2002) 10996–11009.
27. J. B. Witkoskie and J. Cao, Single molecule kinetics. I. Theoretical analysis of indicators, *Journal of Chemical Physics* **121**(13) (2004) 6361–6372.

28. J. B. Witkoskie and J. Cao, Single molecule kinetics. II. Numerical Bayesian approach, *Journal of Chemical Physics* **121**(13) (2004) 6373–6379.
29. J. Cao, Correlations in single molecule photon statistics: Renewal indicator, *Journal of Physical Chemistry B* **110**(38) (2006) 19040–19043.
30. A. Molski, Photon-counting distribution of fluorescence from a blinking molecule, *Chemical Physics Letters* **324**(4) (2000) 301–306.
31. A. Molski, J. Hofkens, T. Gensch, N. Boens and F. De Schryver. Theory of time-resolved single-molecule fluorescence spectroscopy, *Chemical Physics Letters* **318**(4–5) (2000) 325–332.
32. V. Barsegov, V. Chernyak and S. Mukamel, Multitime correlation functions for single molecule kinetics with fluctuating bottlenecks, *Journal of Chemical Physics* **116**(10) (2002) 4240–4251.
33. Y. J. Jung, E. Barkai and R. J. Silbey, Current status of single-molecule spectroscopy: Theoretical aspects. *Journal of Chemical Physics* **117**(24) (2002) 10980–10995.
34. E. Barkai, Y. J. Jung and R. Silbey, Theory of single-molecule spectroscopy: Beyond the ensemble average, *Annual Review of Physical Chemistry* **55**(1) (2004) 457–507.
35. J. Sung and R. J. Silbey, Counting statistics of single molecule reaction events and reaction dynamics of a single molecule, *Chemical Physics Letters* **415**(1–3) (2005) 10–14.
36. O. Flomenbom, J. Klafter and A. Szabo, What can one learn from two-state single-molecule trajectories? *Biophysical Journal* **88**(6) (2005) 3780–3783.
37. F. L. H. Brown, Single-molecule kinetics with time-dependent rates: A generating function approach, *Physical Review Letters* **90**(2) (2003) 28302–1–4.
38. Y. Zheng and F. L. H. Brown, Single-molecule photon counting statistics via generalized optical bloch equations, *Physical Review Letters* **90**(23) (2003) 238305.
39. F. L. H. Brown, Generating function methods in single-molecule spectroscopy, *Accounts of Chemical Research* **39**(6) (2006) 363–373.
40. Z. Wang and D. E. Makarov, Nanosecond dynamics of single polypeptide molecules revealed by photoemission statistics of fluorescence resonance energy transfer: A theoretical study, *Journal of Physical Chemistry B* **107**(23) (2003) 5617–5622.
41. L. P. Watkins and H. Yang, Detection of intensity change points in time-resolved single-molecule measurements, *Journal of Physical Chemistry B* **109**(1) (2005) 617–628.
42. S. C. Kou, X. S. Xie and J. S. Liu, Bayesian analysis of single-molecule experimental data (with discussion), *Journal of Royal Statistical Society Series C* **54**(3) (2005) 469.
43. M. Andrec, R. M. Levy and D. S. Talaga, Direct determination of kinetic rates from single-molecule photon arrival trajectories using hidden Markov models, *Journal of Physical Chemistry A* **107**(38) (2003) 7454–7464.

44. A. M. Berezhkovskii, A. Szabo and G. H. Weiss, Theory of single-molecule fluorescence spectroscopy of two-state systems, *Journal of Chemical Physics* **110**(18) (1999) 9145–9150.
45. A. M. Berezhkovskii, A. Szabo and G. H. Weiss, Theory of the fluorescence of single molecules undergoing multistate conformational dynamics, *Journal of Physical Chemistry B* **104**(16) (2000) 3776–3780.
46. I. V. Gopich and A. Szabo, Single-macromolecule fluorescence resonance energy transfer and free-energy profiles, *Journal of Physical Chemistry B* **107**(21) (2003) 5058–5063.
47. I. V. Gopich and A. Szabo, Statistics of transitions in single molecule kinetics, *Journal of Chemical Physics* **118**(1) (2003) 454–455.
48. I. Gopich and A. Szabo, Theory of photon statistics in single-molecule Förster resonance energy transfer, *Journal of Chemical Physics* **122**(1) (2005) 14707–1–18; I. V. Gopich and A. Szabo, Single-molecule FRET with diffusion and conformational dynamics, *Journal of Physical Chemistry B* **111**(44) (2007) 12925–12932.
49. I. Gopich and A. Szabo, Fluorophore-quencher distance correlation functions from single-molecule photon arrival trajectories, *Journal of Physical Chemistry B* **109**(14) (2005) 6845–6848.
50. I. V. Gopich and A. Szabo, Photon counting histograms for diffusing fluorophores, *Journal of Physical Chemistry B* **109**(37) (2005) 17683–17688.
51. I. V. Gopich and A. Szabo, Theory of the statistics of kinetic transitions with application to single-molecule enzyme catalysis, *Journal of Chemical Physics* **124**(15) (2006) 154712–1–21.
52. Y. Chen, J. D. Muller, P. T. C. So and E. Gratton, The photon counting histogram in fluorescence fluctuation spectroscopy, *Biophysical Journal* **77**(1) (1999) 553–567.
53. P. Kask, K. Palo, D. Ullmann and K. Gall, Fluorescence-intensity distribution analysis and its application in biomolecular detection technology, *Proceedings of the National Academy of Science of the United States of America* **96**(24) (1999) 13756–13761.
54. T. D. Perroud, B. O. Huang and R. N. Zare, Effect of bin time on the photon counting histogram for one-photon excitation, *ChemPhysChem* **6**(5) (2005) 905–912.
55. D. Magde, E. L. Elson and W. W. Webb, Fluorescence correlation spectroscopy. II. An experimental realization, *Biopolymers* **13**(1) (1974) 29–61.
56. J. Mertz, C. Xu and W. W. Webb, Single-molecule detection by two-photon-excited fluorescence, *Optics Letters* **20**(24) (1995) 2532–2534.
57. O. Krichevsky and G. Bonnet, Fluorescence correlation spectroscopy: The technique and its applications, *Reports on Progress in Physics* **65**(2) (2002) 251–297.
58. K. Velonia, O. Flomenbom, D. Loos, S. Masuo, M. Cotlet, Y. Engelborghs, J. Hofkens, A. E. Rowan, J. Klafter and R. J. M. Nolte, Single-enzyme kinetics

of CALB-catalyzed hydrolysis, *Angewandte Chemie International Edition* **44**(4) (2005) 560–564.

59. O. Flomenbom, K. Velonia, D. Loos, S. Masuo, M. Cotlet, Y. Engelborghs, J. Hofkens, A. E. Rowan, R. J. M. Nolte, M. Van der Auweraer, F. C. de Schryver and J. Klafter, Stretched exponential decay and correlations in the catalytic activity of fluctuating single lipase molecules, *Proceedings of the National Academy of Science of the United States of America* **102**(7) (2005) 2368–2372.
60. B. P. English, W. Min, A. M. Oijen, K. T. Lee, G. Luo, H. Sun, B. J. Cherayil, S. C. Kou and X. S. Xie, Ever-fluctuating single enzyme molecules: Michaelis–Menten equation revisited, *Nature Chemical Biology* **2**(2) (2006) 87–94.
61. L. Mandel and E. Wolf, *Optical Coherence and Quantum Optics* (Cambridge U. Press, Cambridge, 1995).
62. R. Short and L. Mandel, Observation of sub-poissonian photon statistics, *Physical Review Letters* **51**(5) (1983) 384–387.
63. D. R. Cox, *Renewal Theory* (Wiley, New York, 1962).
64. D. R. Cox and H. D. Miller, *The Theory of Stochastic Processes* (Wiley, New York, 1965).
65. I. Z. Steinberg and E. Katchalski, Theoretical analysis of the role of diffusion in chemical reactions, fluorescence quenching, and nonradiative energy transfer, *Journal of Chemical Physics* **48**(6) (1968) 2404–2410.
66. M. Tachiya, Theory of diffusion-controlled reactions: Formulation of the bulk reaction rate in terms of the pair probability, *Radiation Physics and Chemistry* **12**(1–2) (1983) 167–175.
67. A. Szabo, Theory of diffusion-influenced fluorescence quenching, *Journal of Physical Chemistry* **93**(19) (1989) 6929–6939.
68. B. J. Berne and R. Pecora, *Dynamic Light Scattering* (Wiley, New York, 1976).
69. G. Wilemski and M. Fixman, Diffusion-controlled intrachain reactions of polymers. I, Theory, *Journal of Chemical Physics* **60**(3) (1974) 866–877.
70. H.-X. Zhou and A. Szabo. Theory and simulation of stochastically-gated diffusion-influenced reactions, *Journal of Physical Chemistry* **100**(7) (1996) 2597–2604.

CHAPTER 7

Memory Effects in Single-Molecule Time Series

Jianshu Cao

Department of Chemistry, Massachusetts Institute of Technology Cambridge, MA 02139

1. Introduction

Single-molecule experiments have become a powerful approach to probe and manipulate liquids, polymers, quantum dots, enzymes, proteins, DNAs, RNAs, and low-temperature glasses.[1–9] Of particular relevance to our research is the observation of single-molecule trajectories, which record dynamic events of individual systems in condensed phases. Motivated by the experimental developments, theorists have analyzed intermittency,[10, 11] interconversion between two conformational states,[12–15] reaction event histograms,[16–24] and photon statistics.[25–36] This chapter will not include an extensive review, but shall focus on memory effects in single-molecule kinetics, including analysis of on–off blinking traces and photon emission traces.

The memory effects observed in single-molecule traces often arise from conformational fluctuations ubiquitous in chemical and biological processes. In macroscopic chemical kinetics, the rate process in a reactive system is usually the slowest timescale such that the bath degrees of freedom are treated as noise, and the reaction dynamics is treated as a Markovian process. Macroscopic chemical kinetics predicts Poisson statistics and exponential population decay, or convolutions of simple rate processes. However, in the presence of slow conformational fluctuations, such as in proteins and glassy systems, the simple exponential decay law breaks down.[37–39] The rate constant is no longer a constant of time but is modulated by evolving conformational configurations, which include solvent interactions, intra-molecular couplings, electrostatic potentials, and topological effects of macromolecules.[40–45] Yet, despite of its ubiquitous presence in chemical processes, such conformational modulation is not reflected in macroscopic chemical kinetics and is often averaged out in bulk measurements. In contrast, single-molecule techniques monitor the time-dependence in the reaction process and reveal the competition between reactions and fluctuations. An important example is the case of the single-molecule enzymatic turnover experiments and subsequent studies, which demonstrate conformational fluctuations in protein environments.[13,46–49] In this chapter, we will address the questions raised by these experiments: how to describe memory effects caused by conformational fluctuations and how to measure these effects quantitatively.

The transfer matrix formalism for the hidden Markovian process provides a theoretical basis for quantifying memory effects and for rigorously applying various theoretical solutions. The transfer matrix solution for a Markovian rate process is standard, but the extension to nonMarkovian processes is not trivial.[50–56] This chapter explores the flexibility of the transfer matrix formalism for several purposes: (i) definition of the probability for a single-molecule trace and, in combination with proper initial conditions, prediction of all single-molecule measurements; (ii) general applicability to various single-molecule time series, including on–off blinking traces and photon emission traces; (iii) rigorous definition of the stochastic rate model via a uniform transformation and a proper initial averaging; (iv) explicit connections to the generating function method; (v) compatibility with Bayesian statistics for data analysis. This chapter is arranged as follows: We present the transfer matrix formalism in the context of the modulated reaction model in Sec. 2, and then apply it to two types of single-molecule time series: on–off blinking traces in Sec. 3 and photon emission traces in Sec. 4. We discuss how to use Bayesian statistics to develop numerical methods to identify and extract memory effects from the noisy data in Sec. 5, and then conclude with a list of future directions.

For related discussions, we encourage readers to consult chapters 1, 2, and 10 for information analysis of single molecule data; chapters 3, 4, and 6 for generating function methods; and chapter 8 for stochastic rate models; and chapter 11 for weak ergodicity breaking. These chapters complement our theoretical treatment of memory effects and, together, they provide a broad overview of existing theoretical and numerical methods for analyzing single molecule kinetic experiments.

2. Modulated reaction model: General formalism

The modulated reaction model captures memory effects in single-molecule kinetics measurements and describes a large class of reaction schemes. We formulate the modulated reaction model in Sec. 2.1 and the transform matrix formalism in Sec. 2.2. Then, we define the time-averaging and event-averaging initial conditions in Sec. 2.3, which are necessary for the explicit evaluation of single-molecule quantities. Finally, using the transform matrix formalism, we introduce the generating function method in Sec. 2.4 and the stochastic rate model in Sec. 2.5.

2.1. *Definitions*

In general, the dynamics of a complex system can often be described with a set of states and transitions between these states. If all the distinct transitions of the system can be resolved in single-molecule experiments, then the single-molecule trace records a trajectory of the Markovian process, which is completely specified by a set of observed waiting time distribution functions.[58] In practice, only a few transitions or states can be optically resolved, posing a major challenge to single-molecule analysis. In the context of conformational fluctuations, the single-molecule time series with unresolved states is described as the modulated reaction model, which is the basis of our analysis. Both the modulated reaction model discussed in this section and the photon statistics model discussed in Sec. 3 are examples of hidden Markovian processes, where the observed transitions are coupled to unresolved transitions.

Formally, we denote the resolved transitions with the reaction rate matrix, K, and the unresolved transitions with the interconversion rate matrix, Γ. Then, the rate equation for the overall population dynamics reads

$$\dot{P}(t) = -(\Gamma + K)P(t), \tag{1}$$

where the rate matrix is explicitly given by

$$\Gamma + K = \begin{pmatrix} \Gamma_a + K_a & -K_{a,b} & -K_{a,c} & \cdots \\ -K_{b,a} & \Gamma_b + K_b & -K_{b,c} & \cdots \\ -K_{c,a} & -K_{c,b} & \Gamma_c + K_c & \cdots \\ \cdots & \cdots & \cdots & \cdots \end{pmatrix}. \tag{2}$$

In Eq. (2), $K_a = \sum_b K_{b,a}$, all the elements are block matrices, a and other English letters denote a resolved state or manifold, and μ and other Latin letters denote an unresolved state, a hidden state, or a conformational channel. As illustrated in Fig. 1, $\Gamma_a = \Gamma_{a\mu,a\nu}$ is the internal conversion rate matrix within a resolved manifold of states, $K_a = K_{a\mu,a\nu}$ is the depletion rate matrix from a resolved

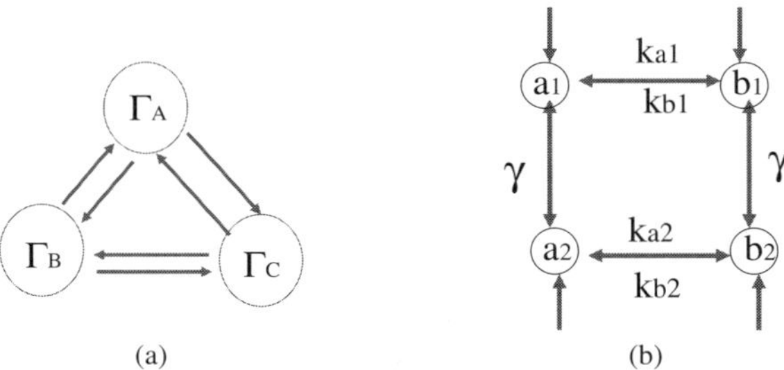

Fig. 1. Two examples of the modulated reaction model: (a) modulated three-state kinetics and (b) multi-conformational-channel on–off two-state kinetics.

manifold, and $K_{a,b} = K_{a\mu,b\nu}$ is the transition rate matrix between two manifolds. Figure 1(a) is a modulated reaction with three resolved states, and Fig. 1(b) is a two-state model with multiple conformational channels.

The general definition in Eq. (2) can be specified to various cases discussed in the literature:

- For discrete chemical states, rate is a matrix, and the distribution function is specified by the state index. A central topic of recent interest is chemical kinetics modulated by conformational fluctuations. The example discussed in Sec. 3 is the on–off blinking time series, which is described by the multiple conformational channel two-state chemical reaction. Another important class of the modulated reaction with discretized states is the photon emission time series discussed in Sec. 4.
- In the continuous limit, a set of discrete states becomes a continuous coordinate, the rate matrix becomes a differential operator (e.g., diffusion operator), and the distribution function depends on the coordinate. An interesting example is diffusion-controlled reactions, where the resolved states are discretized and the unresolved states are continuous. Then, $K(t)$ is a coordinate-dependent rate matrix and $\Gamma(x)$ is a diffusion operator.[16,40]
- If the reaction kinetic scheme is open rather than closed, the modulated reaction model can be used to describe a molecular motor, where the reaction state maps to the ATP reaction loop, and the conformational state maps to the spatial displacement of the motor.[59]
- The model has two obvious limits: In the limit of fast modulation, $\Gamma \to \infty$, the modulated reaction model reduces to a single conformation channel model with an effective rate constant, a result of dynamic narrowing. In the limit of slow modulation, $\Gamma \to 0$, the modulated reaction model reduces to an inhomogeneous average of the hidden states, a result of static disorder. The difficult scenario studied here is the broad parameter range in between these two limits.

2.2. *Transfer matrix method*

The formal solution to Eq. (1),

$$G(t) = e^{-(K+\Gamma)t}, \tag{3}$$

describes the population evolution, and with a proper initial condition, can describe one-time or multiple-time fluorescence correlation functions. The rate matrix in Eq. (2) can be separated to the diagonal part $K_0 = \{K_a\}$, and the off-diagonal part $V = \{K_{a,b}\}$, i.e. $K = K_0 - V$, which is a useful notation for later developments. Then, the Green's function solution in Eq. (1) is rewritten in Laplace space as

$$G(s) = \frac{1}{sI + \Gamma + K} = \frac{1}{sI + \Gamma + K_0 - V} = \frac{1}{1 - G_0(s)V} G_0(s), \tag{4}$$

where $G_0(s) = 1/(sI + \Gamma + K_0)$ is the waiting time distribution function matrix, I is the identity matrix, and all functions of variable s are Laplace transforms. Perturbation expansion of Eq. (4) leads to a series of terms in the order of single-molecule events,[50] giving

$$\begin{aligned} G(s) &= \sum_n G_n(s) \\ &= G_0(s) + G_0(s)VG_0(s) + G_0(s)VG_0(s)VG_0(s) + \cdots, \end{aligned} \tag{5}$$

where the first term represents a single-molecule trace without any observed transitions, the second term represents one transition, the third term two transitions, and so on. In this sense, the single-molecule time series forms a correlated random walk in terms of dwelling times of turnover events. Explicitly, the probability from a_0 to a_m manifolds via m transitions is

$$[G_m(s)]_{a_0,a_m} = \sum_{a_1,\ldots,a_{m-1}} G_{a_m}(s)K_{a_m,a_{m-1}} \cdots G_{a_1}(s)K_{a_1,a_0}G_{a_0}(s), \tag{6}$$

where all repeated indices except for a_0 and a_m and all implicit hidden states are summed over. Hereafter, all G_a with a single subscript is an element of G_0 defined in Eq. (4),

$$[G_a]_{\mu\nu} = [G_0]_{a\mu,a\nu} = \left[\frac{1}{sI + \Gamma + K_0}\right]_{a\mu,a\nu}. \tag{7}$$

Equation (5) demonstrates that population dynamics measured in the bulk is equivalent to the summation of all possible realizations of single-molecule time series. This is exactly the ergodic theorem, which states that ensemble average can be recovered by averaging over dynamic trajectories. The series of individual transition events described by Eq. (6) contains more information than the dynamics of bulk population described by Eq. (3).

A widely used method to analyze single-molecule time traces counts the number of transitions.[17] For this purpose, we introduce the distribution function of the event number

$$P(N,t) = \sum_m \delta(m - N) G_m(t), \tag{8}$$

where N is the total number of transition events that occur along all time series within time t. The moments of the event number can then be computed by integrating over N, giving

$$\int P(N,t) N^l dN = \sum_m m^l G_m(t). \tag{9}$$

With these moments, the Poisson indicator and other related indicators can be computed.

A unique feature of the single-molecule approach is the capability of recording the history of transitions, which yields the statistics and correlation of single-molecule events. The basis for calculating these event-averaged quantities is the transfer matrix expression for the probability of a specific time series, given by

$$G_{a_1,\ldots,a_m}(t_1, t_2, \ldots, t_m) = G_{a_m}(t_m) \cdots K_{a_2,a_1} G_{a_1}(t_1), \tag{10}$$

where the summation is carried over the unresolved states but not over the resolved manifolds. Event-averaged quantities represented by $G_m(t_1, t_2, \ldots, t_m)$ cannot be obtained in bulk experiments and are more informative than time-averaged quantities. The above expressions are not complete without specifying the initial conditions of experimental measurements, which is the subject discussed next.

2.3. *Initial conditions and single-molecule measurements*

2.3.1. *Initial time averaging*

To formulate single-molecule measurements, it is necessary to introduce the initial condition, which depends on the counting method for the origin of a single-molecule trace. Here we consider two types of counting methods and the related initial conditions. If we start randomly from the time series, the initial condition is defined by the steady-state population, $\rho_a = P_{a,ss}/\sum_a P_{a,ss}$, where P_{ss} is solved from Eq. (1) by setting $\dot{P}_{ss} = 0$. As illustrated in Fig. 2(a), this condition is associated with time-averaged quantities, which can be measured in bulk or in single-molecule experiments. For example, the fluorescence correlation function or the occupation correlation

$$C_{aa}(t) = G_{aa}(t)\rho_a = \sum_{\mu\nu} G_{a\mu,a\nu}(t)\rho_{a\nu}, \tag{11}$$

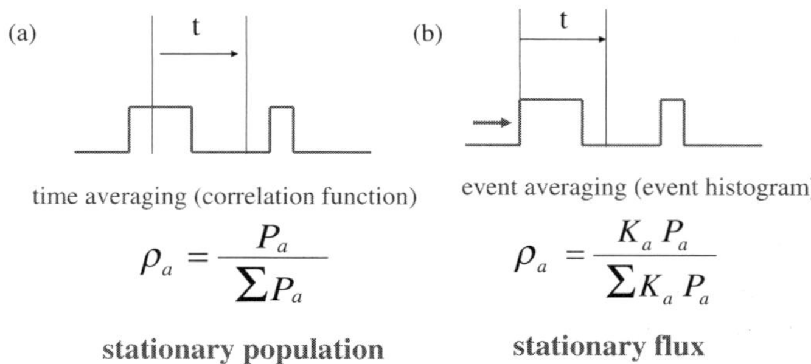

Fig. 2. Two initial conditions for single-molecule on–off traces: (a) time averaging defined by stationary population and (b) event averaging defined by stationary flux.

where the sum represents a summation over all the unresolved states. Note $G_{a\mu,a\nu} = [s + \Gamma + K]_{a\mu,a\nu}$ is a matrix element of G, which differs from G_0 or its matrix element G_a in Eq. (7). Similar expressions can be obtained for multiple times and mixed indices.

2.3.2. *Initial event averaging*

If we start randomly from a transition into the a manifold, we need to define the stationary probability flux,[50]

$$F_a = \sum_b K_{a,b} P_{b,ss}, \tag{12}$$

where a summation is carried over all the states connected to the a manifold. As illustrated in Fig. 2(b), this initial condition is given in terms of the stationary flux vector F_a as $\rho'_a = F_a / \sum F_a$ and is needed for all event-averaged quantities. The resulting distribution function of the dwelling time for the $a \rightarrow b$ transition is

$$f_a(t) = \sum_c K_{c,a} G_a(t) \tilde{\rho}'_a, \tag{13}$$

where the summation is carried over the manifold c as well as the hidden states. The average transition time defined by

$$\langle t_a \rangle = \int_0^\infty f_a(t) t dt \tag{14}$$

determines the rate constant $k_a = 1/\langle t_a \rangle$, used in the phenomenological kinetic description. Higher order moments of the waiting time distribution can also be calculated from $\langle t_a^n \rangle = \int f_a(t) t^n dt$.

2.3.3. *Event correlations*

We can further determine the joint distribution function of two or many events. For example, the joint distribution of two adjacent events labeled as a and b is given by

$$f_{a,b}(t_1, t_2) = \sum_c K_{c,b} G_b(t_2) K_{b,a} G_a(t_1) \rho'_a, \tag{15}$$

where the summation is carried over the c manifold and all the hidden states including those in the initial flux. Both $f(t)$ and $f(t_1, t_2)$ are investigated extensively in the context of single-molecule kinetic echo. Similarly, we consider two events separated by m other events and write

$$f_{\text{ev}}(t_1, t_{m+1}) = \int dt_2 \int dt_3 \cdots \int dt_m \, f_{\text{ev}}(t_1, t_2, \ldots, t_n, t_{m+1}), \tag{16}$$

where the intermediate events can be explicitly integrated. The first moment of the above distribution function is

$$\langle t_1 t_{m+1} \rangle = \int f_{\text{ev}}(t_1, t_{m+1}) t_1 t_{m+1} dt_1 dt_{m+1}. \tag{17}$$

The on-time correlation function[47] is employed to quantitatively measure the correlation in single-molecule time series,

$$\text{Cor}(m) = \frac{\langle t_1 t_{m+1} \rangle - \langle t \rangle^2}{\langle t^2 \rangle - \langle t \rangle^2}, \tag{18}$$

which is useful for inferring the memory function introduced in Sec. 3.4.

2.4. *Generating function method*

A powerful method to evaluate Eq. (4) is the generating function approach, which yields compact forms for those probabilities. Here we present the method in the general framework of the modulated reaction model.

Introducing variable λ_{ab}, we separate the rate matrix as the diagonal and off-diagonal parts, $K = K_0 - V$, and multiply the off-diagonal part with the variable λ. Then, we write the rate equation as

$$\dot{P}(t, \{\lambda\}) = -(\Gamma + K_0 - \lambda V) P(t, \{\lambda\}), \tag{19}$$

which is solved to yield the generating function in Laplace space,

$$G(s, \{\lambda\}) = \frac{1}{sI + \Gamma + K_0 - \lambda V} = \sum_m \lambda^m G_m(s). \tag{20}$$

Here λ is a scalar if transitions are specified and is a tensor if multiple transitions are monitored. Taking mth order derivative of $G(\lambda)$ with respect to λ and then setting $\lambda = 0$, we select out the probability of observing n specific transitions along a single-molecule time series,

$$\frac{1}{m!} \frac{\partial^m G(t, \lambda = 0)}{\partial \lambda^m} = G_m(t), \tag{21}$$

where λ can be a scalar or a tensor depending on the intermediate transitions that are specified. With $\lambda = 1$, we recover the expansion $G(t) = \sum_m G_m(t)$. Taking the derivative of the generating function and then setting $\lambda = 1$, we obtain the moment of transitions

$$\frac{\partial^m G(t, \lambda = 1)}{\partial \lambda^m} = \sum m! G_m(t) = \int N! P(N, t) dN, \tag{22}$$

which is the basis for calculating the Poisson indicator.

2.5. *Stochastic rate model*

An alternative formulation based on the stochastic rate model proves to be insightful. Although the stochastic rate model has appeared long ago in literature, its extension to single-molecule kinetics is recent. Among many related works by other groups, the contributions of this study are a rigorous definition of the stochastic model on the basis of the interaction picture[51] and a proof of Eq. (53) via a combined

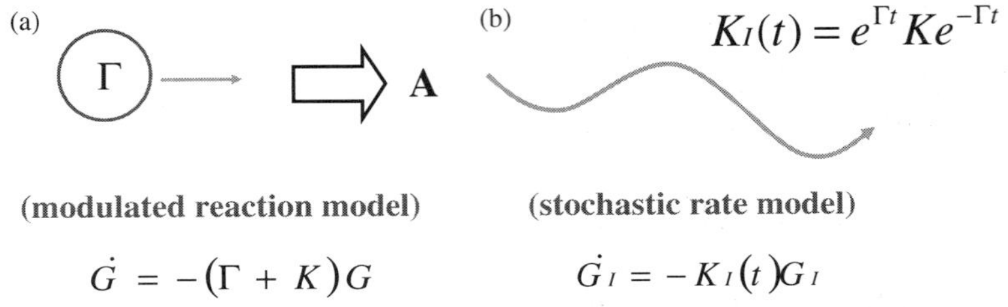

(a) (b) $K_I(t) = e^{\Gamma t} K e^{-\Gamma t}$

Γ A

(modulated reaction model) **(stochastic rate model)**

$\dot{G} = -(\Gamma + K)G$ $\dot{G}_I = -K_I(t)G_I$

Fig. 3. Illustration of (a) the modulated reaction model and (b) the stochastic rate model, and the transformation between the two representations.

application of the stochastic rate model and the generating function method.[52]

As illustrated in Fig. 3, the stochastic model is introduced by defining $P(t) = \exp(-\Gamma t) P_I(t)$ and the time-dependent rate

$$K_I(t) = \exp(\Gamma t) K \exp(-\Gamma t). \tag{23}$$

These definitions are analogous to the interaction picture in quantum mechanics; so we use subscript I to denote the stochastic model. The kinetic equation in the interaction picture becomes $\dot{P}_I(t) = -K_I(t) P_I(t)$, which has Green's function solution

$$G(t) = \exp(-\Gamma t) T \exp\left[-\int_0^t K_I(\tau) d\tau\right], \tag{24}$$

where T denotes time-ordering. An average over the initial condition leads to the survival probability

$$S(t) = \left\langle \exp\left[-\int_0^t K_I(\tau) d\tau\right]\right\rangle, \tag{25}$$

which can be compared to the path integral expression in quantum mechanics. The unresolved states introduce time-dependence into the kinetics of probed states. This expression has appeared in the study of nonPoisson statistics for diffusion-controlled reactions,[10] and has been applied and extended by several groups.[33,34,51,52] Similar to Kubo's stochastic line-shape theory, each realization of the stochastic rate defines a rate process, and single-molecule quantities are the average of all possible realizations of the stochastic rate process. The analogy to line-shape theory is instructive, as the multiple event probability distribution defined in Eq. (10) is analogous to multidimensional spectroscopy.[60] It then follows naturally that a long single molecule trace with a large number of events contains high-dimensional information, which is impossible to be obtained in conventional kinetics.

To account for the probability of events, we introduce a rescaled reaction rate matrix $K(\lambda) = K_0 - \lambda V$, where K_0 is the diagonal part of the matrix and V is the off-diagonal part of the matrix. Then, the generating function is formally given by

$$P_I(t, \lambda) = T \exp\left[- \int_0^t K_I(\tau, \lambda) d\tau \right], \tag{26}$$

which after applying derivatives with respect to λ yields the counting probability, $P(N, t)$.

3. On–off blinking time series

A typical on–off blinking trajectory in Fig. 4 can represent transitions between the dark and bright states in the triplet-state system, the open and close transition in the ion channel, turnover reactions between the reactant and product in enzymatic reactions, or folding transitions between the native state and denatured state in proteins and RNAs. All these trajectories can be treated as a binary code, or more accurately, a two-state time series.[17–19,21–24,50,52,53] First, the kinetic model for the on–off time series is defined in Sec. 3.1, and the phenomenological two-state kinetics is established in Sec. 3.2 with a brief discussion on detailed balance. Then, general formalisms including the transfer matrix approach, generating function method,

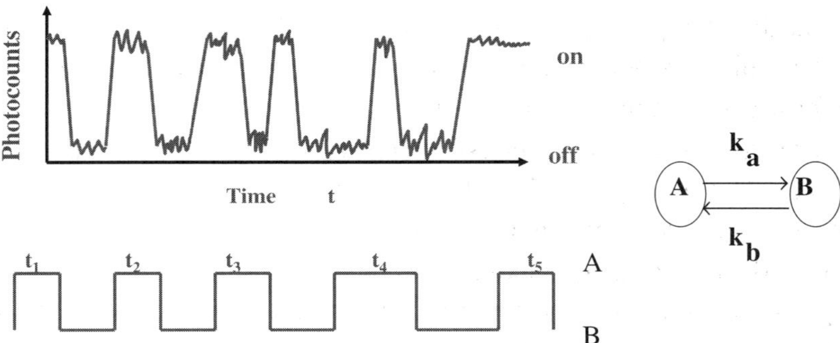

Fig. 4. Photon-counting signal generated by modulated two-state kinetics and the corresponding telegraphic on–off blinking trace.

and stochastic rate description are presented in Sec. 3.3. The memory function is introduced in Sec. 3.4 with the help of cumulant expansion of the stochastic rate. Finally, in Secs. 3.5 and 3.6, the two-event echo, on-time correlation function, and two-event number density are shown to be related to the memory function, quantitatively.

3.1. *Definitions*

The conformational modulation of on–off blinking traces is a special case of the generic nonMarkovian model discussed earlier. Usually, conformational fluctuations imply that (i) Γ matrix is independent of resolved states, $\Gamma_a = \Gamma_b = \cdots$; and (ii) there is no cross-reaction between different conformational channels so that rate matrices, $\{K_a\}$, are always diagonal. Here, we consider on–off blinking traces as a two-state reaction between the bright state a and dark state b, and the more general case of multiple states can be analyzed in a similar fashion. A straightforward way to calculate the on–off time series is the matrix notation, introduced in Sec. 2. The two-state blinking time series (a and b states) is described by the kinetic equation

$$K + \Gamma = \begin{pmatrix} \Gamma_a + K_a & -K_b \\ -K_a & \Gamma_b + K_b \end{pmatrix}, \tag{27}$$

where Γ_a and Γ_b are the internal conversion rates associated with conformational fluctuations, and $K_a = K_{b,a}$ and $K_b = K_{a,b}$ are the reaction rates between the bright and dark states.

Since the two-state kinetics consists of one pair of forward and backward transitions, the transition pathway $a \to b \to a \to b \cdots$ or $b \to a \to b \to a \cdots$ illustrated in Fig. 4 is unique once the initial state is specified, thus bringing significant simplifications: (i) The stationary flux $F_a = K_b \rho_b$ does not involve a summation over the b states: (ii) There is no difference between the specified time series defined in Eq. (10) and the unspecified time series defined in Eq. (6). The time series of on and off events are completely specified by the initial states and the number of transitions. (iii) There is no need for introducing a tensor form for the variable λ since there is only one pair of transitions with a unique sequence of blinking events.

3.2. *Phenomenological chemical kinetics, average rate, and detailed balance*

Macroscopically, the phenomenal two-state kinetics observed in bulk is the average along single-molecule blinking time series, and obeys chemical balance. Specifically, the steady-state solution to Eq. (27) relates ρ_a and ρ_b, and hence leads to the identity,

$$K_a \rho_a = K_b \rho_b = F, \tag{28}$$

which implies the conservation of macroscopic flux for two-state kinetics. With more than two states, the macroscopic flux conservation relation takes a more general form, which imposes zero sum for the total flux entering a macroscopic state. For two-state kinetics, the zero-sum relation implies the flux conservation or, equivalently, the macroscopic chemical balance.

The rate constant in phenomenological chemical kinetics can be interpreted as the average time that the single-molecule system spends in a macroscopic state (i.e. resolved state).[51] To be specific, the average on-time is evaluated from the waiting time distribution function as

$$\langle t_a \rangle = \int_0^\infty t f_a(t) dt = \sum K_a \left(\frac{1}{K_a + \Gamma_a} \right)^2 F_a = \frac{\sum P_{a,ss}}{\sum K_a P_{a,ss}}, \tag{29}$$

where the flux conservation relation, $F_a = F_b$, is used. This relation indicates that the average rate constant is the static average of the rate distribution

$$k_a = \frac{\sum K_a \rho_a}{\sum \rho_a}, \tag{30}$$

which is an inhomogeneous average and does not contain any information about dynamic disorder. A similar result can be obtained for the backward reaction, $k_b = \sum K_b \rho_b / \sum \rho_b$. Then, the ratio of the forward and back rate constants satisfies the phenomenological chemical balance relation

$$k_a \eta_a = k_b \eta_b, \tag{31}$$

where $\eta_a = \sum \rho_a$ and $\eta_b = \sum \rho_b$ are the equilibrium populations in the two resolved states. It should be noted that macroscopic chemical balance does not rule out the possibility of microscopic flux between unresolved states and hence the possibility of detailed balance violations. These types of violations cannot be observed in bulk and can be resolved by analyzing single-molecule time series. Recently, we examined the equilibrium conditions for modulated reactions and identified three signatures for detailed balance violations in one-dimensional and two-dimensional single-molecule histograms.[61]

3.3. *Formalism*

3.3.1. *Transfer matrix methods*

We now apply the transfer matrix formalism to calculate quantities pertinent to on–off blinking trace. The key quantity of several recent studies is the probability of blinking events. Let us assume that counting starts from a flip from the dark to bright state. Using Eqs. (6) or (9), we can then write the probability for a blinking trace with an even number of events as

$$G_{aa,2m}(s) = G_a(s)[K_a G_b(s) K_b G_a(s)]^m, \tag{32}$$

and the probability for an odd number of events as

$$G_{ab,2m+1}(s) = G_a(s) K_a G_b(s)[K_b G_a(s) K_a G_b(s)]^m. \tag{33}$$

Green's function $G_a(s)$ is defined as $G_a(s) = [s + \Gamma_a + K_a]^{-1}$, and Green's function $G_b(s)$ is defined is a similar way.

3.3.2. *Generating functions*

A compact expression for the above probabilities can be obtained through generating functions. Introducing variables λ_a and λ_b, we rewrite the kinetic equation in Eq. (19) as

$$\begin{aligned} \dot{P}_a(t,\lambda) &= -(\Gamma_a + K_a) P_a(t,\lambda) + \lambda_a K_b P_b(t,\lambda), \\ \dot{P}_b(t,\lambda) &= \lambda_b K_a P_a(t,\lambda) - (\Gamma_b + K_b) P_b(t,\lambda). \end{aligned} \tag{34}$$

The solution to the above equation is expressed as the Laplace transform

$$\begin{pmatrix} P_a(s,\lambda) \\ P_b(s,\lambda) \end{pmatrix} = (1 - \lambda_a \lambda_b K_a K_b G_a G_b)^{-1} \times \begin{pmatrix} G_a & \lambda_a K_a G_a G_b \\ \lambda_b K_b G_a G_b & G_b \end{pmatrix} \begin{pmatrix} P_a(0) \\ P_b(0) \end{pmatrix}, \tag{35}$$

which is equivalent to Eqs. (32) and (33). For example, setting $P_a(0) = 1$ and $P_b(0) = 0$, and resuming the probability in Eq. (33), we obtain

$$P_a(s,\lambda) = \sum_{m=0}^{\infty} P_{a,2m}(s)\lambda_a^m \lambda_b^m$$
$$= G_a(s)[1 - G_a(s)K_a\lambda_a G_b(s)K_b\lambda_b]^{-1}, \tag{36}$$

which is exactly the same as the relevant element in Eq. (35).

3.3.3. *Stochastic rate model*

Let us consider a two-state problem and explicitly write the stochastic rate as

$$K(t,\lambda) = \begin{pmatrix} e^{\Gamma_a t}K_a e^{-\Gamma_a t} & -e^{\Gamma_a t}\lambda_b K_b e^{-\Gamma_b t} \\ -e^{\Gamma_b t}\lambda_a K_a e^{-\Gamma_a t} & e^{\Gamma_b t}K_b e^{-\Gamma_b t} \end{pmatrix}, \tag{37}$$

which reduces to

$$K(t,\lambda) = \begin{pmatrix} K_a(t) & -\lambda K_b(t) \\ -\lambda K_a(t) & K_b(t) \end{pmatrix} \tag{38}$$

when $\Gamma_a = \Gamma_b$ and $\lambda_a = \lambda_b$. If the system satisfies a detailed balance, we also have the equilibrium condition $K_a = k_{eq}K_b$. In this case, one can separate the time-dependent part of $K(t,\lambda)$, and the generating function can often be solved analytically.[33,34,51,52]

The difference between the explicit transfer matrix expression and compact time-dependent generating function is mainly numerical: the former involves matrix manipulation, whereas the latter involves finite difference calculations. Establishing the connection between the matrix formalism and the time-dependent generating function requires specific definitions of the stochastic rate and the initial condition for counting, which are demonstrated in Refs. 51 and 52 and in Sec. 2.5

3.4. *Stochastic rate and memory functions*

Memory effects have been measured in single-molecule experiments with various indicators including the fluorescence intensity correlation function, waiting time distribution function, joint probability of two on-time events, and on-time correlation function.[14,17,19,21–23,27,28,34,62] These methods indicate the presence of conformation fluctuations, but do not provide a quantitative measure of memory effects. For this purpose, we introduce the stochastic rate model and define the memory function rigorously.[52] The environmental fluctuations introduce a time-dependence into the reaction rate, which can be treated as a stochastic variable. Similar to the Kubo–Anderson stochastic line-shape theory, each realization of the stochastic rate defines a reaction process, and single-molecule measurements can be obtained as a stochastic average of rate fluctuations. The modulated reaction model is analogous to the Schrodinger picture in quantum mechanics, as the occupancy in each conformational channel changes with time. The stochastic reaction model is analogous to the interaction picture in quantum mechanics, as the rate changes with time. As illustrated in Fig. 3, although different in pictures, the two representations are equivalent.

3.4.1. *Cumulant expansion*

With time-dependent rate, all single-molecule quantities can be evaluated explicitly with the help of cumulant expansion. For example, the average survival probability in the bright state manifold can be written as $S_a = \langle G_a(t) \rangle$, where the average is taken with respect to the equilibrium distribution, $\langle A \rangle = \sum A\rho_a / \sum \rho_a$. Cumulant expansion of $S(t)$ leads to

$$S(t) = \exp\left[\sum_{n=1}^{\infty} \frac{(-1)^n}{n!} \int_0^t d\tau_1 \cdots \int_0^{\infty} d\tau_n \chi_n(\tau_1, \ldots, \tau_n)\right], \quad (39)$$

where $\chi_n(\tau_1, \ldots, \tau_n)$ is the nth order cumulant expansion. Specifically, we have $\chi_1(t) = \langle K_a(t) \rangle = k$ as the average rate constant, and

$$\chi_2(t_1, t_2) = \langle \delta K(t_2)\delta K(t_1)\rangle = \langle \delta K \exp^{-\Gamma(t_2-t_1)} \delta K\rangle, \quad (40)$$

which rigorously defines the memory function for the stochastic rate. The initial value of the memory function is the variance of the rate $\chi(0) = \langle \delta K^2 \rangle$, and the lifetime of the memory function gives a characteristic timescale of memory

$$\gamma = \frac{\chi(0)}{\int_0^\infty \chi(t)dt}. \quad (41)$$

By truncating the expansion in Eq. (39) to second order, we obtain the Gaussian stochastic rate model

$$S(t) = \exp[-kt + M(t)], \quad (42)$$

where $M(t) = \int_0^t (t - \tau)\chi(\tau)d\tau$. Since the survival probability decreases with time, the Gaussian approximation holds when $k > \int_0^\infty (1 - t/\tau)\chi(\tau)d\tau$, which implies a small variance and short memory time. We emphasize that the stochastic rate model and the memory function are completely general, whereas the Gaussian stochastic model is approximate.

3.4.2. *Memory function*

We note that the memory function describes the modulation of the environment on kinetics, but does not directly probe the fluctuation spectrum. Nevertheless, the asymptotic behavior of the memory function reveals the nature of the long-time relaxation of conformational fluctuations. To demonstrate this point, we assume that the relaxation is multi-exponential, so that the memory function can be given as

$$\chi(t) = \sum c_n \exp(-\gamma_n t), \quad (43)$$

where γ_n is the set of exponents, describing the multi-exponential relaxation of the hidden states. The smallest non-zero eigenvalue has the slowest decay and thus dominates the asymptotic behavior,

$$\lim_{t\to\infty} \chi(t) \propto \exp(-\gamma_1 t), \tag{44}$$

which directly probes the fundamental mode of conformational fluctuations. The memory function is a convolution of reactions and conformation fluctuations, but decays asymptotically to the fundamental mode of fluctuations.

3.5. *Two-state two-channel model: Single molecule echo*

For the specific case of on–off time series with two conformational channels, the Green function and probabilities can be solved in closed forms, and the solution leads to the observation of the focal time and two-event echo. Here we summarize the results derived in Refs. 50 and 51. The rate matrix of the two-channel on–off process is defined as

$$\Gamma + K = \begin{pmatrix} \gamma_1 + k_{a1} & -\gamma_2 & -k_{b1} & 0 \\ -\gamma_1 & \gamma_2 + k_{a2} & 0 & -k_{b2} \\ -k_{a1} & 0 & \gamma_1 + k_{b1} & -\gamma_2 \\ 0 & -k_{a2} & -\gamma_1 & \gamma_2 + k_{b2} \end{pmatrix}. \tag{45}$$

To simplify the analysis, we consider the special case of $\gamma_1 = \gamma_2$ and $k_{b1} = k_{b2} = k_b$, so that the stationary fluxes for the two channels are the same. The on-time distribution function is given by the sum of the time-dependent flux functions from the two channels, $f_1(t) = k_{a1}G_{a,11}(t) + k_{a2}G_{a,21}(t)$ and $f_2(t) = k_{a1}G_{a,12}(t) + k_{a2}G_{a,22}(t)$, giving

$$f_a(t) = F_a[f_{a1}(t) + f_{a2}(t)], \tag{46}$$

where $G_{a,\mu\nu}(t)$ is a matrix element of Green's function $G_a(t)$. As a result, the on-time distribution function is found to cross over at the same focal time t_f regardless of the backward reaction rate k_b. Using Eq. (46), the focal time can be found from $f_1(t_f) = f_2(t_f)$

which yields

$$\tanh(t_f\Delta) = \frac{\Delta}{k+\gamma}, \tag{47}$$

where $k = (k_{a1} + k_{a2})/2$. Therefore, at the focal time, the two channels become indistinguishable, the memory effect disappears, and the backward reaction is decoupled from the forward reaction.

A more visual demonstration of the memory effect is the two-dimensional contour plot of $f(t_1, t_{M+1})$ in Fig. 5. If the two events are not correlated, the joint distribution function becomes the product of two single-event distribution functions, i.e. $f(t_1, t_{m+1}) = f(t_1)f(t_{m+1})$. Thus, as a probe of the memory effect, we calculate the difference distribution function

$$\delta(t_1, t_{m+1}) = f(t_1, t_m + 1) - f(t_1)f(t_{m+1}), \tag{48}$$

which vanishes in the absence of memory effects. To see the contour more clearly, we plot the diagonal section of the two-dimensional contour $\delta(t, t)$. The amplitude of the difference increases with the difference of rate constant and the timescale of the interconversion and disappears when $k_1 = k_2$ or $\gamma \to \infty$. Since the memory effect becomes weak as time increases, the difference function decays initially from the peak value at $t = 0$, as expected. Surprisingly, the

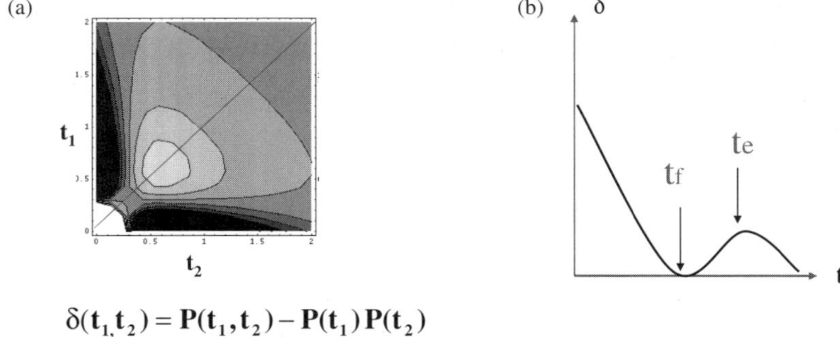

Fig. 5. (a) Two-dimensional contour of the difference function and (b) the diagonal cross-section with the echo time and focal time.

difference function disappears exactly at the focal time and reaches a local maximal at the echo time t_e. The values of t_e and t_f are invariant to the backward reaction rate and the number of separation between events.

To analyze the recurrent behavior, we explicitly evaluate the difference function for the special case of $k_b = \infty$, giving

$$\delta(t_1, t_2) = \chi_1 \chi_2 H(t_1) H(t_2), \tag{49}$$

where $H(t)$ is a function given in Ref. 50 and where the subscript $m = 1$ is implied. The same-time difference function becomes $\delta(t) = \chi_1 \chi_2 H(t)^2 \leq 0$, implying the bunching of the on-time event with the same residence time. Using the extreme condition $\dot{\delta}(t) = 0$, we find the condition for the echo time

$$\tanh(t_e \Delta) = \frac{\Delta}{\gamma + k}. \tag{50}$$

In the limit of slow variation $\gamma \leq k$, $\Delta t \leq 1$, Eqs. (47) and (50) simplify to

$$t_e \approx \frac{2}{\gamma + k_s} \tag{51}$$

and

$$t_f \approx \frac{1}{\gamma + k}, \tag{52}$$

which confirms that $t_e \approx 2t_f$. An interesting outcome of this analysis is the prediction of the two-event echo, the single-molecule analogue of photon echo in nonlinear spectroscopy. Writing $\chi_2(t) \approx \langle \delta K^2 \rangle \exp(-t\gamma)$ allows us to identify $\langle \delta K^2 \rangle$ as the amplitude of the inhomogeneous rate distribution, and γ as the relaxation rate of conformational modulation, i.e. the equivalence of dephasing rate in condensed phase spectroscopy. Thus, the two-event echo is a signature of memory effects, and its amplitude and position can be related to $\langle \delta K^2 \rangle$ and γ, respectively.[51]

3.6. *Direct measurements of memory function*

With the definition of $\chi(t)$ in Eq. (40), we can explore statistical methods to retrieve the memory function from single-molecule traces. In the slow modulation limit, one can show approximately that several quantities including the on-time correlation function are proportional to the memory function.[52] More reliable indicators can be defined based on the event-averaged initial condition, where we randomly select an emission event along the single-molecule time series. For example, in the case of a simple decay process, the two-event number density is exactly equal to the memory function.[52] For the photon emission data, we can define the renewal indicator, which vanishes for a renewal process.[53] The relation between the renewal indicator and the memory function is discussed in Sec. 4.

3.6.1. *On-time correlation function*

The rate correlation function provides an intuitive method to quantify the modulation of conformational fluctuations and can be obtained by analyzing single-molecule time traces. An indicator used in experiments is the on-time correlation function, $\mathrm{Cor}(m)$, defined in Eq. (18), which is shown to be proportional to the memory function

$$\mathrm{Cor}(m) \approx \frac{\chi(2m/k)}{\chi(0)}, \tag{53}$$

where $t_{\mathrm{eff}} = 2m/k$ is the effective separation between $2m$ events. The on-time correlation function is a coarse-grained function averaged over a blinking trajectory and does not suffer from the binning. The relationship in Eq. (53) is valid for slow modulation, i.e. $|K| \ll |\Gamma|$, at a discretized value of integer number m, and therefore does not apply at short-time or for fast modulation.

3.6.2. *Two-event number density*

Several other indicators have been proposed and are shown to relate to the memory function. In particular, we define the two-event number density, $g_{\text{ev}}(t)$, as the probability for finding a single-molecule trace with an event at zero time and another event at time t regardless of the number of events in between. For a photon emission trace governed by $\Gamma + K$, there is a single type of single-molecule events and we can rigorously show

$$P_{\text{ev}}(t) = k\left[1 + \frac{\chi(t)}{k^2}\right], \tag{54}$$

which provides a rigorous measure for the memory function. For an on–off blinking trace, there are two types of events and four event densities, and we can show that

$$P_{\text{ev},aa}(t) \approx \frac{k_a k_b}{k_a + k_b}\left[1 + \frac{\chi_{aa}(t)}{k_a^2}\right], \tag{55}$$

where k_a and k_b are the average forward and backward rate constants. The relationship holds under the general condition of slow modulation. In addition, there is also another general class of indicators introduced for photon statistics, including the Poisson indicator and renewal indicator, discussed in Sec. 4 and in several other chapters of this book.

4. Photon emission time series

Photon emission traces are directly recorded in single-molecule experiments and thus contain the original information about the multiple timescales in complex systems. General definitions useful for photon counting, including number densities, moments, Poisson indicator, and renewal indicator, are formulated in Sec. 4.1. The transfer matrix formalism[50,52] developed for on–off blinking sequences is extended in Sec. 4.2 to define a general waiting distribution function and calculate the Poisson indicator and renewal indicator.[56] The Wilemski–Fixman expression[64] derived for diffusion-controlled reactions is examined in Sec. 4.3 in the context of the relationship between ensemble-averaged survival probabilities and single-molecule photon counting moments.

4.1. *Definitions*

The following definitions are general and independent of the underlying mechanisms or the methods of evaluation.

4.1.1. *Photon densities and moments*

A single-molecule fluorescence trace is represented by a set of emission times, $\{\tau_1, \tau_2, \tau_3, \ldots\}$, which gives the number of emitted photons in the interval $[0, t]$, $g(t) = \int_0^t \sum_i \delta(\tau - \tau_i) d\tau$. Averaging over all realizations of the photon sequences gives the average photon count

$$\langle N(t) \rangle = \int_0^t d\tau \left\langle \sum_i \delta(\tau - \tau_i) \right\rangle = \int_0^t g(\tau) d\tau, \tag{56}$$

where $g(t)$ is number density and $\langle \cdots \rangle$ denotes the stochastic average over all traces.[58] The mean square of the photon count is given by

$$\begin{aligned} \langle N(t)^2 \rangle &= \int_0^t dt_1 \int_0^t dt_2 \left\langle \sum_i \delta(t_1 - t_i) \sum_j \delta(t_2 - t_j) \right\rangle \\ &= \int_0^t g_1(t_1) dt_1 + \int_0^t dt_1 \int_0^t dt_2 g_2(t_1, t_2), \end{aligned} \tag{57}$$

where $g_2(t_1, t_2)$ is similar to the two-event number density discussed in Ref. 52, although the notation used here is different. Similarly, we introduce the three-event number density through

$$\begin{aligned} \langle N(t)^3 \rangle &= \int_0^t dt_1 g_1(t_1) + 3 \int_0^t dt_1 \int_0^t dt_2 g_2(t_1, t_2) \\ &\quad + \int_0^t dt_1 \int_0^t dt_2 \int_0^t dt_3 g_3(t_1, t_2, t_3). \end{aligned} \tag{58}$$

The above definitions can be recast in a compact form as

$$C_m(t) = \langle N(N-1)\cdots(N-m+1)\rangle$$
$$= \left[\prod_i \int_0^t dt_i\right] g_m(t_1,\ldots,t_m), \quad (59)$$

which relates the multiphoton densities (i.e. multi-event number densities), g_m, and photon-counting moments $C_m(t)$. Both densities and moments contain the same information, one in the differential form and the other in the integral form. Photon-counting moments are less sensitive to experimental noise and are more robust in data analysis.

4.1.2. *Poisson indicator and renewal indicator*

As discussed in Sec. 2.3, the evaluation of single-molecule quantities depends on the initial condition of data collection. A photon-emission sequence without correlations is a renewal process, in which photons are emitted from the same state with the same rate constant. Naturally, to measure the deviation from the renewal behavior, one starts to count photons from an emission event, which defines the initial condition for the proposed renewal indicator.

A standard procedure is to start randomly on the time axis along the single-molecule trace, as illustrated in Fig. 6(a), which is the initial condition used in time-averaged quantities. If the single-molecule kinetics is stationary, the photon density is a constant, $g(t) = k = 1/\langle\tau\rangle$, where $\langle\tau\rangle$ is the average waiting time between two adjacent photons. Since the multi-event densities for a Poisson process are constant, the Poisson indicator is usually used to measure deviations from Poisson statistics,

$$\bar{N}Q(t) = \langle N(t)^2\rangle - \langle N(t)\rangle - \langle N(t)\rangle^2 = 2\int_0^t (t-\tau)\delta g(\tau)d\tau, \quad (60)$$

where $\langle N(t)\rangle = \bar{N} = kt$ and $\delta g(\tau) = g_2(\tau) - k^2$. The two-photon density is equivalent to the second-order photon correlation function

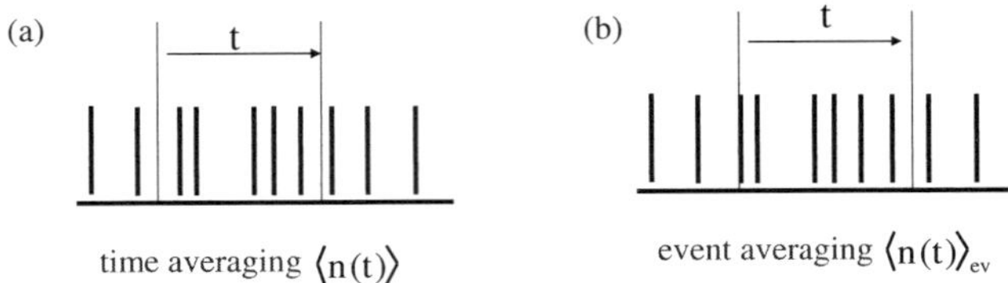

Fig. 6. Two initial conditions for single-molecule photon emission traces: (a) initial time averaging associated with the Poisson indicator and (b) initial event averaging associated with the renewal indicator.

$g(t) = g_2(t)/k^2 = g_{\rm ev}(t)/k$ so that the Poisson indicator in Eq. (60) can be expressed as

$$\bar{N}Q(t) = 2\int_0^t (t-\tau)k^2[g(\tau)-1]d\tau. \tag{61}$$

The long time limit of Eq. (61), $Q_{\rm M} = \lim_{t=\infty} Q(t) = 2\int_0^\infty \delta g(\tau)d\tau/k$, defines Mandel's Q parameter.[63]

As an alternative, we can randomly select an emission event along the single-molecule sequence, as illustrated in Fig. 6(b). This initial condition is a unique feature of single-molecule experiments and was introduced in Ref. 50 for calculating event-averaged quantities. We denote this condition with subscript "ev". Except for the Poisson process, the event-averaged photon density $g_{\rm ev}(t)$ is not constant of time. With the notation of $\delta g_{\rm ev}(t_1, t_2) = g_{2,\rm ev}(t_1, t_2) - g_{\rm ev}(t_1)g_{\rm ev}(t_2 - t_1)$ for $t_2 > t_1$, we define the renewal indicator[56]

$$\begin{aligned}\bar{N}Q_{\rm ev}(t) &= \langle N(t)^2\rangle_{\rm ev} - \langle N(t)\rangle_{\rm ev} - 2\int_0^t \langle N(t-t_1)\rangle_{\rm ev}\langle \dot{N}(t_1)\rangle_{\rm ev}dt_1 \\ &= 2\int_0^t dt_2\int_0^{t_2} dt_1 \delta g_{\rm ev}(t_1, t_2).\end{aligned} \tag{62}$$

The long time limit of $Q_{\rm ev}(t)$ is zero unless the memory of conformational fluctuations persists for infinitely long time. The two initial conditions discussed above are related. The time-averaged photon density associated with the first time variable is always a constant, so that

$$g_m(t_1, t_2, \ldots) = k g_{m-1,\mathrm{ev}}(t_2 - t_1, t_3 - t_1, \ldots), \tag{63}$$

and the event-averaged photon density is proportional to the photon correlation function, $g_{\mathrm{ev}}(t) = g(t)k$. Our notations are different from those used in a recent paper.

4.2. *Transfer matrix formalism*

4.2.1. *Photon-emission time series*

Photon-emission processes can be understood as a special case of the modulated reaction model introduced in Sec. 2, but differ from the on–off blinking process of Sec. 3, in the following aspects: (i) Instead of on and off events with two types of transitions, photon-emission kinetics generates a constant photon intensity baseline interrupted by a single type of emission events. (ii) Photon decay is a nonreversible rate process, often accompanied by nonradiative relaxation. (iii) Even without conformational fluctuations, emission kinetics can be complicated by quantum coherence and intermediate states; however, these effects lead to renewal processes without correlations between emission events. Four examples of photon-emission kinetics are illustrated in Fig. 7: (a) two-state kinetics, (b) triple blinking, (c) quantum two-level-system, and (d) conformational modulation.

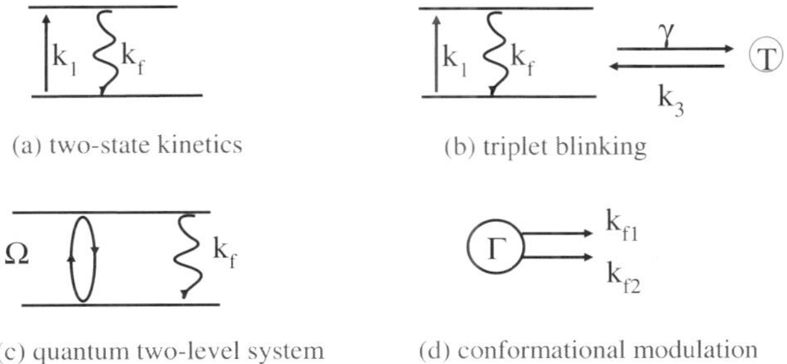

Fig. 7. Examples of photon-emission kinetics with k_f the emission rate.

Several methods including generating functional methods and numerical simulations have been employed to evaluate Poisson indicators and other statistical measures. Since the photon-emission process is a special case of the modulated reaction model defined in Eq. (2), the same transform matrix approach can be applied. Here, we demonstrate that the transfer matrix formalism of Sec. 2.2 provides a simple and robust approach to evaluate photon statistics, and note that quantum effects can be easily treated as a multistate rate process with quantum coherence described on the level of the Bloch equation.

4.2.2. *Transfer matrix and photon number density*

The central quantity is the interphoton distribution tensor, the conditional probability density of emitting another photon at time t after an emission event at time zero. After emitting a photon, the system assumes the ground state and reaches the fluorescent state after time t, giving the probability of emitting another photon[56]

$$\psi(s) = K_{\mathrm{f}} \frac{1}{sI + \Gamma + K} = K_{\mathrm{f}} \rho_{\mathrm{f}}(s), \tag{64}$$

where $\rho_{\mathrm{f}}(t)$ is the population evolution at the fluorescent state after time t, $\dot{\rho} = -(\Gamma + K)\rho$. Here, Γ represents conformational fluctuations, K represents the rate process including fluorescence emission, K_{f} represents the fluorescence emission rate, similar to V in Eq. (4). Both K_{f} and ρ_{f} are understood as vectors for discretized fluorescent states and functions for continuous states. The long time limit defines the stationary flux $\lim_{t\to\infty} \psi(t) = K_{\mathrm{f}} \rho_{\mathrm{f}}$, where ρ_{f} is the equilibrium population distribution at the fluorescence state. The interphoton distribution function, $\psi(t)$ is a scalar for a renewal process and becomes a tensor in the presence of conformational modulation. Once $\psi(t)$ is obtained, the emission trace in Fig. 8 is completely specified as

$$g_{\mathrm{ev}}(t_1, t_2, \ldots, t_m) = \langle \psi(t_m) \cdots \psi(t_2) \psi(t_1) K_{\mathrm{f}} \rangle, \tag{65}$$

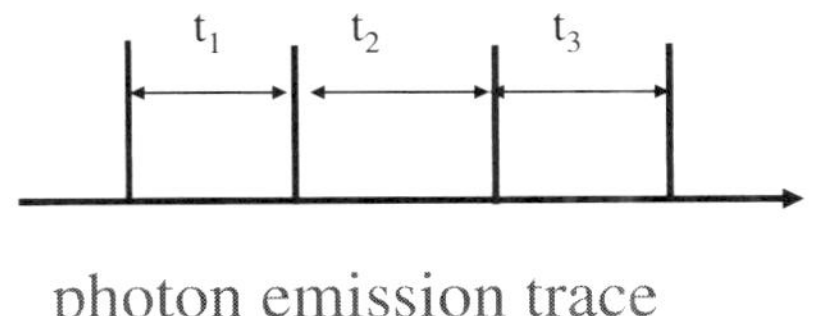

F ⇒ K —K_f→ K —K_f→ K —K_f→

$$g_{\mathrm{ev}}(t_1, t_2, \ldots, t_m) = \langle \psi(t_m) \cdots \psi(t_2)\psi(t_1)K_{\mathrm{f}} \rangle$$

Fig. 8. A segment of single photon emission trace and its probability in terms of the waiting time distribution matrix $\psi(t)$.

and we can calculate the the number density g_m via Eq. (63). Similar results have been obtained with the generating function method.[57]

4.2.3. *Renewal processes*

We first consider photon statistics of a single-channel emission process, where the photon is emitted from a single fluorescence state. Then the emission process is a renewal process, and the renewal indicator is zero, whereas the Poisson indicator is generally non-zero. For a renewal process, $\psi(t)$ is a scalar, and is the same as the phonon density, $\psi(t)$. Formally, as shown in Eq. (64), $\psi(t)$ is solved from the time-dependent population of the fluorescence emission state with the initial condition at the ground state, $\psi(t) = k_{\mathrm{f}}\rho_{\mathrm{f}}(t)$, with k_{f} the fluorescence rate. The Poisson indicator is given explicitly as

$$\mathcal{L}[\bar{N}Q] = 2k\frac{\psi(s)}{s^2} - \frac{2k^2}{s^3}, \tag{66}$$

which leads to the asymptotic limit of the Q parameter $\lim_{t\to\infty} Q(t) = [\langle\tau^2\rangle - 2\langle\tau\rangle^2]/\langle\tau\rangle^2$. In this chapter, functions of the s variable are implicitly Laplace transforms and $\mathcal{L}$ denotes Laplace transformation. Equation (66) is essentially the same as Eq. (61), but with explicit definitions for $g(t)$.

4.2.4. *Non-renewal processes*

Next, we consider single-molecule photon-emission processes modulated by conformational fluctuations. The generic case of the multichannel fluorescence emission process is defined with a set of emission states, $\rho_{\rm f}$, and emission rate constants, $K_{\rm f}$. The single channel correlation function $\psi(t)$ can be generalized to a matrix $\psi(t)$ in Eq. (64), where element $[\psi(t)]_{\mu\nu}$ is the photon correlation function of an emission event from channel μ and another emission event from channel ν. The time-averaged initial condition is determined by equilibrium populations of fluorescence states associated with each channel, $\{\rho_{\rm f}\}$. The event-averaged initial condition is determined by a set of stationary fluxes associated with each channel, $\{k_{\rm f}\rho_{\rm f}\}$. As a result, we arrive at[56]

$$\mathcal{L}[\bar{N}Q] = \frac{2}{s^2}\langle\psi(s)K_f\rangle - \frac{2}{s^3}\langle K_{\rm f}\rangle^2, \tag{67}$$

and similarly

$$\mathcal{L}[\bar{N}Q_{\rm ev}] = \frac{2}{s}\frac{\langle\psi(s)\psi(s)K_{\rm f}\rangle}{\langle K_{\rm f}\rangle} - \frac{2}{s}\left[\frac{\langle\psi(s)K_{\rm f}\rangle}{\langle K_{\rm f}\rangle}\right]^2, \tag{68}$$

where $\bar{N} = kt = \langle K_{\rm f}\rangle t$ and $\mathcal{L}$ denotes Laplace transforms. The conformational average is explicitly defined over the equilibrium distribution of the fluorescence state $\langle A\rangle = \sum_{\mu\nu} A_{\nu\mu}\rho_{\mu,{\rm f}}$, where ν and μ denote conformational channels.

We now consider a simple case of the multiple-channel fluorescence emission process, where the conformation modulation is homogeneous along the reaction, i.e. the interconversion kinetics Γ is independent of the chemical state. Then the photon correlation function matrix has the same functional form

$$\psi(k, s) \rightarrow \psi(K, s + \Gamma), \tag{69}$$

but with the rate matrix K and the modulation matrix Γ as the new variables.

4.3. *Diffusion-controlled reactions and Wilemski–Fixman expression*

A typical bulk experiment measures the profile of fluorescence decay, equivalent to the probability of the first photon arrival time along the single-molecule trace. Hence, the ensemble-averaged survival probability can be expressed in terms of single-molecule photon counts as

$$S(t) = 1 - \langle N(t)\rangle + \frac{1}{2!}\langle N^2(t) - N(t)\rangle \cdots$$
$$= \sum \frac{(-1)^m}{m!} C_m(t), \tag{70}$$

which is an expansion of the identity $S(t) = \langle \delta_{N,0}\rangle = \langle (1-x)^N\rangle_{x=1}$. The survival probability is measured in fluorescence-quenching experiments and discussed in the context of diffusion-controlled reactions. Here we repeat the derivation of the Wilemski–Fixman (WF) expression and show its equivalence to renewal theory.

We now explicitly evaluate the ensemble-averaged survival probability using single-molecule quantities. The fluorescence intensity decay described by the survival probability $S(t)$ in Eq. (70) is equivalent to the probability of finding the first photon from a random time on the single-molecule sequence. For a renewal process, the photon density distribution is

$$g_m(s_1, \ldots, s_m) = (k/s_1)\prod_{i=2}^{m} \psi(s_i), \tag{71}$$

and the photon counting moments in Eq. (70) becomes $C_m(s) = m!k\psi^{m-1}(s)/s^2$. Re-summation of the moments in Eq. (70) yields the survival probability

$$S(s) = \frac{1}{s}\left[1 - \frac{k}{s(1+\psi)}\right] = \frac{1 + k\chi(s)}{k + s[1 + k\chi(s)]}, \tag{72}$$

where the memory function $\chi(t)$ is identified as $\tilde{g}(t) = [1 + \chi(t)]$. Equation (72) is exactly the WF expression[64] for diffusion-controlled reactions. The average fluorescence decay time is

$$\langle t \rangle = \int_0^\infty S(t)dt = 1/k + \chi(0), \tag{73}$$

where the average memory time $\chi(s = 0)$ is related to Mandel's Q parameter through $Q_M = 2k\chi$. Thus, the WF theory holds exactly for renewal processes and relates the ensemble-averaged quantities to the second-order correlators and Mandel's parameter.[64]

In general, Eq. (70) not only leads to the WF expression for renewal processes but also forms the basis for deriving high-order corrections to the WF expression for nonrenewal processes. For details, we refer to a generalized WF expression shown in Refs. 65 and 66.

5. Data analysis

Single-molecule time series contain rich information, but retrieval of the information from single-molecule time series proves to be difficult because only certain signals of the system are observed in experiments. Data collected from experiments are inherently noisy because of the background contributions, shot noise, and fluctuations of the dynamic system. In addition to intrinsic noise, single-molecule time series also suffer from finite duration and experimental uncertainty. These factors make it difficult to directly read information from single-molecule time series and necessitate the applications of robust data analysis techniques. Several groups have explored information theory to analyze single-molecule time series.[54,55,67–69] Here, two methods are discussed based on Bayesian statistics, one for complete trajectory analysis and the other for histogram analysis.

The question addressed is: Given a single-molecule time series, what is the likelihood that a proposed model reproduces the complete time series? The basic idea of Bayesian statistics can be formulated as

$$P(M|D) \propto P(D|M)P(M), \tag{74}$$

where $P(M|D)$ is the probability of the model given the data, $P(D|M)$ is the probability of the data given the model, and $P(M)$ is *a priori* probability of the model. Equation (74) suggests a robust Monte Carlo algorithm, which explores the parameters of the proposed model as a random walk in kinetic space. The Monte Carlo algorithm does not give one model with one set of parameters, but provides the probability distribution function of parameters. We recently applied the Bayesian approach to analyze the photon emission measurements on a single protein and calculated the relaxation spectrum which is consistent with the maximal entropy fitting and physical pictures for protein relaxation.[71] The difficulty of the Bayesian method is the computational cost of carrying out the complete time series analysis and the exhaustive search in kinetic space.

To circumvent the computation cost of the full time series analysis, we can apply Bayesian statistics to two-dimensional binned histograms.[55] Consistent with the indicator analysis, this approach does not necessarily determine a kinetic model as in the full time series analysis but characterizes the basic features of the underlying physical process. We illustrate this idea with the two-event distribution function. The experimental time series generates the two-event histogram h_{ij}, which presents the best fitting $P_{0,ij} = h_{ij}/N$ without any restrictive features. Here, N is the number of data points on the two-dimensional histogram, so that the original data defines the best fit. A property of the underlying process such as the memory effects yields a restrictive model for the time series, P_{ij}. The probability of realizing the two-dimensional histogram with the restrictive model is

$$P(D|M) = \sum_{ij} (P_{ij})^{h_{ij}}. \tag{75}$$

Following the principle of Bayesian statistics, the existence of the particular feature is quantified by the difference function in the logarithm, giving $\sum_{ij} h_{ij}[\ln(P_{ij}) - \ln(P_{0,ij})]$. The test for a particular feature is specified through the restrictive representation of the two-dimensional histogram P_{ij}. The Bayesian analysis of two-dimensional histograms examines various features of the time series quantitatively, thus avoiding the ambiguity of inspection, and is advantageous for analyzing limited data points.

To establish a reliable model for a single-molecule experiment, we apply a set of standard tests for the basic kinetic features, including memory effects, detailed balance violations, and aging correlations. We then extract the correlation spectrum using the maximal entropy method and devise a model consistent with the basic physical properties and the spectrum. To test the validity of the model, we simulate complete time series with the Baysian-based Monte Carlo method and collect the distribution of the fitting parameters of the model. The procedure allows us to extract reliable information from the data within the limitation of its quality and length.

6. Concluding remarks

The chapter summarizes our efforts in describing memory effects in single-molecule time series, such as on–off blinking traces and photon emission traces. In addition to the two types of signals, single-molecule FRET and other fluorescence measurements record several levels or continuous variations of intensities, which suggest that the signal is generated from multiple states or from a continuous coordinate. These scenarios can be formulated within the framework of the transfer matrix formalism introduced in Sec. 2 and are discussed in the other chapters of the book. Issues to be addressed in further studies include: (i) Our discussion in this chapter is limited to rate processes and will be extended to general waiting time distribution functions. (ii) Unimolecular linear rate processes are a special case of chemical kinetics. A broader class of nonlinear kinetics problems needs to be studied for a number of fluctuation effects. (iii) The detailed balance condition will be established not only for the on–off blinking processes but also for more complex kinetic schemes, where detailed balance violations lead to subtle single-molecule signatures, which can be related to nonequilibrium steady-state thermodynamics.

Acknowledgments

The research was supported by the NSF Career Award and the Camille Dreyfus Teacher-Scholar Award. The author thanks Jim Witkoskie and Shilony Yang for their contributions.

References

1. W. E. Moerner, Examining nanoenvironments in solids on the scale of a single, isolated impurity molecule, *Science* **265** (1994) 46.
2. T. Bache, W. E. Moerner, M. Orrit and U. P. Wild, *Single-Molecule Optical Detection, Imaging and Spectroscopy* (VCH, 1996).
3. S. Nie and R. N. Zare, Optical detection of single molecules, *Annual Reviews of Biophysics and Biomolecular Structure* **26** (1996) 567.
4. X. S. Xie and J. K. Trautman, Optical studies of single molecules at room temperature, *Annual Reviews of Physical Chemistry* **49** (1998) 441.
5. S. Weiss, Fluorescence spectroscopy of single biomolecules, *Science* **283** (1999) 1676.
6. Y. Jia, D. S. Talaga, W. L. Lau, H. S. M. Lu, W. F. DeGrado and R. M. Hochstrasser, Folding dynamics of single gcn4 peptides by fluorescence resonant energy transfer confocal microscopy, *Chemical Physics* **247** (1999) 69.
7. W. E. Moerner and M. Orrit, Illuminating single molecules in condensed matter, *Science* **283** (1999) 1670.
8. X. Zhuang, L. E. Bartley, H. P. Babcock, R. Russell, T. Ha, D. Herschlag and S. Chu, A single-molecule study of RNA catalysis and folding, *Science* **288** (2000) 2048.
9. C. Bustamante, Z. Bryant and S. B. Smith, Ten years of tension: Single-molecule DNA mechanics, *Nature* **421** (2003) 423.
10. J. Wang and P. Wolynes, Intermittency of single molecule reaction dynamics in fluctuating environments, *Physical Review Letters* **74** (1995) 4317.
11. I. Chung, J. Witkoskie, J. Cao and M. Bawendi, Fluorescence intensity time traces of collections of CdSe nanocrystals QDs, *Phys. Rev. E* **73** (2006) 011106.
12. Y. Jia, A. Sytnik, L. Li, S. Vladimirov, B. S. Cooperman and R. M. Hochstrasser, Nonexponential kinetics of a single tRNA^{phe} molecule under physiological conditions, *Proceedings of the Natural Acadamy of Sciences USA* **94** (1997) 7932.

13. L. Edman, U. Mets and R. Rigler, Conformational transitions monitored for single molecules in solution. *Proceeding of the Natural Academy of Sciences USA* **93** (1996) 6710.
14. E. Geva and J. L. Skinner, Two-state dynamics of single biomolecules in solution, *Chemical Physics Letters* **288** (1998) 225.
15. A. M. Berezhkovskii, A. Szabo and G. H. Weiss, Theory of single-molecule fluorescence spectroscopy of two-state systems, *Journal of Chemical Physics* **110** (1999) 9145.
16. N. Agmon, Conformational cycle of a single working enzyme, *Journal of Physical Chemistry B* **104** (2000) 7830.
17. I. V. Gopich and A. Szabo, Statistics of transitions in single molecule kinetics, *Journal of Chemical Physics* **118** (2003) 454.
18. I. V. Gopich and A. Szabo, Single-macromolecule fluorescence resonance energy transfer and free-energy profiles, *Journal of Physical Chemistry B* **107** (2003) 5058.
19. I. V. Gopich and A. Szabo, Theory of photon statistics in single-molecule forster resonance energy transfer, *Journal of Chemical Physics* **122** (2005) 014707.
20. I. V. Gopich and A. Szabo, Photon counting histgrams for diffusing fluorophores, *Journal of Physical Chemistry B* **109** (2005) 17683.
21. O. Flomenbom, K. Velonia, D. Loos, S. Masuo, M. Cotlet, Y. Engelborghs, J. Hofkens, A. E. Rowan, R. J. M. Nolte, M. van der Auweraer, F. C. de Schryver and J. Klafter, Stretched exponential decay and correlations in the catalytic activity of fluctuating single lipase molecules, *Proceeding of the Natural Academy Sciences USA* **102** (2005) 2368.
22. O. Flomenbom, J. Klafter and A. Szabo, What can one learn from two-state single-molecule trajectories, *Biophysical Journal* **88** (2005) 3780.
23. H. P. Lerch, R. Rigler and A. S. Mikhailov, Functional conformational motions in the turnover cycle of cholesterol oxidase, *Proceeding of the Natural Academy Sciences USA* **102** (2005) 10807.
24. M. O. Vlad, F. Moran, F. W. Schneider and J. Ross, Memory effects and oscillations in single-molecule kinetics, *PNAS* **99** (2002) 12548.
25. A. Moski, J. Hopkens, T. Gensch, N. Boens and F. De Schryver, Theory of time-resolved single-molecule fluorescence spectroscopy, *Chemical Physical Letters* **318** (2000) 325.
26. D. Makarov and H. Metiu, Control, with an RF field, of photon emission times by a single molecule and its connection to laser-induced localization of an electron in a double well, *Journal of Chemical Physics* **115** (2001) 5989.
27. V. Chernyak, M. Schuls and S. Mukamel, Stochastic-trajectories and non-Poisson kinetics in single-molecule spectroscopy, *Journal of Chemical Physics* **111** (1999) 7416.

28. V. Barsegov, V. Chernyak and S. Mukamel, Multitime correlation functions for single molecule kinetics with fluctuating bottlenecks, *Journal of Chemical Physics* **116** (2002) 4240.
29. F. Sanda and S. Mukamel, Liouville-space pathways for spectral diffusion in photon statistics from single molecules. *Physical Review A* **71** (2005) 033807.
30. E. Barkai, Y. Jung and R. Silbey, Time-dependent fluctuations in single molecule spectroscopy: A generalized Wiener-Khintchine approach, *Physical Review Letters* **87** (2001) 207403.
31. Y. Jung, E. Barkai and R. Silbey, A stochastic theory of single molecule spectroscopy, *Advances in Chemical Physics* **123** (2002) 199.
32. J. Sung and R. J. Silbey, Counting statistics of single-molecule reaction events and reaction dynamics of a single molecule, *Chemical Physics Letters* **415** (2005) 10.
33. Y. He and E. Barkai, Theory of single photon control from a two level system source, *Physical Review Letters* **93** (2004) 068302.
34. J. Cao and R. J. Silbey, Generic schemes for single molecule kinetics: Self-consistent pathway solutions for renewal processes (submitted)
35. F. Kulzer and M. Orrit, Single-molecule optics, *Annual Reviews of Physical Chemistry* **55** (2004) 585.
36. M. Lippitz, F. Kulzer and M. Orrit, Statistical evalutation of single nano-object fluorescence, *Chem Phys Chem* **6** (2005) 770.
37. R. F. Grote and J. T. Hynes, The stable states picture of chemical reactions: II. Rate constants for condensed and gas-phase reaction models, *Journal of Chemical Physics* **73** (1980) 2715.
38. J. E. Straub, M. Borkovec and B. J. Berne, Calculation of dynamic friction on intramolecular degrees of freedom, *Journal of Physical Chemistry* **91** (1987) 4995.
39. P. Hanggi, P. Talkner and M. Borkovec, Fifty years after Kramers, *Reviews of Modern Physics* **62** (1990) 251.
40. N. Agmon and J. J. Hopfield, Co-binding to heme proteins: A model for barrier height distributions and slow conformational changes, *Journal of Chemical Physics* **79** (1983) 2042.
41. R. Zwanzig, Rate processes with dynamical disorder, *Accounts of Chemical Research* **23** (1990) 148.
42. R. Zwanzig, Dynamical disorder: Passage through a fluctuating bottleneck, *Journal of Chemical Physics* **97** (1992) 3587.
43. J. Wang and P. G. Wolynes, Passage through fluctuating geometrical bottlenecks: The generating Gaussian fluctuating case, *Chemical Physics Letters* **212** (1993) 427.
44. D. J. Bicout and A. Szabo, Escape through a bottleneck undergoing non-Markovian fluctuations, *Journal of Chemical Physics* **108** (1998) 5491.

45. N. A. Baker and J. A. McCammon, Non-Boltzmann rate distributions in stochastically gated reactions, *Journal of Physical Chemistry B* **103** (1999) 615.
46. L. Edman, Z. Foldes-Papp, S. Wennmalm and R. Rigler, The fluctuating enzyme: A single molecule approach, *Chemical Physics* **247** (1996) 11.
47. H. P. Lu, L. Xun and X. S. Xie, Single-molecule enzymatic dynamics, *Science* **282** (1998) 1877.
48. H. Yang, G. Luo, P. Karnchanaphanurach, T.-M. Louie, I. Rech, S. Cova, L. Xun and X. S. Xie, Protein conformational dynamics probed by single-molecule electron transfer, *Science* **302** (2003) 262.
49. K. Velonia, O. Flomenbom, D. Loos, S. Masuo, M. Cotlet, Y. Engelborghs, J. Hofkens, A. E. Rowan, J. Klafter, R. J. M. Nolte and F. C. de Schryver, Single enzyme kinetics of calb catalyzed hydrolysis, *Angewaridte Chemie International Edition* **44** (2005) 560.
50. J. Cao, Event-averaged measurements of single molecule kinetics, *Chemical Physics Letters* **327** (2000) 38.
51. S. Yang and J. Cao, Two-event echos in single molecule kinetics: A signature of conformational fluctuations, *Journal of Physical Chemistry B* **105** (2001) 6535.
52. S. Yang and J. Cao, Direct measurements of memory effects in single molecule kinetics, *Journal of Chemical Physics* **117** (2002) 10996.
53. J. Witkoskie and J. Cao, Single molecule kinetics I. Analysis of indicators, *Journal of Chemical Physics* **121** (2004) 6361.
54. J. Witkoskie and J. Cao, Single molecule kinetics II. Baysian approach, *Journal of Chemical Physics* **121** (2004) 6372.
55. J. B. Witkoskie and J. Cao, Testing for renewal and detailed balance violations in single molecule blinking processes, *Journal of Physical Chemistry B* **110** (2006) 19009.
56. J. Cao, Correlation in single molecule photon statistics: Renewal indicator, *Journal of Physical Chemistry B* **110** (2006) 19040.
57. I. V. Gopich and A. Szabo, Theory of the statistics of kinetic transitions with application to single molecule enzyme catalysis, *Journal of Physical Chemistry* **124** (2006) 1.
58. N. G. van Kampen, *Stochastic Processes in Physics and Chemistry* (Elsevier Science, New York, 1992).
59. M. E. Fisher and A. B. Kolomeisky, The force exerted by a molecular motor, *Proceeding of the Natural Academy of Sciences USA* **96** (1999) 6567.
60. S. Mukamel, *The Principles of Nonlinear Optical Spectroscopy* (Oxford University Press, London, 1995).
61. J. Witkoskie and J. Cao, Signatures of detailed balance violations in single molecule kinetics, submitted (2008).

62. E. Barkai, CTRW pathways to the fractional diffusion equation, *Chemical Physics* **284** (2002) 13.
63. L. Mandel and E. Wolf, *Optical Coherence and Quantum Optics* (Cambridge University Press, New York, 1995).
64. G. Wilemski and M. Fixman, Diffusion-controlled interchain reactions of polymers, *Journal of Chemical Physics* **60** (1974) 866–876.
65. S. Yang and J. Cao, Theoretical analysis and computer simulation of fluorescence lifetime measurements: I. Kinetic regimes and experimental timescales, *Journal of Chemical Physics* **120** (2004) 562.
66. S. Yang and J. Cao, Theoretical analysis and computer simulation of fluorescence lifetime measurements: II. Contour-length dependence in a polymer chain, *Journal of Chemical Physics* **120** (2004) 572.
67. H. Yang and X. S. Xie, Probing single molecule dynamic photon by photon, *Journal of Chemical Physics* **117** (2002) 10965.
68. L. P. Watkins and H. Yang, Information bounds and optimal analysis of dynamic single molecule measurements, *Biophysical Journal* **86** (2004) 4015.
69. S. C. Kou, X. S. Xie and J. S. Liu, Bayesian analysis of single molecule experimental data, *Journal of Royal Statistical Society C* **54** (2005) 469.
70. E. A. Donley and T. Plakhotnik, Statistics for single-molecule data, *Single molecules* **2** (2001) 23–30.
71. J. Witkoskie and J. Cao, Analysis of the entire sequence of a single photon experiment on a flavin protein, *Journal of Physical Chemistry B* **112** (2008) 5988–5996.

CHAPTER 8

Analysis of Experimental Observables and Oscillations in Single-Molecule Kinetics

Marcel O. Vlad*,† and John Ross*

*Department of Chemistry, Stanford University
Stanford CA 94305-5080 USA
†Institute of Mathematical Statistics and Applied Mathematics
Casa Academiei Romane, Calea 13 Septembrie 13, Bucharest
050711 Romania

1. Introduction

This chapter deals with the theoretical analysis of different types of oscillations which may occur in single-molecule kinetics and their influence on different experimental observables. The main difference between a single-molecule system and a macroscopic kinetic system is that for a macroscopic system the details of intramolecular dynamics are lost due to the overlapping of a large number of signals produced by the different molecules present in the system, whereas for a single-molecule system these microscopic signals have a direct influence on the experimental observables. In particular, oscillations of observables due to intramolecular dynamics, which in macroscopic experiments are smoothed out and do not show up in the observed data, can be observed directly in single-molecule experiments.

In this chapter we do not intend to develop a general theory of oscillations in single-molecule kinetics; instead we consider a few models, which can be investigated in detail. We intend to show the reader how to build different stochastic models for single-molecule kinetics, with special reference to oscillations. Experiments in single-molecule kinetics[1–7] consist in studying the chemical changes of one large molecule, such as a protein or an enzyme, immobilized on a support; the process may involve either a large molecule alone or a large molecule interacting with smaller molecules; intramolecular and molecular fluctuations are large and thus the deterministic mass action laws of chemical kinetics do not hold and are replaced by probabilistic laws. Although it would be desirable to develop a microscopic description based on nonequilibrium statistical mechanics, this is an extremely difficult task and thus the approaches in single-molecule kinetics are based on a stochastic, mesoscopic description involving two different types of stochastic processes. In the following we use variations of the basic model: (a) A single-molecule can exist in different chemical states $u = 1, 2, \ldots$ and the random transitions from one chemical state to another can be described by a local, Markovian master equation with time-dependent transition rates, $k_{uu'} = k_{uu'}(t)$. (b) Due to the conformational and other (energy) fluctuations in the single molecule, the rate coefficients $k_{uu'}(t)$ themselves are random functions of time. Some approaches consider directly the fluctuations of the rate coefficients, whereas other approaches assume that the stochastic properties of the rate coefficients can be represented in terms of a set of control parameters,[3,7–9] such as total energy of the molecule or the energy corresponding to a given degree of freedom; in this chapter we use both approaches, (a) and (b). In both cases we can write a Markovian master equation with random rate coefficients for the probability $P_u(t)$ that the molecule is in the chemical state u at time t

$$\frac{\partial}{\partial t} P_u(t) = \sum_{u' \neq u} P_{u'}(t) k_{u'u}(t) - P_u(t) \sum_{u'' \neq u} k_{uu''}(t), \tag{1}$$

where $k_{uu'}(t)$ is the rate of transition (rate coefficient) from the state u to the state u' at time t. The rate coefficients are random functions of time; thus, we need additional stochastic equations for describing the fluctuations of the rate coefficients $k_{uu'}(t)$. The simplest approximation is to neglect the fluctuations of the rate coefficients altogether and describe the kinetics of the process in terms of a master equation with constant coefficients; such an approach is similar to traditional chemical kinetics and is used as a first approximation. More sophisticated approaches make certain assumptions regarding the fluctuations of the rate coefficients based on theoretical models or experimental observations. In this chapter, we assume that the fluctuations of the rate coefficients can be described in terms of known characteristic functionals.

Among the different states $u = 1, 2, \ldots$ of the molecule studied, some are fluorescent and some are not. The molecule undergoes a random walk among these states, resulting in random variations of the fluorescent signal. The direct, raw experimental observable in a single-molecule experiment is the fluorescent signal $I(t)$ as a function of time, collected from the molecule studied; since the experiments are usually carried out in a time-independent regime, the time series describing the evolution of $I(t)$ is stationary. The most commonly used approach of data analysis is based on the computation of the correlation functions[10] of the fluorescence signal at times $t_1, \ldots, t_m$

$$C_m = \langle (I(t_1) - \langle I(t_1) \rangle) \cdots (I(t_m) - \langle I(t_m) \rangle) \rangle. \tag{2}$$

The second type of observables include the on/off time distributions,[6,7] that is, the distributions of the time intervals for which the fluorescent signal is on or off, respectively.

The third type of observables which can be extracted from the experimental data include the statistical properties of the numbers of reaction events,[11] that is, the numbers of occurrences of different reactions occurring in a given time interval. The models used in this paper are variations of the basic model based on the master equation with random coefficients (1) supplemented by suitable descriptions for the fluctuations of the rate coefficients. In the following sections we focus on using these models for investigating the connections among the three types of experimental observables mentioned earlier, and the possible occurrence of oscillations in single-molecule kinetics.

Our analysis is related to other aspects of single-molecule kinetics presented in this book. For details about data processing in single-molecule kinetics the readers may consult Chapters 1, 2 and 10. The presentation of generating function methods and kinetic models in Chapters 3, 4, 6 and 7 supplements our discussion.

2. Correlation functions and oscillations

Following Ref. 9 we start out by considering a special class of systems for which the fluctuating rate coefficients obey a separability condition $k_{uu'} = k^0_{uu'}\chi(\mathbf{s})$, that is, they are made up of the multiplicative contributions of two factors: (a) a universal factor, $\chi(\mathbf{s})$ which is fluctuating and is the same for all interaction processes and (b) process-dependent factors, $k^0_{uu'}$ which depend on the initial and final chemical states of the molecule but are not random. This separability condition makes it possible to introduce an intrinsic timescale and use the method of characteristic functionals for computing the correlation functions of the fluorescent signal. The separability condition is consistent with the condition of detailed balance for a system with a unique equilibrium state and is automatically fulfilled by a system with two chemical states and a unique equilibrium.

In Ref. 9, the theory was developed for the general case when the single molecule has an arbitrary number of chemical states and general expressions were derived for correlation functions of all orders. However, for simplicity, we begin by considering a system with two different chemical states. By applying the general theory developed in the study of Ref. 9, we obtain the following formula for the second-order correlation function:

$$\langle \Delta I(t)\Delta I(t+\tau)\rangle = \frac{K}{(K+1)^2}\mathscr{I}(\tau), \tag{3}$$

where

$$K = k_+(t)/k_+(t) \text{ independent of } t \tag{4}$$

is the equilibrium constant of the process, which, for a single chemical equilibrium is time-independent and not random. The term

$$\mathscr{I}(\Delta t) = \left\langle \exp\left(- \int_0^{\Delta t} k_\Sigma(t')dt' \right) \right\rangle \tag{5}$$

is a dynamic damping factor and

$$k_\Sigma(t) = k_+(t) + k_-(t) \tag{6}$$

is a total fluctuating rate coefficient, which is the sum of forward and backward reaction rates $k_+(t)$ and $k_-(t)$, respectively, and $\langle ... \rangle$ denotes a dynamic average over all possible values of the total fluctuating rate coefficient $k_\Sigma(t)$. For a system with two chemical states the separability condition mentioned earlier is automatically fulfilled and Eq. (3) is not subjected to any restriction.

If we assume that the cumulants $\langle\langle k_\Sigma(t)\rangle\rangle$, $\langle\langle k_\Sigma(t_1)k_\Sigma(t_2)\rangle\rangle \ldots$ of the total rate coefficient exist and are finite, the damping factor can be expressed by a cumulant expansion. We arrive at

$$\mathscr{I}(\Delta t) = \exp\left\{ \sum_{m=1}^{\infty} \frac{(-1)^m}{m!} \right. \\ \left. \times \int_0^{\Delta t} \cdots \int_0^{\Delta t} \langle\langle k_\Sigma(t_1) \cdots k_\Sigma(t_m)\rangle\rangle dt_1 \cdots dt_m \right\}. \tag{7}$$

The data can be analyzed in terms of the effective decay rate

$$\begin{aligned} k_{\mathrm{eff}}(\Delta t) &= -\frac{\partial}{\partial \Delta t} \ln \mathscr{I}(\Delta t) \\ &= \langle k_\Sigma \rangle + \frac{\partial}{\partial \Delta t} \sum_{m=2}^{\infty} \frac{(-1)^{m-1}}{m!} \\ &\quad \times \int_0^{\Delta t} \cdots \int_0^{\Delta t} \langle\langle k_\Sigma(t_1) \cdots k_\Sigma(t_m)\rangle\rangle \, dt_1 \cdots dt_m. \end{aligned} \tag{8}$$

The effective decay rate bears information about the nature of intramolecular fluctuations. If the fluctuations of the rate of change are of short range in time (Markovian or independent fluctuations), then in the long run, the effective decay rate is independent of the time difference Δt

$$k_{\rm eff}(\Delta t) = \text{independent of } \Delta t \text{ as } \Delta t \to \infty. \tag{9}$$

If condition (9) is fulfilled by the experimental data, then the intramolecular fluctuations are of short range in time. If $k_{\rm eff}(\Delta t)$ varies with Δt for large time differences Δt, then the intramolecular fluctuations are of long range. In addition, we notice that the effective rate is a better function for identifying the existence of oscillations in single-molecule kinetics than the correlation function of the fluorescent signal. In order to evaluate $k_{\rm eff}(\Delta t)$, however, accurate measurements are necessary.

In order to express the contribution of intramolecular fluctuations to the effective decay rate, we evaluate the difference

$$\begin{aligned}\Delta k_{\rm eff}(\Delta t) &= k_{\rm eff}(\Delta t) - \langle k_\Sigma(t)\rangle \\ &= \frac{\partial}{\partial \Delta t}\sum_{m=2}^{\infty}\frac{(-1)^{m-1}}{m!} \\ &\quad\times \int_0^{\Delta t}\cdots\int_0^{\Delta t}\langle\langle k_\Sigma(t_1)\cdots k_\Sigma(t_m)\rangle\rangle dt_1\cdots dt_m. \end{aligned}\tag{10}$$

In the particular case of Gaussian fluctuations of the total rate coefficient, all cumulants of order higher than 2 vanish and the difference $\Delta k_{\rm eff}(\Delta t)$ is simply given by

$$\begin{aligned}\Delta k_{\rm eff}(\Delta t) &= -\frac{1}{2}\frac{\partial}{\partial \Delta t}\int_0^{\Delta t}\int_0^{\Delta t}\langle\langle k_\Sigma(t_1)k_\Sigma(t_2)\rangle\rangle dt_1 dt_2 \\ &= -k_\Sigma\frac{\partial}{\partial \Delta t}\int_0^{\Delta t}(\Delta t - x)g(x)dx, \end{aligned}\tag{11}$$

where

$$g(|t_2 - t_1|) = \frac{1}{\langle k_\Sigma\rangle^2}\langle\langle k_\Sigma(t_1)k_\Sigma(t_2)\rangle\rangle \tag{12}$$

is the relative value of the correlation function of the total rate coefficient. The relative correlation function can be evaluated from experimental data by solving Eq. (11) for $g(\Delta t)$, resulting in

$$g(\Delta t) = -\frac{1}{k_{\Sigma}} \frac{\partial}{\partial \Delta t} \Delta k_{\text{eff}}(\Delta t). \tag{13}$$

This simplified model makes it possible to discuss the possible existence of damped oscillations in single-molecule kinetics. We start out by analyzing the oscillations due to the intramolecular fluctuations. For simplicity we limit ourselves to the case of Gaussian fluctuations. In this case it is easy to show that if the relative correlation function of the rate of change, $g(\Delta t)$, which expresses the intramolecular fluctuations, displays damped oscillations, then damped oscillations may also occur in the correlation functions of the fluorescent signal. According to the normal mode theory,[12] the function $g(\Delta t)$ can be expressed as

$$g(|t_1 - t_2|) = \sum_q c_q \exp[-\varepsilon_q |t_1 - t_2|] + \int_q c(q) \exp[-\varepsilon(q) |t_1 - t_2|] dq. \tag{14}$$

In this equation, c_q, $c(q)$, ε_q, $\text{Re}(\varepsilon_q) > 0$ and $\varepsilon(q)$, $\text{Re}(\varepsilon(q)) > 0$ are amplitude and frequency factors attached to the different normal modes. Since, in general, both c_q and ε_q are complex, their values must be chosen in such a way that the corresponding Gaussian process is physically consistent. For a purely discrete mode spectrum in Eq. (14) the integral term is missing, and the stochastic process, even though generally nonMarkovian, has short memory. The Markovian memory corresponds to a single exponential, that is, to a single mode. For a discrete spectrum, the Markovian approximation is accurate for large time differences, because in this case the main contribution to the sum in Eq. (14) is given by a single exponential which corresponds to the frequency with the smallest absolute value. If the mode spectrum has a continuum branch, then the tail of the correlation function may obey a scaling law of the inverse power type and the system may display long memory. Our analysis in this chapter is limited to the case of short-range fluctuations, for which Eq. (14) contains only the contribution of the discrete spectrum.

We consider the following physical constraints:

(1) For a time difference equal to zero, the autocorrelation function is equal to the dispersion of the total relative rate, $\nu(t) = k_\Sigma(t)/\langle k_\Sigma(t)\rangle$ at time t, $\langle\langle \nu^2(t)\rangle\rangle$, which, by definition, must be non-negative.
(2) Since the characteristic frequency is a real function of time, the modes with complex frequencies ε_q must occur in conjugated pairs.
(3) For large times the autocorrelation function of the relative total rate must decay to zero.

We keep in Eq. (14) only the contribution of the discrete spectrum and express the contribution of real eigenvalues $\varepsilon_q^{(\mathrm{real})}$ and of complex eigenvalues $\varepsilon_q^{(\mathrm{compl})} = \mu_q \pm i\sigma_q$. After some calculations, we obtain

$$
\begin{aligned}
g(|t_1 - t_2|) &= \sum_{\text{real values}} c_q^{(\mathrm{real})} \exp[-\varepsilon_q^{(\mathrm{real})}|t_1 - t_2|] \\
&\quad + \sum_{\text{complex values}} c_q^{(\mathrm{compl})} \exp[-\varepsilon_q^{(\mathrm{compl})}|t_1 - t_2|] \\
&= \sum_{\text{real values}} c_q^{(\mathrm{real})} \exp[-\varepsilon_q^{(\mathrm{real})}|t_1 - t_2|] \\
&\quad + \sum_{\text{complex values}} 2\{a_q \cos[\sigma_q|t_1 - t_2|] \\
&\quad + b_q \sin[\sigma_q|t_1 - t_2|]\} \exp[-\mu_q|t_1 - t_2|], \qquad (15)
\end{aligned}
$$

where a_q and b_q are the real and imaginary parts of the complex amplitude factors, and $c_q^{(\mathrm{compl})} = a_q \pm ib_q$. In order that the constraints (1)–(3) be valid we introduce the following restrictions for the parameters in Eq. (15):

$$c_q^{(\text{real})},\ a_q > 0, \quad b_q > 0 \tag{16}$$

$$\varepsilon_q^{(\text{real})},\ \mu_q > 0. \tag{17}$$

The restrictions (16) ensure that the dispersion of the characteristic frequency is non-negative, whereas the restrictions (17) are necessary in order that the autocorrelation function tends to zero for large time differences. The damping factor of the correlation function $\mathscr{I}(\Delta t)$ can be expressed as

$$\mathscr{I}(\Delta t) = \exp[-k_\Sigma \Delta t + k_\Sigma \Theta(\Delta t)], \tag{18}$$

where $\Theta(\Delta t)$ is a phase factor given by

$$\begin{aligned}
\Theta(\Delta t) = \Delta t & \left[\sum_{\text{real values}} \frac{c_q^{(\text{real})}}{\varepsilon_q} + \sum_{\text{complex values}} \frac{2(a_q\mu_q + b_q\sigma_q)}{(\mu_q)^2 + (\sigma_q)^2} \right] \\
& - \sum_{\text{real values}} \frac{c_q^{(\text{real})}}{(\varepsilon_q)^2} \\
& + 2 \sum_{\text{complex values}} \left[\frac{a_q[(\sigma_q)^2 - (\mu_q)^2] - 2b_q\mu_q\sigma_q}{[(\mu_q)^2 + (\sigma_q)^2]^2} \right] \\
& + \sum_{\text{real values}} \frac{c_q^{(\text{real})}}{(\nu_q)^2} \exp(-\varepsilon_q \Delta t) + \frac{2}{[(\mu_q)^2 + (\sigma_q)^2]^2} \\
& \times \sum_{\text{complex values}} \{a_q[(\mu_q)^2 - (\sigma_q)^2]\cos(\sigma_q\Delta t) + 2b_q\mu_q\sigma_q\} \\
& \times \exp(-\mu_q\Delta t) + \frac{2}{[(\mu_q)^2 + (\sigma_q)^2]^2} \\
& \times \sum_{\text{complex values}} \{b_q[(\mu_q)^2 - (\sigma_q)^2]\sin(\sigma_q\Delta t) - 2a_q\mu_q\sigma_q\} \\
& \times \exp(-\mu_q\Delta t).
\end{aligned} \tag{19}$$

The phase factor $\Theta(\Delta t)$ has the following asymptotic behavior:

$$\Theta(\Delta t) \sim \begin{cases} \mathscr{M}\Delta t^2 & \text{as } \Delta t \to 0 \\ \mathfrak{S}\Delta t & \text{as } \Delta t \to \infty \end{cases}, \tag{20}$$

where the proportionality factors $\mathscr{M}$ and $\mathfrak{S}$ are given by

$$\mathscr{M} = \frac{1}{2} \sum_{\text{real values}} c_q^{(\text{real})} + \sum_{\text{complex values}} a_q > 0, \tag{21}$$

$$\mathfrak{S} = \sum_{\text{real values}} \frac{c_q^{(\text{real})}}{\varepsilon_q} + \sum_{\text{complex values}} \frac{2(a_q\mu_q + b_q\sigma_q)}{(\mu_q)^2 + (\sigma_q)^2} > 0. \tag{22}$$

In this case the variation of the effective rate coefficient has the asymptotic behavior

$$\Delta k_{\text{eff}}(\Delta t) \sim k_\Sigma \begin{cases} -2\mathscr{M}\Delta t & \text{as } \Delta t \to 0 \\ -\mathfrak{S} & \text{as } \Delta t \to \infty \end{cases}. \tag{23}$$

For large as well as short time differences, the variation $\Delta k_{\text{eff}}(\Delta t)$ is negative: as expected, for large time differences the variation is constant, as it should be for short-range intramolecular fluctuations.

According to Eq. (19) the phase factor $\Theta(\Delta t)$ may display damped oscillations in the time difference Δt. From Eq. (18) it follows that the same type of damped oscillation must be displayed by the damping factor $\mathscr{I}(\Delta t)$. We did a numerical study of the possible occurrence of damped oscillations in the correlation functions of the fluorescent signal, due to the presence of damped oscillations in intramolecular dynamics, represented by the complex eigenmodes in Eq. (19). In order for the damped oscillations to show up in the correlation functions on the fluorescent signal, it is necessary that the timescale of the chemical process be of the same order of magnitude as the timescale of intramolecular dynamics. Figure 1 shows such a damped oscillating behavior for the second-order correlation function of the fluorescent signal $C_2(\tau) = \langle \Delta I(t+\tau)\Delta I(t)\rangle$, which is similar to the oscillations observed in the experiments of Edman and Rigler.[10] Similar behavior is displayed by the correlation functions of higher order.

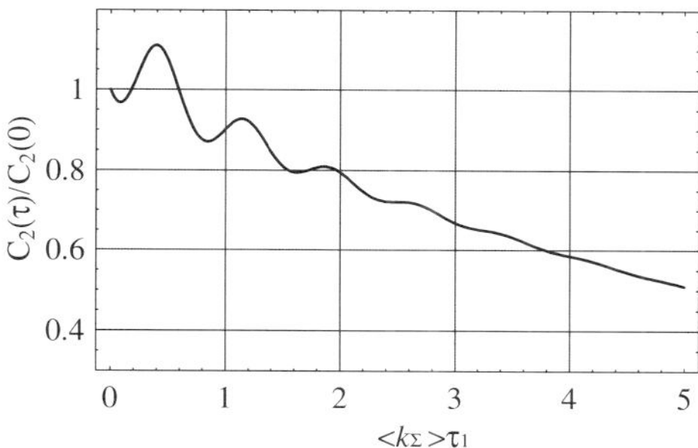

Fig. 1. Graphical representation of the absolute value of the second-order correlation function $C_2(\tau)$ versus the time difference $\tau = |t_2 - t_1|$ for a two-state Gaussian model. (The relative correlation function $g(\Delta t)$ is represented by a linear combination of complex exponential terms.)

It has been suggested that the damped oscillations of the correlation functions of high order can be used for the characterization of the nonMarkovian nature of the two-state fluorescent process.[10] A non-Markovian function (NMF) has been defined in terms of the second- and third-order correlation functions of the fluorescent signal:

$$\mathrm{NMF}(\tau_1, \tau_2) = p_f \left[\frac{C_3(\tau_1, \tau_2)}{C_2(\tau_2)} - C_2(\tau_1) \right], \tag{24}$$

where p_f is the stationary probability that the molecule is in a fluorescent state. Figure 2 shows a typical memory landscape for the NMF computed by applying our approach from Eqs. (3)–(5), (18), and (19), and the expression

$$\langle \Delta I(t) \Delta I(t + \tau_1) \Delta I(t + \tau_1 + \tau_2) \rangle = \mathscr{I}(\tau_1 + \tau_2) \frac{K(K-1)}{(K+1)^3}, \tag{25}$$

for the third-order correlation function, computed by applying our theory presented in Ref. 9. The computed landscape displays the same type of damped oscillations as the ones observed in the experiments of Edman and Rigler.[10]

Another possible cause for the occurrence of the damped oscillations of the correlation functions is the interaction between chemical kinetics and intramolecular dynamics. This cause was suggested by

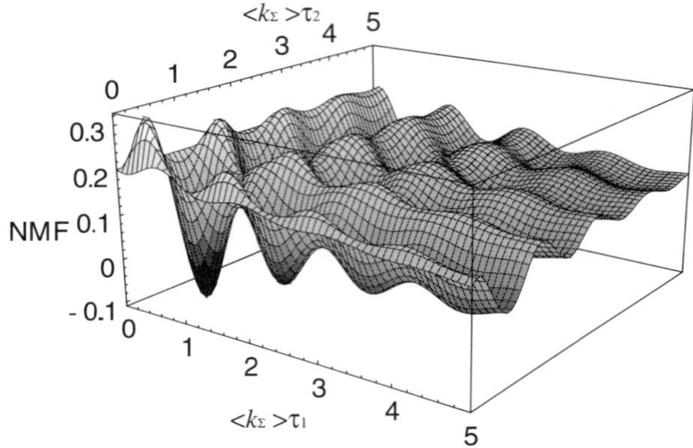

Fig. 2. Graphical representation of the non-Makovian function NMF(τ_1, τ_2) of Edman and Rigler[10] versus the time differences τ_1 and τ_2 for a two-state Gaussian model. (The relative correlation function $g(\Delta t)$ is represented by a linear combination of complex exponential terms.)

Edman and Rigler[10] in order to explain their experimental data on the oxidation reaction involving a single molecule of immobilized horseradish peroxidase. These authors neglected the random fluctuations of the rate coefficients and assumed that the kinetics of the process can be described by a simplified form of the master equation (1), where the rate coefficients $k_{u'u}$ are constant. They have chosen sets of rates $k_{u'u}$, which correspond to closed loops of states and violate detailed balance. It has been theoretically proven that a master equation with rates $k_{u'u}$, which violate the detailed balance, are capable of producing damped oscillations.[9] The model of Edman and Rigler[10] seems to contradict the principles of statistical mechanics, because they evaluate time invariant, equilibrium correlation functions by using a model which violates the detailed balance. However, we show shortly that this is not necessarily the case. A single molecule is not a macroscopic system; therefore, it does not have to obey equilibrium statistical mechanics; however, a single molecule is not isolated, but connected to its environment, and in most experiments, the ensemble molecule plus environment are at statistical equilibrium. Actually, most experimental studies of single-molecule kinetics involve the measurement of the regression of the equilibrium fluctuations of the fluorescent signal.

We consider a different approach, which shows that the model of Edman and Rigler[10] may be correct and do not violate detailed balance. We assume that the intramolecular dynamics, expressed in terms of the control parameters $\mathbf{s}(t) = (s_1(t), s_2(t), \ldots)$, can be described by a Markovian stochastic process. We denote by $\mathscr{R}(\mathbf{s}; t)d\mathbf{s}$ the probability that at time t the vector of control parameters is between $\mathbf{s}$ and $\mathbf{s} + d\mathbf{s}$ and assume that its time evolution is described by a linear evolution equation:

$$\frac{\partial}{\partial t}\mathscr{R}(\mathbf{s}; t) = \mathbb{L}\mathscr{R}(\mathbf{s}; t), \tag{26}$$

where $\mathbb{L}$ is a Markovian operator of the Fokker–Planck, master or Liouville type. We introduce the joint probability density $B_u(\mathbf{s}; t)$ for the chemical state of the molecule u and the control vector $\mathbf{s}$. This joint probability density is the solution of a compound stochastic Liouville equation[13,14]:

$$\begin{aligned}\frac{\partial}{\partial t}B_u(\mathbf{s}; t) = \mathbb{L}B_u(\mathbf{s}; t) + \sum_{u' \neq u} B_{u'}(\mathbf{s}; t)k_{u'u}(\mathbf{s}) \\ - B_u(\mathbf{s}; t)\sum_{u' \neq u} k_{uu'}(\mathbf{s}).\end{aligned} \tag{27}$$

We are interested in the evaluation of the marginal probability

$$P_u(t) = \int B_u(\mathbf{s}; t)d\mathbf{s}, \tag{28}$$

in terms of which we can compute the experimental observables, the correlation functions of the fluorescence signal. A simple way would be to derive an approximate equation for the marginal probability $P_u(t)$ by eliminating the stochastic vector $\mathbf{s}$ from Eq. (27). This is a standard topic in statistical physics,[13,14] which is usually referred to as the "renormalization of stochastic evolution equations." In quantum field theory, a "bare" particle interacting with a field is replaced by a "dressed" particle with renormalized parameters, which take into account the contribution of the field. By analogy with quantum field theory, we start out with a set of "bare" rate coefficients, $k_{u'u}(\mathbf{s})$, which depend on the fluctuations of the control parameters and then introduce a set of "dressed", renormalized rate coefficients $\check{k}_{u'u}$, which express the contribution of the fluctuations of the control parameters. If the intramolecular fluctuations, expressed in terms of the control parameters, have a short correlation time compared to chemical dynamics, the renormalized rate coefficients can be evaluated by using Van Kampen's renormalization method based on a cumulant expansion.[13,14] We get a "dressed" master equation for the marginal probability $P_u(t)$

$$\frac{\partial}{\partial t} P_u(t) = \sum_{u' \neq u} P_{u'}(t)\check{k}_{u'u} - P_u(t) \sum_{u' \neq u} \check{k}_{uu'}, \tag{29}$$

where the renormalized rate coefficients $\check{k}_{u'u}$ are expressed by fluctuation–dissipation relations as integrals of functional transformations of the correlation functions of the control variables. These expressions can be derived by using a cumulant expansion; in order to save space they are not given here. The important thing is that Eq. (29) is of the type used by Edman and Rigler[10] in their analysis. By starting out from a set of "bare" rate coefficients $k_{u'u}(\mathbf{s})$, which obey detailed balance, we end up with a set of renormalized, "dressed" rate coefficients $\check{k}_{u'u}$, which do not have to obey a similar condition of detailed balance. In general the renormalized rate coefficients $\check{k}_{u'u}$ are different from the average values of the "bare" rate coefficients, $\langle k_{u'u}(\mathbf{s})\rangle$. The differences

$$\Delta\check{k}_{u'u} = \check{k}_{u'u} - \langle k_{u'u}(\mathbf{s})\rangle, \tag{30}$$

measure the contribution of intramolecular fluctuations to the values of the renormalized rate coefficients.

The model of Edman and Rigler is valid if the following conditions are fulfilled:

(a) The intramolecular fluctuations have a short correlation time compared to the chemical timescale. If this condition is fulfilled, then the stochastic Liouville equation (27) can be replaced by the time homogeneous Markovian equation (29).
(b) Although short, the correlation time of intramolecular fluctuations is long enough so that the renormalized rate coefficients $\check{k}_{u'u}$ are different from the average values of the "bare" rate coefficients, $\langle k_{u'u}(\mathbf{s})\rangle$, that is, $\Delta\check{k}_{u'u} \neq 0$. If this constraint is fulfilled, it is possible that the renormalized rate coefficients do not obey detailed balance even though the bare coefficients do.
(c) The chemical states of the system are connected with at least a loop[15]; for this condition to be fulfilled there must be at least three chemical states.

An interesting problem pointed out by a referee is to identify the conditions for which "bare" kinetic equations obeying detailed balance would lead to "renormalized" kinetic equations which violate detailed balance. In general, this is still an open problem. Based on the method of projection operators, recently we have identified a few particular cases for which "bare" detailed balance leads to "renormalized" detailed balance.

It is interesting to compare the two mechanisms for damped chemical oscillations of correlation functions discussed in this paper. Although both mechanisms involve the intramolecular fluctuations expressed in terms of the random variations of control variables, their role is different in the two cases. In the first case, the timescale of the intramolecular fluctuations must be of the same order of magnitude as the chemical dynamics, and the damped oscillations of the correlation functions of the fluorescent signal are a direct result of damped oscillations at the intramolecular level. Here chemical dynamics plays a marginal role in the occurrence of damped oscillations; the oscillations may emerge even if there are only two chemical states. In the second case the intramolecular fluctuations play an indirect role: their job is to produce "dressed" rate coefficients which may violate detailed balance. In this case the intramolecular fluctuations are characterized by a characteristic time, which is smaller than the timescale of chemical kinetics. The chemistry plays a major role in the generation of the oscillations: they are produced by feedback loops involving at least three chemical states, which may violate detailed balance. In contrast to the first case, damped oscillations may exist even if the intramolecular dynamics do not have an oscillatory component.

There is a unified approach which can be used for describing both types of oscillations, intramolecular and chemical, respectively. The starting point is the stochastic Liouville equation (27). Instead of using approximate renormalization group methods for the elimination of the control parameters, we use an exact method, based on the use of an additional variable, the age a[16,17] of a given chemical state of the single molecule. We introduce a joint probability density $\mathscr{B}_u(a, \mathbf{s}; t)$ for the state u of the molecule, the age a of the state u, and the vector $\mathbf{s}$ of the control parameters, all evaluated at time t and the marginal probability density:

$$\mathscr{R}_u(a; t) = \int_{\mathbf{s}} \mathscr{B}_u(a, \mathbf{s}; t) d\mathbf{s} \tag{31}$$

of the state u of the molecule and the age a of the state u at time t. $\mathscr{B}_u(a, \mathbf{s}; t)$ is the solution of a system of age-dependent stochastic Liouville equation:

$$\left(\frac{\partial}{\partial t} + \frac{\partial}{\partial a}\right) \mathscr{B}_u(a, \mathbf{s}; t) = \mathbb{L}\mathscr{B}_u(a, \mathbf{s}; t) - \mathscr{B}_u(a, \mathbf{s}; t) \sum_{u' \neq u} k_{uu'}(\mathbf{s}), \tag{32}$$

$$\mathscr{B}_u(0, \mathbf{s}; t) = \sum_{u' \neq u} k_{u'u}(\mathbf{s}) \int_0^\infty \mathscr{B}_{u'}(a', \mathbf{s}; t) da'. \tag{33}$$

Equations (32) and (33) have a special structure which makes it possible to eliminate formally the vector $\mathbf{s}$ of control variables; such elimination methods were developed in connection with the description of heavy-ion collisions in nuclear physics.[17] The elimination leads to closed systems of equations for the marginal probability $\mathscr{R}_u(a; t)$:

$$\left(\frac{\partial}{\partial t} + \frac{\partial}{\partial a}\right) \mathscr{R}_u(a; t) = -\mathscr{R}_u(a; t) \sum_{u' \neq u} \kappa_{uu'}(a, t), \tag{34}$$

$$\mathscr{R}_u(0; t) = \sum_{u' \neq u} \kappa_{u'u}(a, t) \int_0^\infty \mathscr{R}_{u'}(a'; t) da', \tag{35}$$

where $\kappa_{u'u}(a, t)$ are renormalized rate coefficients which are complicated functionals of the Green functions $\mathfrak{S}_u(\mathbf{s}, t|\mathbf{s}_0, t_0)$ attached to the operator $\mathbb{L}$, which are the solutions of the equation $(\partial_t - \mathbb{L})\mathfrak{S}_u(\mathbf{s}, t|\mathbf{s}_0, t_0) = \delta(\mathbf{s} - \mathbf{s}_0)\delta(t - t_0)$. The evaluation of $\kappa_{u'u}(a, t)$ requires the numerical evaluation of the Green functions $\mathfrak{S}_u(\mathbf{s}, t|\mathbf{s}_0, t_0)$. In general the renormalized rates $\kappa_{uu'}(a, t)$ depend both on the age a of the state u and on the current time t. If the fluctuations of the control parameters are stationary, then the renormalized rate coefficients depend only on age, not on time, $\kappa_{uu'}(a, t) = \kappa_{uu'}(a)$ independent of t. Unlike the renormalized master equations (29), Eqs. (34) and (35) are exact; they describe both intramolecular as well as chemical fluctuations. If the renormalized rate coefficients are independent of time, then Eqs. (34) and (35) can be transformed into a generalized master equation (GME)[16,18]:

$$\begin{aligned}\frac{\partial}{\partial t} P_u(t) = \int_0^t \sum_{u'} [&P_{u'}(t - \Delta t)\omega_{u'u}(\Delta t) \\ &- P_u(t - \Delta t)\omega_{uu'}(\Delta t)] d\Delta t,\end{aligned} \tag{36}$$

where $\omega_{u'u}(\Delta t)$ are acceleration coefficients which can be computed in terms of the renormalized rate coefficients. We have

$$\begin{aligned}\omega_{uu'}(\Delta t) = \mathscr{L}^{-1}_{(\Delta t, s)} \\ \times \left\{ \frac{\mathscr{L}_{(s,a)} \left[\kappa_{uu'}(a) \exp\left(-\sum_{u''} \int_0^a \kappa_{uu''}(a'') da''\right)\right]}{\mathscr{L}_{(s,a)} \exp\left(-\sum_{u''} \int_0^a \kappa_{uu''}(a'') da''\right)} \right\},\end{aligned} \tag{37}$$

where $\mathscr{L}_{(s,a)}$ and $\mathscr{L}^{-1}_{(\Delta t,s)}$ denote the direct and inverse Laplace transformations, respectively.

Among the three sets of renormalized rate parameters, $\check{k}_{uu'}$, $\kappa_{uu'}(a)$, and $\omega_{uu'}(\Delta t)$, the following asymptotic relations are obtained:

$$\check{k}_{uu'} = \lim_{a\to\infty} \kappa_{uu'}(a) = \int_0^\infty \omega_{uu'}(\Delta t) d\Delta t. \tag{38}$$

The dependence of $\kappa_{uu'}(a)$ and $\omega_{uu'}(\Delta t)$ on a and Δt is due to intramolecular fluctuations. The approximate renormalized equations (29) do not contain detailed information about intramolecular dynamics, and thus they are able to describe only chemical oscillations. In contrast, the exact renormalized equations (34) and (35) and (36) are formally equivalent to the stochastic Liouville equations (32) and (33), and describe both intramolecular as well as chemical kinetics, therefore, they are able to describe both types of oscillations.

In conclusion, in single-molecule kinetics, both intramolecular as well as chemical processes can produce oscillations. There are no simple recipes to distinguish between these two types of oscillations; intramolecular oscillations tend to be faster than the chemical oscillations but there can be situations for which this is not true. Regarding the description of oscillation displayed by correlation functions, there are different types of models available. (1) Global techniques based on the use of characteristic functionals; (2) Local methods based on the use of stochastic Liouville equations, combined with the use of age-dependent and generalized master equations and the renormalization group approach.

3. On–off time distributions and oscillations

The analysis of on–off time distributions in single-molecule kinetics and their connections with chemical and intramolecular oscillations can be carried out by using an approach based on age-dependent master equations,[19] similar to Eqs. (32)–(35) combined with the method of characteristic functionals.[9] We consider a given realization of the random rate coefficients $\mathbf{k} = \mathbf{k}(t)$ with $\mathbf{k} = [k_{uu'}]$, $k_{uu} = 0$; for this realization we introduce the joint probability

$$\mathscr{P}_u(a; t, \mathbf{k}(t'))d\tau = \mathscr{P}_u(a; t)d\tau, \tag{39}$$

for the chemical state u of the molecule and for the age a of the state u at time t. In the following considerations we describe the fluctuations of the rates directly, without the use of any control parameters. This joint probability obeys the normalization condition $\sum_u \int_0^\infty \mathscr{P}_u(a; t)da = 1$. In terms of $\mathscr{P}_u(a; t)da$ we can introduce the probability density $\gamma(a|u; t)$ of the age (lifetime) of a given chemical state u, which obeys the normalization condition, $\int_0^\infty \gamma(a|u; t)da = 1$, and can be expressed in terms of the joint probability $\mathscr{P}_u(a; t)da$ as

$$\gamma(a|u; t)da = \frac{\mathscr{P}_u(a; t)da}{\int_0^\infty \mathscr{P}_u(a; t)da} = \frac{\mathscr{P}_u(a; t)da}{P_u(t)}, \tag{40}$$

where $P_u(t) = \int_0^\infty \mathscr{P}_u(a; t)da$ is a state probability which is the solution of the master equation (1). The conditional probability $\gamma(\tau|u; t)$ is the solution of a system of modified age-dependent master equations[19]:

$$\left(\frac{\partial}{\partial t} + \frac{\partial}{\partial a}\right)\gamma(a|u; t) = -\gamma(a|u; t)\tilde{\Theta}_u(t), \tag{41}$$

$$\gamma(0|u; t) = \sum_{u' \neq u} \tilde{k}_{u'u}(t) \int_0^\infty \gamma(a|u'; t)d\tau = \sum_{u' \neq u} \tilde{k}_{u'u}(t) = \tilde{\Theta}_u(t), \tag{42}$$

where

$$\tilde{k}_{u'u}(t) = k_{u'u}(t)P_{u'}(t)/P_u(t) \tag{43}$$

are adjoint rate coefficients and

$$\tilde{\Theta}_u(t) = \sum_{u' \neq u} \tilde{k}_{u'u}(t) \tag{44}$$

is a total (fluctuating) adjoint transition (decay) rate attached to the chemical state u. Equation (41) has an exact formal solution. We assume that the fluctuations of the rate coefficients are stationary, a hypothesis which is justified by the fact that usually single-molecule kinetic experiments are carried out at statistical equilibrium. For this reason, without loss of generality, we can push the initial condition to minus infinity, $t_0 \to -\infty$, and can represent this formal solution as

$$\gamma(a|u;t) = \tilde{\Theta}_u(t-a)\exp\left[-\int_{t-a}^{t} \tilde{\Theta}_u(t')dt'\right]. \tag{45}$$

An experimental observable is the averaged distribution of the lifetime over all possible values of the rate coefficients $\varphi(a|u) = \langle \gamma(a|u;t)\rangle$. We have

$$\varphi(a|u) = \left\langle \tilde{\Theta}_u(t-a)\exp\left[-\int_{t-a}^{t} \tilde{\Theta}_u(t')dt'\right]\right\rangle, \tag{46}$$

where the dynamic average $\langle ... \rangle$ is taken over all possible fluctuations of the rate coefficients. In Eq. (46) we take into account that the fluctuations of the rate coefficients are stationary, and as a result the average probability density of the lifetimes, $\varphi(a|u)$, depends on the lifetime a and is independent of the current time, t. For a given chemical state u, this dynamical average can be expressed in terms of a single random variable, the total adjoint rate $\tilde{\Theta}_u(t)$. By assuming that the cumulants of the total adjoint rate $\tilde{\Theta}_u(t)$ exist and are finite, the experimental observable $\varphi(a|u)$ can be easily evaluated. We have

$$\varphi(a|u) = \tilde{\Theta}_u^{\text{eff}}(a)\exp\left[-\int_{0}^{a} \tilde{\Theta}_u^{\text{eff}}(a')da'\right], \tag{47}$$

where $\tilde{\Theta}_u^{\text{eff}}(a)$ is the total adjoint effective decay rate attached to the state u, which is given by a cumulant expansion similar to Eq. (8)

from Sec. 2:

$$\tilde{\Theta}_u^{\text{eff}}(a) = \langle \tilde{\Theta}_u \rangle + \frac{\partial}{\partial a} \sum_{m=2}^{\infty} \frac{(-1)^{m-1}}{m!} \times \int_0^a \cdots \int_0^a \langle\langle \tilde{\Theta}_u(t_1) \cdots \tilde{\Theta}_u(t_m) \rangle\rangle dt_1 \cdots dt_m. \quad (48)$$

The identical structures of Eqs. (8) and (48) for $k_{\text{eff}}(\Delta t)$ and $\tilde{\Theta}_u^{\text{eff}}(a)$, respectively, make it possible to bring many of the results about correlation functions to the study of on/off time distributions, and vice versa. In particular, the survival function of a state u,

$$\ell(a|u) = \int_a^{\infty} \varphi(a|u) da = \exp\left[- \int_0^a \tilde{\Theta}_u^{\text{eff}}(a') da' \right], \quad (49)$$

plays the same role as the two-point correlation function of the fluorescent signal. We notice however that the rate coefficients involved in the two cases are different. They are direct rates for correlation functions and adjoint rates for on/off time distributions. In addition, we notice that the results about correlation functions are valid only for separable models, whereas the results for on/off times are more general; they apply to any master equation (1) with fluctuating rate coefficients. As far as we know, no experimental data about oscillations and their possible connections with on/off statistics have been published in the literature. Anyway, the same approaches as the ones used for correlation functions can be applied for the analysis of the oscillations. In particular, the analysis of the effective rates $\tilde{\Theta}_u^{\text{eff}}(a)$ is more suitable for the identification and analysis of oscillation, rather than the distribution $\varphi(a|u)$ or the survival function $\ell(a|u)$.

4. Reaction event statistics and oscillations

Of all observables in single-molecule kinetics, the reaction events have the longest history; they were studied decades before single-molecule kinetic experiments were possible. The numbers of reaction events as stochastic variables in time were first discussed over

80 years ago in Jean Perrin's celebrated book, *The Atoms*.[20] In 1974, Milan Solc[21–23] computed the probability distribution of the reaction events for a monomolecular reaction. In connection with a problem of nuclear physics, Vlad and Pop[24] developed a systematic Lippmann–Schwinger expansion for the solution of a master equation for the joint probability density of the state and the number of transition events for Markovian process in continuous time and discrete state space.[24] A more general expansion approach, which in particular can be applied to chemical reactions, was developed by Vlad and Ross.[25] Two different Lippmann–Schwinger expansions are developed: the first one produces exact expressions for the probabilities of the reaction events, and the second one produces exact expressions for the factorial moments and cumulants of the reaction events.

Regarding the applications to single-molecule kinetics, a serious mathematical difficulty arises: the average over the fluctuations of the rate coefficients is very hard to compute analytically or numerically. The simplest approach is to ignore the fluctuations of rate coefficients.[26,27] Although the method of Lippmann–Schwinger expansions can be extended to fluctuating rate coefficients, its application is numerically intensive; it requires repeated numerical integration of the master equations for different realizations of the rate coefficients, storing the resulting Green functions in a database, followed by the evaluation of dynamic averages.

The basic theory starts from a master equation for the joint probability $F_u(\mathbf{q}; t)$ of the state u and of the matrix $\mathbf{q} = [q_{uu'}]$ of the numbers $q_{uu'}$ of reaction events $u \to u'$:

$$\frac{\partial}{\partial t} F_u(\mathbf{q}; t) = \sum_{u' \neq u} F_u(\cdots q_{uu'} - 1; \ldots; t) k_{u'u}(t) - F_u(\mathbf{q}; t) \times \sum_{u'' \neq u} k_{uu''}(t). \tag{50}$$

If the fluctuations of the rate coefficients are neglected, $k_{u'u}(t) = k_{u'u}$, independent of t, and $F_u(\mathbf{q}; t)$ as well as the moments and cumulants of $q_{uu'}$ can be easily evaluated by deriving from Eq. (50) an equation for the partial generating function of $F_u(\mathbf{q}; t)$, $G_u(s; t) = \sum_{\mathbf{q}} \prod_{uu'} (s_{uu'})^{q_{uu'}} F_u(\mathbf{q}; t)$ with $|s_{uu'}| < 1$, followed by the use of Lippmann–Schwinger expansions. (For details, see Ref. 25). In order to take the flucuations of the rate coefficients into account, the method should be applied repeatedly for different realizations of the rate coefficients $k_{u'u}(t)$, followed by the evaluation of dynamic averages.

For fluctuating rate coefficients, a simpler procedure is based on the use of age-dependent renormalization technique mentioned in Sec. 2. For simplicity we take the fluctuating rates themselves as control parameters, $\mathbf{k} = \mathbf{s}$, and write a system of stochastic Liouville, age-dependent equations for the joint probability density $Q_u(a, \mathbf{k}, \mathbf{q}; t)$ of the state u of the molecule, its age a, its matrix $\mathbf{k} = \mathbf{s}$ of the rate coefficients (control parameters), and its matrix $\mathbf{q}$ of the numbers of reaction events

$$\left(\frac{\partial}{\partial t} + \frac{\partial}{\partial a}\right) Q_u(a, \mathbf{k}, \mathbf{q}; t) = \mathbb{L} Q_u(a, \mathbf{k}, \mathbf{q}; t) - Q_u(a, \mathbf{k}, \mathbf{q}; t) \sum_{u' \neq u} k_{uu'}, \tag{51}$$

$$Q_u(0, \mathbf{k}, \mathbf{q}; t) = \sum_{u' \neq u} k_{u'u} \int_0^\infty Q_u(a', \mathbf{k}, \ldots, q_{u'u} - 1; \ldots; t) da'. \tag{52}$$

The elimination of the fluctuation rate coefficients in Eqs. (51) and (52) leads to a system of renormalized age-dependent master equation for the joint probability density $\mathscr{F}_u(a, \mathbf{q}; t)$ of the state u of the molecule, its age a, and its matrix $\mathbf{q}$ of the numbers of reaction events

$$\left(\frac{\partial}{\partial t} + \frac{\partial}{\partial a}\right) \mathscr{F}_u(a, \mathbf{q}; t) = -\mathscr{F}_u(a, \mathbf{q}; t) \sum_{u' \neq u} \kappa_{uu'}(a, t), \tag{53}$$

$$\mathscr{F}_u(0, \mathbf{q}; t) = \sum_{u' \neq u} \kappa_{u'u}(a, t) \int_0^\infty \mathscr{F}_{u'}(a', \ldots, q_{u'u} - 1; \ldots; t) da', \tag{54}$$

where $\kappa_{u'u}(a,t)$ are the same renormalized rate coefficients as the ones from Eqs. (34) and (35). In particular, if the fluctuations of the rate coefficients (control parameters) are stationary, then the renormalized rate coefficients depend only on age, and not on time $\kappa_{uu'}(a,t) = \kappa_{uu'}(a)$, independent of t. In this case too, it is possible to use a generating function transformation, $\mathscr{G}_u(a,s;t) = \sum_{\mathbf{q}} \prod_{uu'} (s_{uu'})^{q_{uu'}} \mathscr{F}_u(a,\mathbf{q};t)$ with $|s_{uu'}| < 1$, followed by the use of Lippmann–Schwinger expansions.

The application of these approaches in the general case to experimental data is rather complicated and involves the use of advanced numerical techniques. For separable systems, the average over the fluctuations of the rate coefficients can be carried out analytically. Although complicated, the theory leads to analytical results, which are similar to the ones obtained for the correlation functions of the fluorescent signal. In particular, for a system with two states, one fluorescent and the other one non-fluorescent, the second-order correlation function of the "out" reaction events, attached to each state, can be expressed in terms of the damping factor $\mathscr{I}(\Delta t)$, given by Eq. (7). It follows that the considerations account for the relations between oscillations and correlation functions can be easily adapted to the study of the relations between oscillations and the statistics of the reaction events.

5. Conclusions

In this chapter we tried to get the reader acquainted with the process of building stochastic models for single-molecule kinetics, with special reference to oscillations. The emphasis was on model building, rather than on model classification or model solving. We have illustrated model building based on various techniques from nonequilibrum statistical physics, including both global as well as local methods: global characteristic functionals, stochastic Liouville equations, age-dependent and generalized master equations, and different stochastic renormalization techniques.

We tried to show how the models can be used for describing intramolecular as well as chemical oscillations in single-molecule kinetics. Based on theoretical considerations, we also tried to develop criteria for distinguishing between these two types of oscillations. The difficulty of the analysis of experimental data depends on the type of experimental observable used. On/off time data statistics are the easiest to process. Moreover, the theory imposes few restrictions for the models, which describe on/off time distributions; in particular, no separability conditions are needed. Building models for the other two observables, the correlation functions of the fluorescent signal and the statistics of the reaction events, is more complicated. These models can be solved analytically only in a few cases (separable models, deterministic rate coefficients). In general the use of these models require advanced numerical computations.

Acknowledgments

This research has been supported in part by the National Science Foundation and by the CEEX-M1-C2-3004/2006 Grant of the Romanian Ministry of Research and Education.

References

1. R. M. Dickson, A. B. Cubitt, R. Y. Tsien and E. W. Moerner, *Nature* **388** (1997) 355.
2. P. H. Lu and X. S. Xie, *Journal of Physical Chemistry B* **10** (1997) 2753.
3. G. K. Schenter, H. P. Lu and X. S. Xie, *Journal of Physical Chemistry A* **103** (1999) 10477.
4. E. J. Peterman, S. Brasselet and E. W. Moerner, *Journal of Physical Chemistry A* **103** (1999) 10553.
5. A. D. Mehta, M. Rief, J. A. Spudich, D. A. Smith and R. M. Simmons, *Science* **283** (1999) 1689.
6. W. E. Morner and M. Orrit, *Science* **283** (1999) 1670.
7. W. E. Moerner, *Journal of Physical Chemistry B* **106** (2002) 910.
8. J. Wang and P. Wolynes, *Journal of Chemical Physics A* **110** (1999) 4812.

9. M. O. Vlad, F. Moran, F. W. Schneider and J. Ross, *Proceedings of the National Academy of Sciences* **99** (2002) 12548.
10. L. Edman and R. Rigler, *Proceedings of the National Academy of Sciences* **97** (2000) 8266.
11. F. L. H. Brown, *Physical Review Letters* **90** (2003) 028302.
12. J. P. Boon and S. Yip, *Molecular Hydrodynamics* (McGraw-Hill, New York, 1980).
13. N. G. Van Kampen, *Stochastic Processes in Physics and Chemistry*, 2nd edn. (North Holland, Amsterdam, 1992), Chapter XVI.
14. M. O. Vlad, J. Ross and M. C. Mackey, *Journal of Mathematical Physics* **37** (1996) 803.
15. B. L. Clarke, *Advances in Chemical Physics* **43** (1980) 1.
16. M. O. Vlad and A. Pop, *Physica A* **155** (1989) 276.
17. M. O. Vlad and A. Pop, *Journal of Physics A: Mathematical General* **22** (1989) 3945–3957.
18. U. Landman, E. W. Montroll and M. F. Shlesinger, *Proceedings of the National Academy of Sciences* **74** (1977) 430.
19. M. O. Vlad, F. Moran and J. Ross, *Chemical Physics (Elsevier)* **287** (2003) 83.
20. J. Perrin, *Atoms* (translated; Original — French (1939)) (Ox Bow Press, Woodbridge, CT, 1990).
21. M. Solc, *Zeitschrift fur Physikalische Chemie, Neue Folge* **92** (1974) 1.
22. M. Solc, *Collection of Czechoslovak Chemical Communications* **39** (1974) 197.
23. M. Solc, *Collection of Czechoslovak Chemical Communications* **46** (1981) 1217.
24. M. O. Vlad and A. Pop, *Czechoslovak Journal of Physics* **40** (1990) 9.
25. M. O. Vlad and J. Ross, *Journal of Physical Chemistry A* **104** (2000) 3159.
26. F. L. H. Brown, *Physical Review Letters* **90** (2003) 028302.
27. I. V. Gopich and A. Szabo, *Journal of Chemical Physics* **118** (2003) 454.

CHAPTER 9

Discrete Stochastic Models of Single-Molecule Motor Proteins Dynamics

Anatoly B. Kolomeisky

Department of Chemistry, Rice University, Houston, TX 77005, USA

Individual motor proteins, or molecular motors, are enzymatic molecules that utilize chemical energy in producing mechanical work important for cellular functions. Motor proteins typically step unidirectionally along linear tracks, provided by microtubules, actin filaments, and DNA and RNA molecules, and they play a crucial role in many biological transport processes, cellular organization, and functioning. In this chapter we review some theoretical approaches of analyzing motor protein dynamics in light of current experimental methods that measure the biochemical and biophysical properties at the single-molecule level. Our main focus is on discrete kinetic and stochastic models that yield exact and explicit expressions for dynamical properties of single motor proteins. It is shown that this approach provides a satisfactory description of all current experimental observations, and it might provide a convenient theoretical framework for understanding the mechanisms of functioning of these enzymatic molecules.

1. Introduction

Biological cells are complex inhomogeneous systems that undergo many dynamic processes such as gene replication, transcription, translation, transport of vesicles and organelles, and segregation of chromosomes during cell division.[1,2] The ability of cells to support these processes in a fast and effective way strongly depend on several classes of protein molecules, generally called motor proteins or molecular motors.[1–4]

There are many different types of motor proteins that are currently known, such as kinesins, dyneins, myosins, DNA and RNA polymerases, helicases, etc., and new molecular motor species are constantly discovered. It is widely believed that all of them utilize the same general mechanism: motor proteins function by converting a chemical energy into a mechanical work. The most common sources of the chemical energy in motor proteins are the hydrolysis of ATP (adenosine triphosphate) or related compounds, and polymerization of DNA, RNA and protein molecules. These transformations of chemical energy into mechanical work typically involve a complex network of biochemical reactions and physical processes, and they take place on very fast timescales with a high efficiency.[3,4] However, mechanisms of the mechanochemical coupling in motor proteins are still not well understood.[1–4]

Motor proteins can be viewed as submicroscopic nanometer-sized motors[3,4] that consume fuel (from the chemical processes) to produce a mechanical motion. However, in contrast to macroscopic engines, molecular motors mainly function at the single-molecule level in nonequilibrium but isothermal conditions. At these conditions the state of the local environment and thermal fluctuations play a critical role is dynamical properties of single molecules. A successful theoretical description of motor protein mechanisms should take into consideration the existence of multiple protein conformations, account for the complexity of biochemical and biophysical processes involved, and explain their efficiency.

In recent years a tremendous progress in experimental studies of motor proteins has been reported.[5–28] It is now possible to monitor and control the motion of single motor protein molecules at different external conditions with high spatial and time resolution.

These single-molecule experiments have uncovered many previously unknown microscopic details of motor proteins. These investigations also provided a significant amount of quantitative information that stimulated various theoretical discussions of the mechanisms that control the dynamics of molecular motors.[4, 30–55]

In this chapter, after briefly reviewing some single-molecule experimental observations, we present several theoretical developments in the field of motor proteins. Although our theoretical approaches are mainly developed for linear motor proteins, the analysis can be easily extended and applied for several important classes of molecular motors that rotate.[3, 4, 7, 9, 22, 26, 27, 33, 40] We focus theoretically on discrete stochastic (or chemical kinetic) models that currently seem to provide the most appropriate framework for understanding motor protein dynamics.[4]

2. Single-molecule experiments

A significant amount of information on mechanisms of motor proteins have recently come from the single-molecule experiments, including optical-trap spectrometry, magnetic tweezers, Förster Resonance Energy Transfer (FRET), dynamic force microscopy, fluorescent imaging, and many other techniques (see also Chapters 1 and 8).[5–27] These methods, that allow to passively monitor and to actively influence the dynamics of single protein molecules, currently provide a very powerful tool for uncovering details of molecular motor dynamics.

One of the most successful and widely used single-molecule experimental techniques is optical-trap spectrometry.[5, 8, 11–16, 20, 21, 23, 25, 29] In this method a single protein molecule is chemically attached to a polystyrene bead that is held by controlled external laser field. The bead follows the motion of the protein molecule as it moves along the filament track. Because the external electromagnetic field is nonuniform, the bead is trapped in the focus region where the light is most intense. Any motion of the motor molecules leads to the bead displacement from the focus, and it produces a returning force that is proportional to the displacement. Thus, the optical-trap system works like an electronic version of the harmonic potential that can be calibrated with a high precision.

Closely related to optical-trap spectroscopy is the magnetic tweezers method.[10,18,22] In this approach a motor protein molecule is chemically connected to a magnetic bead, that is maintained under tension by the field gradient of permanent magnets, and to the surface perpendicular to the field. The distance, z, from the bead to the surface and the magnitude of transverse fluctuations of the bead, $\langle \delta x^2 \rangle$, are recorded. From these measurements the force exerted on the motor protein is calculated using the equipartition theorem, $F = k_B Tz/\langle \delta x^2 \rangle$. Magnetic tweezers experiments are especially suitable for investigations of motor proteins that unwind, untangle, and remove supercoiling in double-stranded DNA molecules, such as different types of topoisomerases and helicases.[10,18]

A different experimental approach is developed by Selvin and coworkers.[17,24] The method provides a fluorescent imaging with one-nanometer accuracy, called FIONA, and it allows to track the position of a single dye molecule covalently attached to a specific position on a motor protein molecule with nanometer accuracy and sub-second resolution. Although the fluorescent image has a diffraction-limited size of several hundred nanometers, the brightest part of the image, which directly corresponds to the position of the dye, can be determined accurately with the precision up to 1 nm. Using this technique, it has been proved that single two-headed motor proteins, such as kinesins and myosins V and VI, move in hand-over-hand fashion, i.e. the motor domain heads alternate their positions along the molecular tracks.

3. Theoretical models

Successes in single-molecule experiments have stimulated multiple efforts to describe motor proteins theoretically.[30–55] In the development of a theoretical framework for the motor protein dynamics it is important to note that a successful theoretical model should respect basic laws of chemistry and physics and it should satisfy several requirements on symmetry, such as periodicity, polarity, and chirality.

The central goal of all theoretical models for molecular motors is the connection between biochemical transformations and mechanical motions. It is a fundamental result that all biochemical transitions are reversible, even if the available experimental data may not provide direct evidence. The reverse transitions might be very slow but they should not be totally neglected in the comprehensive theoretical analysis because it might lead to unphysical conclusions and wrong assumptions on the mechanism.[36] This observation implies that under some conditions, the motor proteins, that typically hydrolyze ATP molecules when moving forward, can also synthesize ATP. This fact is experimentally shown for the rotary molecular motor F_0–F_1-ATP synthase.[22,27] For the processive motor proteins for a long time the existence of backward transitions has been completely neglected. However, recent experiments on mitotic kinesins[69,72] suggest that ATP can also be produced by linear molecular motors.

All currently existing theoretical approaches can be divided into two main groups: continuum ratchet models[31–34,39,45–47,54] and discrete stochastic (or chemical kinetic) methods.[30,35–37,41–44,48–53,55] Also, there are several theoretical works that combine both approaches by using continuum and discrete descriptions in different parts of their models.[81]

3.1. *Continuum ratchet models*

In this physics-oriented theoretical approach a motor protein molecule is viewed as moving along several spatially parallel, periodic but generally asymmetric free-energy potential surfaces.[31–34,39,45–47,54] The particle can stochastically switch between different potentials that describe different biochemical states. Evolution of the system can be described by a system of coupled

Fokker–Planck equations. The sustained unidirectional motion of the particle requires a constant supply of the chemical energy.

These chemically driven ratchets,[32,39,47] which are also called Markov–Fokker–Planck models,[54] provide a simple and physically reasonable picture of the motor protein's dynamics with a small number of parameters that can be handled by well-established analytical tools. However, there are several aspects of ratchet models that complicate their application for modeling motor protein dynamics. With the exception of a few oversimplified and mostly unrealistic potentials, general analytical results cannot be obtained. The necessary numerical computations with many assumed parameters are quite demanding. Furthermore, it is almost impossible to derive the realistic potentials from the available data on protein structures, and approximate potentials have to be used in the computations of molecular motor dynamic properties. As a consequence, although successful fits to some experimental data have been obtained,[54] it is hard to estimate the reliability and applicability of these methods for real motor proteins. Thus, for the present, we believe that ratchet continuum models can be reasonably used only for description of some qualitative features of motor protein dynamics.

3.2. *Discrete stochastic models*

A rather different approach to describe the motion of single motor proteins adapts discrete stochastic models of traditional chemical kinetics.[30,35–37,41–44,48–53,55] In the simplest model it is assumed that during the enzymatic cycle the molecule jumps from the binding site l to the site $l+1$ through the sequence of N intermediate biochemical states that might have different spatial positions, as shown in Fig. 1. Two neighboring binding sites are separated by distance d. The step size d for molecular motors depends on the structure of linear track. It is equal to 8.2 nm for kinesins and dyneins traveling along the microtubules, while for myosins V and VI that move along actin filaments, the step size is larger, $d \approx 36$ nm. The motor protein in

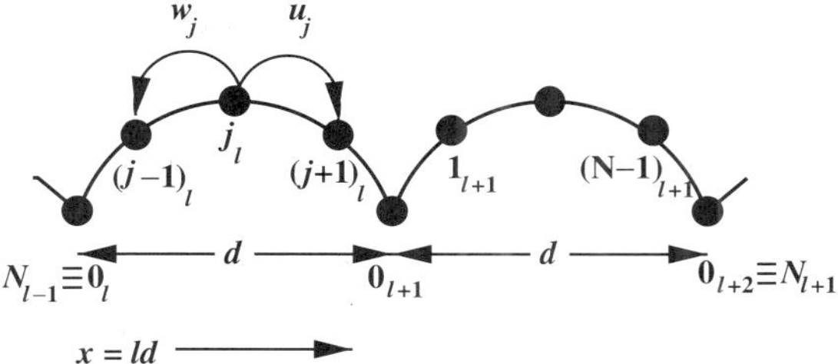

Fig. 1. Schematic picture of a linear sequential discrete stochastic model for the motion of single motor proteins. Transition rates u_j and w_j describe forward and backward steps from the state j.

the mechanochemical state j_l ($j = 0, 1, \ldots, N-1$) moves forward to the state $(j+1)_l$ at a rate u_j, or it can step backward to the state $(j-1)_l$ at a rate w_j. It is assumed that 0_l corresponds to the state when the motor protein is strongly bound to the molecular track, awaiting the arrival of a fuel molecule. By moving from the state 0_l to 0_{l+1}, the motor protein molecule catalyzes the hydrolysis of one fuel molecule. Note that reverse transitions are explicitly taken into account in this approach, in agreement with the experimental observations on backward steps.[12,15,25]

The dynamics of motor proteins is governed by Master equations,

$$\frac{dP_j(l,t)}{dt} = u_{j-1}P_{j-1}(l,t) + w_{j+1}P_{j+1}(l,t) - (u_j + w_j)P_j(l,t), \tag{1}$$

where $P_j(l,t)$ is a probability to find the molecule in the state j_l at time t. It can be shown that these equations also describe the hopping of a single particle along the infinite one-dimensional periodic lattice (with a period of size N). This observation allows one to utilize the mathematical formalism, developed by Derrida[70] in 1983, to obtain exact and explicit expressions for the asymptotic mean velocity

$$V = V(\{u_j, w_j\}) = \lim_{t\to\infty} \frac{d\langle x(t)\rangle}{dt}, \tag{2}$$

and for the dispersion (or effective diffusion constant)

$$D = D(\{u_j, w_j\}) = \frac{1}{2} \lim_{t \to \infty} \frac{d}{dt} [\langle x^2(t) \rangle - \langle x(t) \rangle^2], \tag{3}$$

where $x(t)$ is the position of the motor protein molecule along the linear track at time t. These expressions directly relate the rates u_j and w_j, typically measured in bulk chemical kinetic experiments, with the biophysical properties (V and D) of the motor proteins measured in the single-molecule experiments. For the simplest nontrivial model with $N = 2$ states, the theory gives the following expression for the mean velocity and dispersion:

$$V = d \frac{u_0 u_1 - w_0 w_1}{u_0 + w_0 + u_1 + w_1}, \qquad D = \frac{d^2}{2} \frac{(u_0 u_1 + w_0 w_1) - 2(V/d)^2}{u_0 + w_0 + u_1 + w_1}. \tag{4}$$

It is important to note that simultaneous knowledge of both mean velocity and dispersion provides a valuable microscopic information on the number of intermediate states N. It follows from the fact that a dimensionless parameter, $r = \frac{2D}{dV}$, called *randomness*, which is a measure of dynamical fluctuations of a motor, has a lower bound of $1/N$.[71] Indeed, if all backward rates are equal to zero, i.e. $w_j \equiv 0$ for all j, and forward rates are equal ($u_j = u$) we obtain $r = 1/N$. For all other parameters randomness increases. At saturating ATP concentrations, $r \simeq 0.39$ has been observed for kinesins.[5] This means that $N \geq 3$ [ATP]-independent intermediate transitions significantly contribute to the motor protein dynamics. This conclusion is in agreement with the current biochemical view on the process of ATP hydrolysis with the help of kinesin motor proteins that assume $N = 4$ basic states:

$$\begin{aligned} \mathrm{M \cdot K + ATP} &\underset{w_1}{\overset{u_0}{\rightleftharpoons}} \mathrm{M \cdot K \cdot ATP} \underset{w_1}{\overset{u_1}{\rightleftharpoons}} \mathrm{M \cdot K \cdot ADP \cdot P_i} \\ &\underset{w_3}{\overset{u_2}{\rightleftharpoons}} \mathrm{M \cdot K \cdot ADP} \underset{w_4}{\overset{u_3}{\rightleftharpoons}} \mathrm{M \cdot K}, \end{aligned} \tag{5}$$

where K and M denote kinesin and microtubule, while ADP and P_i represent adenosine diphosphate and inorganic phosphate, respectively. Note that the first forward rate is proportional to ATP concentration, $u_0 = k_0$ [ATP].

A motor protein molecule, that constantly hydrolyzes fuel molecules while moving along the linear track, is exerting a driving force that can be easily analyzed in discrete stochastic models. It was shown that this force for the simplest sequential chemical kinetic model in Fig. 4 is given by

$$F = \frac{k_B T}{d} \ln \prod_{j=0}^{N-1} \frac{u_j}{w_j}. \tag{6}$$

This result can be easily understood by using the standard physical–chemical arguments.[36,37] Let us define a function $K = \prod_{j=0}^{N-1} \frac{u_j}{w_j}$ that corresponds to an effective equilibrium constant for the process of moving the motor protein from the binding site l to the next binding site $l + 1$. Then, the expression $\Delta G = -k_B T \ln K$ gives the free energy difference between two neighboring binding states of the motor protein. It is assumed that all this free energy is converted into mechanical work that moves the molecule over the step-size distance d, thus exerting the force given in Eq. (6). Note again, if one of the backward rates w_j is assumed to be equal to zero then the predicted stall force diverges, leading to unphysical result.

A unique property of single-molecule experiments is the ability to impose a measured force, F, that can be applied directly to a single motor protein. In discrete stochastic models the effect of external forces can be easily incorporated[36,37,44] by introducing *load distribution factors*, θ_j^+ and θ_j^- (for $j = 0, 1, \ldots, N-1$). These parameters describe how the work performed by external forces is distributed between various forward and backward transitions. Assuming that the external force acts parallel to the filament track, it produces a work Fd on a system. Then the original free-energy surfaces are modified by changing the relative heights of transition-state barriers, and the transition rates become force-dependent. The standard reaction-rate theories[3] suggest that

$$\begin{aligned} u_j(F) &= u_j(0)\exp(-\theta_j^+ Fd/k_B T), \\ w_j(F) &= w_j(0)\exp(\theta_j^- Fd/k_B T), \end{aligned} \tag{7}$$

with the additional reasonable requirement that

$$\sum_{j=0}^{N-1} (\theta_j^+ + \theta_j^-) = 1. \tag{8}$$

The load distribution factors provide a significant information on mechanochemical transitions in motor proteins since they effectively describe the free-energy landscapes for the particles. In addition, the products $\theta_j^{\pm} d$ correspond to projections of free-energy landscape extrema along the reaction coordinate, defining the substeps for the motion of motor proteins.[36,37,44]

The substitution of the load-dependent rates from Eq. (7) into expressions for velocities and dispersions provides a formal way to analyze theoretically the effect of external forces on the dynamics of motor proteins.[36,37,44] The applied external force that completely stops the motor protein molecule, i.e. when $V(F) = 0$, is called a *stall force*. For the sequential discrete stochastic models it corresponds to an equilibrium state between internal chemical forces of the motor protein and external fields. The stall force is given analytically by Eq. (6),[36,37,44] and it directly depends on the ATP concentration.

Let us illustrate our method in application to the analysis of the single-molecule data on kinesin motor proteins.[5,8] Because the fluctuation analysis indicates that there are, at least, $N = 4$ important biochemical states, we utilized explicit expressions for velocity V and randomness r to fit experimental observations. The results are shown in Figs. 2 and 3. In fitting procedure the rate constants and load-distribution factors have been used, and the resulting parameters agree well with the available estimates of the rate constants for kinesin motor proteins from the bulk chemical-kinetic measurements.[44]

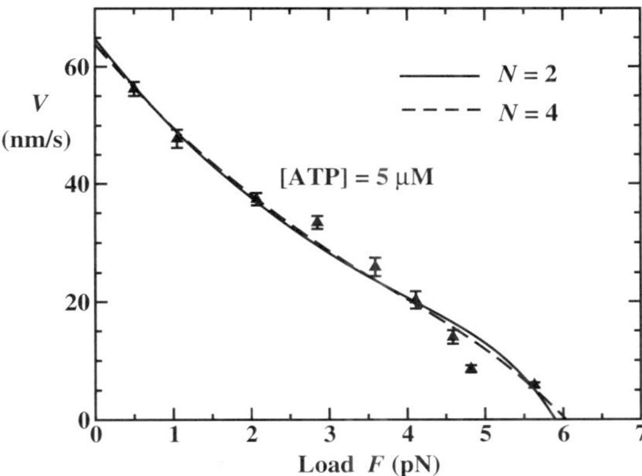

Fig. 2. Theoretical fits of experimental data on kinesin motor proteins for force–velocity curves from Refs. 5 and 8. Solid lines correspond to $N = 2$ model with nonexponential waiting-time distributions, while dashed lines describe chemical kinetic model with $N = 4$. (The figure is adopted and modified from Ref. 44.)

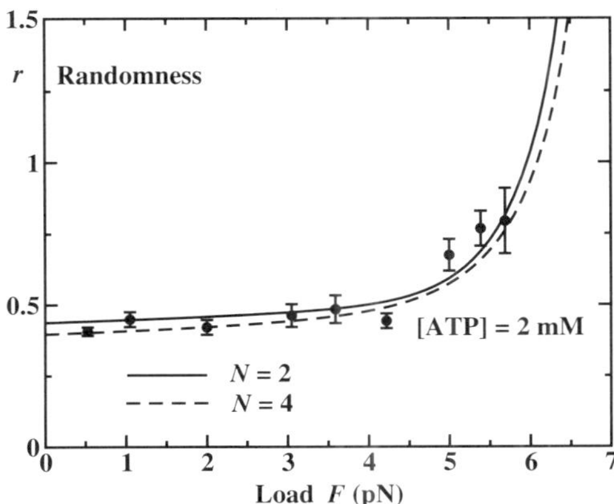

Fig. 3. Theoretical fits of experimental data on kinesin motor proteins for randomness as a function of the external resisting force from Refs. 5 and 8. Solid lines correspond to $N = 2$ model with nonexponential waiting-time distributions, while dashed lines describe chemical kinetic model with $N = 4$. (The figure is adopted and modified from Ref. 44.)

A major advantage of the discrete stochastic models is their ability to easily handle more complex biochemical reactions than the linear sequence. Biochemical experiments indicate that many motor proteins do not follow a single linear sequence of states that connects the corresponding binding states. The more realistic picture of biochemical transitions includes multiple parallel pathways, loops, branched states that do not lead to directed motion, and irreversible detachments of molecular motors, as shown in Figs. 4–6. Experiments on processivity of single-headed kinesins[74] and RNA polymerases[75] clearly demonstrate the possibility of multiple parallel biochemical pathways and branched states. The original theoretical analysis, derived for a single pathway,[36,37,44] can be extended to obtain analytical results for the dynamic properties of motor protein in periodic sequential chemical kinetic models with branches (Fig. 6)[41] and for parallel chemical kinetic models (Fig. 5).[41,43] It is found that the presence of branches and/or parallel channels generally increases fluctuations and decreases the drift velocities of the single particles, as compared with the unmodified simple sequential models. However, the presence of the branched states does not change the value of the stall force, while for the systems with parallel pathways the

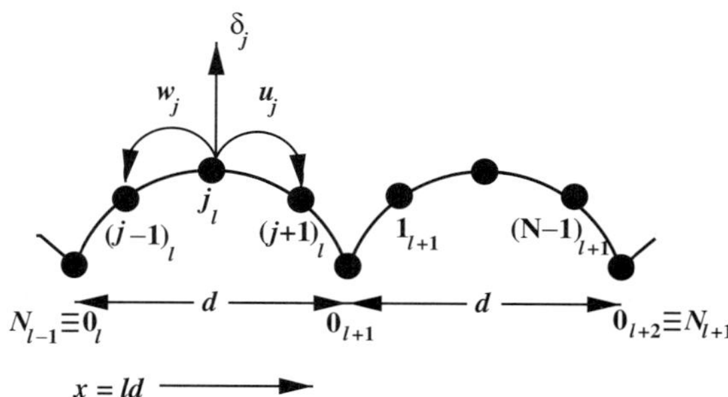

Fig. 4. Schematic picture of a linear sequential discrete stochastic model with irreversible detachments for the motion of single motor proteins. Transition rates u_j and w_j describe forward and backward steps from the state j, while δ_j is a detachment rate from the state j.

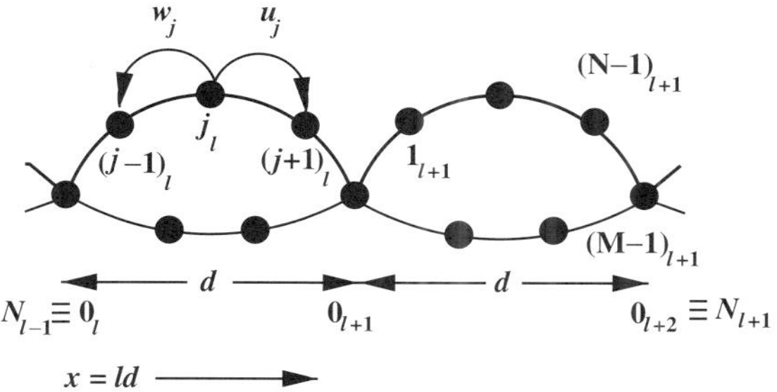

Fig. 5. Schematic picture of a discrete stochastic model with two parallel pathways for the motion of single motor proteins. Upper channel has N intermediate states, while the lower channel has M discrete states.

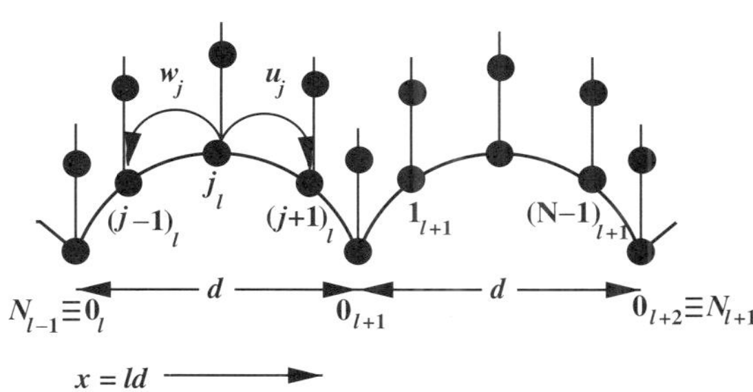

Fig. 6. Schematic picture of a discrete stochastic model with branched states for the motion of single motor proteins.

stall force can change significantly when there is a breaking of the detailed balance conditions.[43]

The active motor protein molecule cannot stay forever on the molecular track and it will eventually dissociate, as illustrated in Fig. 4. Since the probability of the motor protein rebinding under the single-molecule experimental conditions[5,8] is low, these dissociations can be regarded as effectively irreversible. The effect of such irreversible detachments on the dynamic properties of molecular motors can be investigated in the discrete stochastic models.[41,44] It was shown that exact results for velocities and dispersions can be obtained at stationary-state conditions by mapping the discrete model with detachments onto an effective sequential model without detachments. These irreversible dissociations strongly influence run lengths of the molecular motors, while the effect on the velocities and dispersions is typically smaller.[8,44] There is an interesting theoretical prediction for some range of parameters that suggests that irreversible detachments can accelerate the motor particles. This is due to the fact that slower moving particles spend more time in the states from which they can dissociate with higher rates, while fast-moving molecules survive.

The central idea of discrete stochastic models is that the motion of motor proteins consists of multiple transformations between different protein conformations, i.e. the molecule undergoes a chemical transition from one biochemical state to another one, and different conformations have generally different spatial positions. Chemical kinetics supposes that these transitions are fully independent of each other, and the time intervals between these events are distributed according to a Poisson statistics. It means that the waiting-time distribution functions for time intervals between transitions are exponential functions of time with coefficients equal to the overall transition rates from the given state. For example, we can define the waiting-time distribution function $\psi_j^+(t)dt$ as the probability of jumping one step forward from the state j in the time interval from t to $t + dt$ after arriving at state j, while $\psi_j^-(t)dt$ is the corresponding probability of moving backward. For pure chemical kinetic models it can be shown that $\psi_j^{\pm}(t) \propto \exp[-(u_j + w_j)t]$. However, high mechanical efficiency of motor proteins suggest that some transitions might have more mechanical than chemical nature, and the waiting-time distributions for such processes could be nonexponential.[42] In addition, the presence of hidden intermediate states can also result in deviations from the Poisson statistics in motor proteins. It is interesting to consider a more general class of discrete stochastic models with arbitrary waiting-time distributions. Exact analytic expressions for the dynamic properties for these models have been obtained.[42] In this case, the theoretical analysis is also based on the application of Derrida's method to generalized Master equations. It was found that deviations from the exponential waiting-time distributions, that describe the standard chemical kinetic processes, do not affect the mean velocities. However, there are significant changes in the expressions for dispersions. It was also suggested that the inclusion of the mechanical processes might provide a more effective and economic (with less number of parameters) description of motor protein dynamics.[42,44]

Experiments on motor proteins determine single-molecule time trajectories that consist of forward and backward steps and pauses.[5,12,25] Measured fractions and dwell time distributions of forward and backward steps and detachments provide a valuable information on the mechanisms of motor proteins. Although these dynamic properties of molecular motors are related to mean velocities and dispersions, they provide a different dynamical information on molecular motors since they describe the so-called first-passage time processes.[76] Neglecting these differences can lead to erroneous conclusions on the mechanisms of motor proteins.[12,49]

In recent single-molecule experiments,[12,25] dwell times and fractions have been measured for single kinesin molecules at different ATP concentrations and for different external forces. It was found that dwell times before the forward and backward steps and before irreversible dissociations are the same for all experimental conditions. Based on these observations a theoretical model was presented with the conclusion that ATP is hydrolyzed during the forward *and* the backward steps. However, it contradicted earlier single-molecule and bulk biochemical experiments,[77–79] and it could not explain the detachment processes. Using the method of first-passage times, these experimental data on mechanochemical coupling in kinesins have been analyzed in terms of the discrete sequential stochastic models.[48,49] It was assumed that ATP is hydrolyzed only in the forward steps, while the backward steps correspond to ATP hydrolysis. Analytical calculations[49] led to surprising results — that dwell times for the kinesin to move one step forward or backward, or to dissociate irreversibly, are the same, although the probabilities for these events are quite different. Experimental observations confirm these theoretical predictions fairly well.[12,25] As a result, interpretations suggesting that ATP is hydrolyzed on back steps, or not synthesized, are not convincing.[49,51]

The explicit results for first-passage time properties of motor proteins[48,49] correspond to a full mechanochemical cycle when a single major forward or backward step is taken. The resulting expressions can be applied to the analysis of experimental observations *only if* these data allow one to identify precisely the beginning and/or the end of each cycle, i.e. when the molecular motor can be found in the states 0_l (see Fig. 1). However, experimental single-molecule trajectories are quite noisy and some substeps can escape detection, so that only the major transitions are reported. More important is the fact that a molecular motor may execute a major step without actually finishing a full enzymatic cycle from a state 0_l to $0_{l\pm1}$. In such cases the analytical method[49] should be modified to include the contributions from the hidden substeps.[80]

The foundation of discrete stochastic models is the dynamics of motor proteins along free-energy potential surfaces. In the simplest discrete stochastic models, an effective one-dimensional free-energy picture is assumed, which is not realistic for the description of motor protein dynamics. In addition, in the latest single-molecule optical-trap experiments[14,15] perpendicular and sideways controlled forces have been imposed on the moving motor protein molecules. To fully understand the mechanisms of molecular motors a full three-dimensional free-energy landscape is required. The original discrete stochastic models can be extended by explicitly implementing a free-energy landscape picture.[51,52] Using these results to analyze the motion of single kinesins it is suggested that after binding with the ATP the molecular motor moves closer to the microtubule by 0.5–0.7 nm, while advancing forward by only 0.1–0.2 nm. After the hydrolysis of ATP the motor protein jumps $\simeq$7.8 nm to complete the mechanochemical cycle. Also predicted but not yet verified is the strong sensitivity to bead size in the optical trap under assisting loads. At the same time, theoretical analysis suggested a nonmonotonous dynamical behavior at superstall loads that have been observed experimentally.[25]

Many motor proteins consist of several domains that individually also exhibit enzymatic activity.[2,3,19] For example, a RecBCD helicase motor protein is made of three domains, and two of them, RecB and RecD, as single proteins behave like helicases by consuming ATP and unwinding DNA molecules.[19] Single-molecule measurements of the speeds of helicase molecular motors indicate that the protein cluster moves faster than the individual motor proteins.[19] To explain these surprising observations again a discrete stochastic model, in which two parts of the protein molecule interact through some energy potential, has been used.[50] Interaction between motor proteins effect the free-energy landscapes of each molecular motor, leading to different dynamical properties. It was shown that for some sets of parameters the cluster motor protein can be more efficient (by moving faster and fluctuating less) than individual motor particles. Theoretical model suggested that dynamics of RecBCD helicase molecules can be explained by assuming that domains interact with energy of the order of 6 $k_{\mathrm{B}}T$. Groups of molecular motors and multi-motor complexes also produce many interesting phenomena of collective behavior, such as flagellar beating and chromosome and spindle oscillations, that are important for biological systems.[82]

4. Conclusions

It was shown that discrete stochastic models compute exact and explicit expressions for the dynamical properties of molecular motors, and they can account for all available experimental observations on motor proteins. It is suggested then that the discrete stochastic models provide a flexible theoretical framework for understanding motor protein mechanisms. There are many problems concerning motor proteins and their dynamics, and new biological nanomachines are constantly discovered. One might anticipate that theory might yield predictions and conceptual insights in the mechanisms of these biological systems.

At this point it is interesting to speculate about what makes an optimal motor protein from a biological perspective. Although the answer depends on specific functions of these enzymes, one might think (and there are few indications from known molecular motor systems[50]) that the nature tuned them to produce the maximal speed with maximal efficiency and minimum diffusion constant. The presented theoretical method allows to analyze these questions from theoretical point of view, but it will be very important to find the answers also in experimental studies.

References

1. H. A. Lodish, A. Berk, S. L. Zipursky and P. Matsudaira, *Molecular Cell Biology*, 4th edn. (Scientific American Books, New York, 1999), p. 1084.
2. D. Bray, *Cell Movements: From Molecules to Motility*, 2nd edn. (Garland Publishing, New York, 2001), p. 372.
3. J. Howard, *Mechanics of Motor Proteins and the Cytoskeleton* (Sinauer Associates, Sunderland, MA, 2001), p. 367.
4. A. B. Kolomeisky and M. E. Fisher, Molecular motors: A theorist's perspective, *Annual Reviews of Physical Chemistry* **58** (2007) 675–695.
5. K. Visscher, M. J. Schnitzer and S. M. Block, Single kinesin molecules studied with a molecular force clamp, *Nature* **400** (1999) 184–189.
6. S. Rice, A. W. Lin, D. Safer, C. L. Hart, N. Naber, B. O. Carragher, S. M. Cain, E. Pechatnikova, E. W. Wilson-Kubalek, M. Whittaker, E. Pate, R. Cooke, E. W. Taylor, R. A. Milligan and R. D. Vale, A structural change in the kinesin motor protein that drives motility, *Nature* **402** (1999) 778–784.
7. R. M. Berry and J. P. Armitage, The bacterial flagella motor, *Advances in Microbial Physiology* **41** (1999) 292–337.
8. M. J. Schnitzer, K. Visscher and S. M. Block, Force production by single kinesin motors, *Nature Cell Biology* **2** (2000) 718–723.
9. K. Adachi, R. Yasuda, H. Noji, H. Itoh, Y. Harada, M. Yoshida and K. Kinosita, Stepping rotation of F_1-ATPase visualized through angle-resolved single-fluorophore imaging, *Proceedings of the Natural Academy of Sciences USA* **97** (2000) 7243–7247.

10. T. R. Strick, J. F. Allemand and D. Bensimon, Stress-induced structural transitions in DNA and proteins, *Annual Review of Biophysics and Bimolecular Structure* **29** (2000) 523–543.
11. D. E. Smith, S. J. Tans, S. B. Smith, S. Grimes, D. L. Anderson and C. Bustamante, The bacteriophage ϕ29 portal motor can package DNA against a large internal force, *Nature* **413** (2001) 748–752.
12. M. Nishiyama, H. Higuchi and T. Yanagida, Chemomechanical coupling of the forward and backward steps of single kinesin molecules, *Nature Cell Biology* **4** (2002) 790–797.
13. K. Kaseda, H. Higuchi and K. Hirose, Coordination of kinesin's two heads studied with mutant heterodimers, *Proceedings of the National Academy of Sciences USA* **99** (2002) 16058–16063.
14. M. J. Lang, C. L. Asbury, J. W. Shaevitz and S. M. Block, An automated two-dimensional optical force clamp for single molecule studies, *Biophysical Journal* **83** (2002) 491–501.
15. S. M. Block, C. L. Asbury, J. W. Shaevitz and M. J. Lang, Probing the kinesin reaction cycle with a 2D optical force clamp, *Proceedings of the National Academy of Sciences USA* **100** (2003) 2351–2356.
16. C. L. Asbury, A. N. Fehr and S. M. Block, Kinesin moves by an asymmetric hand-over-hand mechanism, *Science* **302** (2003) 2130–2134.
17. A. Yildiz, M. Tomishige, R. D. Vale and P. R. Selvin, Kinesin walks hand-over-hand, *Science* **302** (2003) 676–678.
18. G. Charvin, D. Bensimon and V. Croquette, Single-molecule study of DNA unlinking by eukaryoric and prokaryotic type-II topoisomerases, *Proceedings of the National Academy of Sciences USA* **100** (2003) 9820–9825.
19. A. F. Taylor and G. R. Smith, RecBCD enzyme is a DNA helicase with fast and slow motors of opposite polarity, *Nature* **423** (2003) 889–893.
20. T. T. Perkins, H. W. Li, R. V. Dalal, J. Gelles and S. M. Block, Forward and reverse motion of RecBCD molecules on DNA, *Biophysical Journal* **86** (2004) 1640–1648.
21. B. Maier, M. Koomey and M. P. Sheetz, A force-dependent switch reverses type IV pilus retraction, *Proceedings of the National Academy of Sciences USA* **101** (2004) 10961–10966.
22. H. Itoh, A. Takahashi, K. Adachi, H. Noji, R. Yasuda, M. Yoshida and K. Kinosita, Mechanically driven ATP synthesis by F1-ATPase, *Nature* **427** (2004) 465–468.
23. R. Mallik, B. C. Carter, S. A. Lex, S. J. King and S. P. Gross, Cytoplasmic dynein functions as a gear in response to load, *Nature* **427** (2004) 649–652.
24. G. E. Snyder, T. Sakamoto, J. A. Hammer, J. R. Sellers and P. R. Selvin, Nanometer localization of single green fluorescent proteins: Evidence that myosin V walks hand-over-hand via telemark configuration, *Biophysical Journal* **87** (2004) 1776–1783.

25. N. J. Carter and R. A. Cross, Mechanics of kinesin step, *Nature* **435** (2005) 308–312.
26. Y. Sowa, A. D. Rowe, M. C. Leake, T. Yakushi, M. Homma, A. Ishijima and R. M. Berry, Direct observation of steps in rotation of the bacterial flagellar motor, *Nature* **437** (2005) 916–919.
27. Y. Rondelez, G. Tresset, T. Nakashima, Y. Kato-Yamada, H. Fujita, S. Takeuchi and H. Noji, Highly coupled ATP synthesis by F_1-ATPase single molecules, *Nature* **433** (2005) 774–777.
28. M. T. Valentine, P. M. Fordyce, T. C. Krzysiak, S. P. Gilbert and S. M. Block, Individual dimers of the mitotic kinesin motor Eg5 step processively and support substantial loads *in vitro*, *Nature Cell Biology* **8** (2006) 470–476.
29. S. Toba, T. M. Watanabe, L. Yamaguchi, Okimoto, Y. Y. Toyoshima and H. Higuchi, Overlapping hand-over-hand mechanism of single molecular motility of cytoplasmic dynein, *Proceedings of the National Academy of Sciences USA* **103** (2006) 5741–5745.
30. S. Leibler and D. A. Huse, Porters versus rowers: A unified stochastic model of motor proteins, *Journal of Cell Biology* **121** (1993) 1356–1368.
31. C. S. Peskin and G. Oster, Coordinated hydrolysis explains the mechanical behavior of kinesin, *Biophysical Journal* **68** (1995) 202s–211s.
32. F. Jülicher, A. Ajdari and J. Prost, Modeling molecular motors, *Review of Modern Physics* **69** (1997) 1269–1281.
33. T. C. Elston and G. Oster, Protein turbines I: The bacterial flagellar motor, *Biophysical Journal* **73** (1997) 703–721.
34. H. Y. Wang, T. Elston, A. Mogilner and G. Oster, Force generation in RNA polymerase, *Biophysical Journal* **74** (1998) 1186–1202.
35. A. B. Kolomeisky and B. Widom, A simplified "ratchet" model of molecular motors, *Journal of Statistical Physics* **93** (1998) 633–645.
36. M. E. Fisher and A. B. Kolomeisky, The force exerted by a molecular motor, *Proceedings of the National Academy of Sciences USA* **96** (1999) 6597–6602.
37. M. E. Fisher and A. B. Kolomeisky, Molecular motors and the forces they exert, *Physica A* **274** (1999) 241–266.
38. R. Lipowsky, Universal aspects of the chemomechanical coupling for molecular motors, *Physical Review Letters* **85** (2000) 4401–4404.
39. D. Keller and C. Bustamante, The mechanochemistry of molecular motors, *Biophysical Journal* **78** (2000) 541–556.
40. R. M. Berry, Theories of rotary motors, *Philosophical Transations of the Royal Society of London B* **355** (2000) 503–509.
41. A. B. Kolomeisky and M. E. Fisher, Periodic sequential kinetic models with jumping, branching and deaths, *Physica A* **279** (2000) 1–20.
42. A. B. Kolomeisky and M. E. Fisher, Extended kinetic models with waiting-time distributions: Exact results, *Journal of Chemical Physics* **113** (2000) 10867–10877.

43. A. B. Kolomeisky, Exact results for parallel-chain kinetic models of biological transport, *Journal of Chemical Physics* **115** (2000) 7253–7259.
44. M. E. Fisher and A. B. Kolomeisky, Simple mechanochemistry describes the dynamics of kinesin molecules, *Proceedings of the National Academy of Sciences USA* **98** (2001) 7748–7753.
45. C. Bustamante, D. Keller and G. Oster, The physics of molecular motors, *Accounts of Chemical Research* **34** (2001) 412–420.
46. A. Mogilner, A. J. Fisher and R. J. Baskin, Structural changes in the neck linker of kinesin explain the load dependence of the motor's mechanical cycle, *Journal of Theoretical Biology* **211** (2001) 143–157.
47. P. Reimann, Brownian motors: Noisy transport far from equilibrium, *Physics Reports* **361** (2002) 57–265.
48. A. B. Kolomeisky and M. E. Fisher, A simple kinetic model describes the processivity of myosin-V, *Biophysical Journal* **84** (2003) 1642–1650.
49. A. B. Kolomeisky, E. B. Stukalin and A. A. Popov, Understanding mechanochemical coupling in kinesins using first-passage-time processes, *Physical Review E* **71** (2005) 031902.
50. E. B. Stukalin, H. Phillips and A. B. Kolomeisky, Coupling of two motor proteins: A new motor can move faster, *Physical Review Letters* **94** (2005) 238101.
51. M. E. Fisher and Y. C. Kim, Kinesin crouches to sprint but resists pushing, *Proceedings of the National Academy of Sciences USA* **102** (2005) 16209–16214.
52. Y. C. Kim and M. E. Fisher, Vectorial loading of processive motor proteins, *Journal of Physics: Condensed Matter* **17** (2005) S3821–S3838.
53. H. Qian, Cycle kinetics, steady state thermodynamics and motors — A paradigm for living matter physics, *Journal of Physics: Condensed Matter* **17** (2005) S3783–S3794.
54. J. Xing, J. C. Liao and G. Oster, Making ATP, *Proceedings of the National Academy of Sciences USA* **102** (2005) 16536–16546.
55. E. B. Stukalin and A. B. Kolomeisky, Transport of single molecules along the periodic parallel lattices with coupling, *Journal of Chemical Physics* **124** (2006) 204901.
56. R. D. Vale and R. J. Fletterick, The design plan of kinesin motors, *Annual Review of Cell and Developmental Biology* **13** (1997) 745–777.
57. J. Gelles and R. Landick, RNA polymerase as a molecular motor, *Cell* **93** (1998) 13–16.
58. A. Mehta, Myosin learns to walk, *Journal of Cell Science* **114** (2001) 1981–1998.
59. K. Oiwa and H. Sakakibara, Recent progress in dynein structure and mechanism, *Current Opinion in Cell Biology* **17** (2005) 98–103.

60. F. Kozielski, S. Sack, A. Marx, M. Thormählen, Schönbrunn, V. Biou, A. Thompson, E. M. Mandelkow and E. Mandelkow, The crystal structure of dimeric kinesin and implications for microtubule-dependent motility, *Cell* **91** (1997) 985–994.
61. M. Tomishige, D. R. Klopfenstein and R. D. Vale, Conversion of Unc104/KIF1A kinesin into a processive motor after dimerization, *Science* **297** (2002) 2263–2267.
62. E. M. De La Cruz and E. M. Ostap, Relating biochemistry and function in the myosin superfamily, *Current Opinion in Cell Biology* **16** (2004) 1–7.
63. A. Sadhu and E. W. Taylor, A kinetic study of the kinesin ATPase, *Journal of Biological Chemistry* **267** (1992) 11352–11359.
64. Y. Z. Ma and E. W. Taylor, Mechanism of microtubule kinesin ATPase, *Biochemistry* **34** (1995) 13242–13251.
65. M. L. Moyer, S. P. Gilbert and K. A. Johnson, Pathway of ATP hydrolysis by monomeric and dimeric kinesin, *Biochemistry* **37** 800–813.
66. E. M. De La Cruz, A. L. Wells, S. S. Rosenfeld, E. M. Ostap and H. L. Sweeney, The kinetic mechanism of myosin V, *Proceedings of the National Academy of Sciences USA* **96** (1999) 13726–13731.
67. E. M. De La Cruz, E. M. Ostap and H. L. Sweeney, Kinetic mechanism and regulation of myosin VI, *Journal of Biological Chemistry* **276** (2001) 32373–32381.
68. A. L. Lucius and T. L. Lohman, Effects of temperature and ATP on the kinetic mechanism and kinetic step-size for *E. coli* RecBCD helicase-catalyzed DNA unwinding, *Journal of Molecular Biology* **339** (2004) 751–771.
69. J. C. Cochran, J. E. Gatial, T. M. Kapoor and S. P. Gilbert, Monastrol inhibition of the mitotic kinesin Eg5, *Journal of Biological Chemistry* **280** (2005) 12658–12667.
70. B. Derrida, Velocity and diffusion constant of a periodic one-dimensional hopping model, *Journal of Statistical Physics* **31** (1983) 433–450.
71. Z. Koza, General relation between drift velocity and dispersion of a molecular motor, *Acta Physica Polonica B* **33** (2002) 1025–1030.
72. D. D. Hackney, The tethered motor domain of a kinesin-microtubule complex catalyzes reversible synthesis of bound ATP, *Proceedings of the National Academy of Sciences USA* **102** (2005) 18338–18343.
73. K. C. Neuman, O. A. Saleh, T. Lionnet, G. Lia, J.-F. Allemand, D. Bensimon and V. Croquette, Statistical determination of the step size of molecular motors, *Journal of Physics: Condensed Matter* **17** (2005) S3811–S3820.
74. Y. Okada and N. Hirokawa, Mechanism of the single-headed processivity: Diffusional anchoring between the K-loop of kinesin and the C terminus of tubulin, *Proceedings of the National Academy of Sciences USA* **97** (2000) 640–645.

75. M. Guthold, X. Zhu, C. Rivetti, G. Yang, N. H. Thomson, S. Kasas, H. G. Hansma, B. Smith, P. K. Hansma and C. Bustamante, Direct observation of one-dimensional diffusion and transcription by *Escherichia coli* RNA polymerase, *Biophysical Journal* (1999) 2284–2294
76. N. G. van Kampen, *Stochastic Processes in Physics and Chemistry*, Elsevier (New York, 1992), p. 465.
77. M. J. Schnitzer and S. M. Block, Kinesin hydrolyzes one ATP per 8-nm step, *Nature* **388** (1997) 386–390.
78. W. Hua, E. C. Young, M. L. Fleming and J. Gelles, Coupling of kinesin steps to ATP hydrolysis, *Nature* **388** (1997) 390–393.
79. D. L. Coy, M. Wagenbach and J. Howard, Kinesin takes one 8-nm step for each ATP that it hydrolyzes, *Journal of Biological Chemistry* **274** (1999) 3667–3671.
80. D. Tsygankov, M. Linden and M. E. Fisher, Back-stepping, hidden substeps, and conditional dwell times in molecular motors, *Physical Review E* **75** (2007) 021909.
81. K. Nishinari, Y. Okada, Schadschneider and D. Chowdhury, Intracellular transport of single-headed molecular motors KIF1A, *Physical Review Letters* **95** (2005) 118101.
82. S. W. Grill, K. Kruse and F. Jülicher, Theory of mitotic spindle oscillations, *Physical Review Letters* **94** (2005) 108104.

CHAPTER 10

Unique Mechanisms From Finite Two-State Trajectories

Ophir Flomenbom and Robert J. Silbey

Department of Chemistry, Massachusetts Institute of Technology, Cambridge, MA 02139

1. Introduction

Single-molecule experiments are diverse, ranging from the passage of ions and biopolymers through individual channels,[3–5] activity and conformational changes of biopolymers,[1–2,6–17] diffusion of molecules,[18–21] and blinking of nanocrystals.[22–25] The monitored signal in the measurements of the processes mentioned is different from system to system, and includes photon recordings, flux of ions recordings, and force recordings. In all single molecule experiments, the raw data is noisy, and so the basic analysis should address issues of the interpretation of noisy data (which is specific to the type of measurement). There is a vast literature dealing with noise in single molecule experiments — see chapters 1 and 6 of this book and the references therein. In this chapter, we address a different issue in the analysis of single molecule data: the assignment of a mechanism to the measured process. We assume that the required noise analysis has already been applied to the data, so that we are given a noiseless two-state trajectory (see Fig. 1(A)). The mechanism of the observed process is usually described by a multi-substate *on–off* Markovian kinetic scheme (KS),[26–40] (Fig. 1(B)); see Refs. 41–56

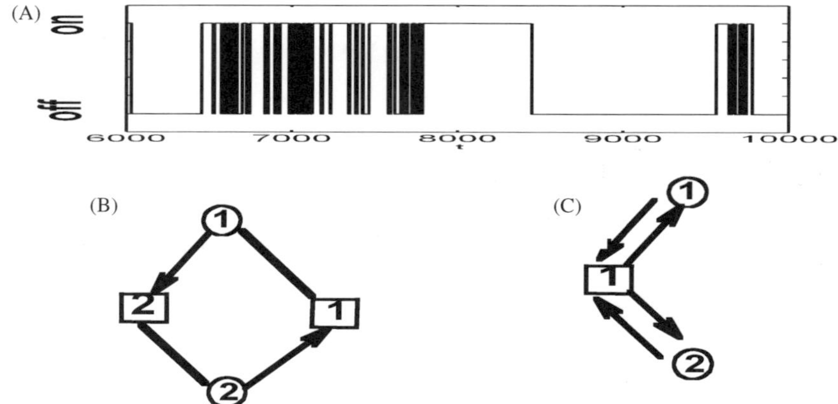

Fig. 1. A two-state trajectory (A), the KS (B), and the RD form (C). (The KS (B) has $L_{on} = 2$ (squared substates), $L_{off} = 2$ (circled substates), irreversible transitions, and $N_{on} = M_{on} = 2$ and $N_{off} = M_{off} = 2$. For generating the data, we take the following transition rate values (k_{ji} connects substates $i \to j$): $k_{1_{off}1_{on}} = 0.3$, $k_{2_{off}2_{on}} = 0.02$, $k_{1_{on}1_{off}} = 0.425$, $k_{2_{on}1_{off}} = 0.075$, $k_{1_{on}2_{off}} = 0.0085$, $k_{2_{on}2_{off}} = 0.0015$ (with arbitrary units). The equality of the ratios, $k_{2_{on}2_{off}}/k_{1_{on}2_{off}} = k_{2_{on}1_{off}}/k_{1_{on}1_{off}} \equiv p_R/p_L$ ($p_L + p_R = 1$), means that the KS is symmetric in the sense that the ranks of the 2D WT-PDFs of successive x, y (= *on, off*) events are all equal one, except $R_{on,off}$ which equal two. As a result, the RD form (C) has one *on* substate and two *off* substates. The RD form has also direction-dependent WT-PDFs for the *on* to *off* connections, $\varphi_{on,11}(t) = p_L k_{1_{off}1_{on}} e^{-k_{1_{off}1_{on}}t}$, $\varphi_{on,21}(t) = p_R k_{2_{off}1_{on}} e^{-k_{2_{off}1_{on}}t}$, and $\varphi_{off,11} = k_{1_{off}} e^{-k_{1_{off}}t}$, $\varphi_{off,12}(t) = k_{2_{off}} e^{-k_{2_{off}}t}$. Here, $k_i = \sum_j k_{ji}$, and $\varphi_{x,ij}(t)$ is the WT-PDF that connects substates $x_j \to y_i$ in the RD form.

for related descriptions. The KS describes a discrete conformational energy landscape of a biomolecule, chemical kinetics with (or without) conformational changes, or environmental changes, stands for quantum states, etc. The underlying stochastic dynamics of the process in the multi-substate *on–off* KS is thus encoded in the two-state trajectory (the stochastic signal changes value only when transitions between substates of different states in the KS take place). From single-molecule experiments, we wish to learn as much as possible about the underlying KS, much more than the averaged properties obtained from bulk measurements.

However, determining the KS from the two-state trajectory is difficult since the number of the substates in each of the states, L_x ($x = on, off$), is usually large, and the connectivity among the substates is usually complex. A more fundamental difficulty in finding the correct KS arises from the equivalent of KSs, i.e. there are KSs that lead to the same two-state trajectories in a statistical sense.[29–31,33–37] A way to deal with this issue is to use canonical (unique) forms rather than KSs[33–36]: the space of KSs is mapped into a space of canonical forms. A given KS is mapped to a unique canonical form but many KSs can be mapped to the same canonical form. Underlying KSs with the same canonical form are equivalent to each other, and cannot be discriminated based on the information in a single infinitely long two-state trajectory.

In this chapter, we present our canonical forms of reduced dimensions (RD),[34] (Fig. 1(C)). RD forms have many advantages over other canonical forms that were previously suggested for this problem. RD forms can handle *any* KS, i.e. KSs with irreversible connections and/or symmetry (symmetry means, e.g., that the spectrum of the waiting time probability density function (WT-PDF) for the single x ($= on, off$) period is degenerate), and constitute a powerful tool in discriminating among KSs, because the mapping of a KS into an RD form is based on the KS's *on–off* connectivity and therefore can be done, to a large extent, without actual calculations. Using this property, we give an ensemble of relationships between the data, the topology of the RD forms, and properties of KSs. These relationships are useful in discriminating KSs and in the analysis of the data. With respect to the actual analysis of the data, we give a simple procedure for finding the RD form from finite data. It turns out that RD forms can be constructed from data sets fairly accurately, and, importantly, much more accurately than other mechanisms (at least in the cases studied by us which represent commonly encountered cases for this problem.)

This chapter is laid out as follows: Section 2 gives the mathematical formulations of the system in terms of the master equation and the path representation. Section 3 introduces the canonical forms of reduced dimensions. RD forms stem from the path representation. Section 4 builds an RD form from finite data, and Sec. 5 concludes.

2. Mathematical formulations

In this section, we express the WT-PDFs for single periods, $\phi_x(t), x = on, off$, and for joint successive periods, $\phi_{x,y}(t_1, t_2), x, y = on, off$, in terms of both the master equation and the path representation. The relationship between the two representations is made. *On–off* KSs are commonly described in terms of the master equation, but our canonical forms are naturally related to the path representation. This will be studied in Sec. 3.

2.1. *Matrix formulation of the system*

The problem at hand is formulated in terms of a (single) random walker in an *on–off* KS. The time-dependent occupancy probabilities of state x for the coupled *on–off* process, $\mathbf{P}_x(t)$, with the elements $(\mathbf{P}_x(t))_i = P_{x,i}(t)$ for $i = 1, \ldots, L_x$ (L_x is the number of substates in state x), obey the reversible master equation,

$$\frac{\partial}{\partial t}\begin{pmatrix}\mathbf{P}_{on}(t)\\ \mathbf{P}_{off}(t)\end{pmatrix} = \begin{pmatrix}\mathbf{K}_{on} & \mathbf{V}_{off}\\ \mathbf{V}_{on} & \mathbf{K}_{off}\end{pmatrix}\begin{pmatrix}\mathbf{P}_{on}(t)\\ \mathbf{P}_{off}(t)\end{pmatrix}. \tag{1}$$

In Eq. (1), matrix $\mathbf{K}_x$, with dimensions $[\mathbf{K}_x] = L_x, L_x$, contains transition rates among substates in state x, and "irreversible" transition rates from substates in state x to substates in state y. (The "irreversible" transition rates are given on the diagonal, and are called irreversible because matrix $\mathbf{K}_x$ does not contain the back transition rates from state y to state x.) Matrix $\mathbf{V}_x$, with dimensions $[\mathbf{V}_x] = L_y, L_x$, contains transition rates between states $x \to y$, where $(\mathbf{V}_x)_{ji}$ is the transition rate between substates $i_x \to j_y$. $\mathbf{P}_x(ss)$ is the vector of occupancy probabilities in state x at steady state ($t \to \infty$). $\mathbf{P}_x(ss)$ is found from Eq. (1) for vanishing time derivative. $\phi_x(t)$ and

$\phi_{x,y}(t_1, t_2)$ are given in terms of the matrices in Eq. (1),

$$\phi_x(t) = \mathbf{1}_y^{\mathrm{T}}\mathbf{V}_x\mathbf{G}_x(t)\mathbf{V}_y\mathbf{P}_y(ss)/N_x, \tag{2}$$

and

$$\phi_{x,y}(t_1, t_2) = \mathbf{1}_x^{\mathrm{T}}\mathbf{V}_y\mathbf{G}_y(t_2)\mathbf{V}_x\mathbf{G}_x(t_1)\mathbf{V}_y\mathbf{P}_y(ss)/N_x, \tag{3}$$

where $N_x = \mathbf{1}_x^{\mathrm{T}}\mathbf{V}_y\mathbf{P}_y(ss)$ and $\mathbf{1}_x^{\mathrm{T}}$ is the summation row vector of 1, L_x dimensions, $[\mathbf{1}_x^{\mathrm{T}}] = 1, L_x$. (The expression for $\phi_{x,x}(t_1, t_2)$ is obtained from Eq. (3) by inserting the factor $\mathbf{V}_y\overline{\mathbf{G}}_y(0)$, $\phi_{x,x}(t_1, t_2) = \mathbf{1}_y^{\mathrm{T}}\mathbf{V}_x\mathbf{G}_x(t_2)\mathbf{V}_y\overline{\mathbf{G}}_y(0)\mathbf{V}_x\mathbf{G}_x(t_1)\mathbf{V}_y\mathbf{P}_y(ss)/N_x$. Here, $\overline{\mathbf{G}}_y(0)$ is the Laplace transform of $\mathbf{G}_y(t)$, $\overline{\mathbf{G}}_y(s) = \int_0^\infty \mathbf{G}_y(t)e^{-st}dt$, at $s = 0$.) $\mathbf{G}_x(t)$ in Eqs. (2) and (3) is the Green's function of state x for the irreversible process, $\partial\mathbf{G}_x(t)/\partial t = \mathbf{K}_x\mathbf{G}_x(t)$, with the solution,

$$\mathbf{G}_x(t) = \exp(\mathbf{K}_x t) = \mathbf{X}\exp(\lambda_x t)\mathbf{X}^{-1}. \tag{4}$$

The second equality in Eq. (4) follows from a similarity transformation $\lambda_x = \mathbf{X}^{-1}\mathbf{K}_x\mathbf{X}$, and all the matrices in Eq. (4) have dimensions L_x, L_x. Nonsymmetric KSs must have nondegenerate eigenvalue matrices λ_xs.

2.2. *Path representation of the WT-PDFs*

Our approach is based on expressing the WT-PDFs in Eqs. (2) and (3) in path representation that utilizes the *on–off* connectivity of the KS. The *on–off* process is separated into two irreversible processes that occur sequentially, and we have for $\phi_{x,y}(t_1, t_2)$ $(x \neq y)$,

$$\begin{aligned}\phi_{x,y}(t_1, t_2) &= \sum_{n_y=1}^{N_y}\left(\sum_{n_x=1}^{N_x} W_{n_x} f_{n_y n_x}(t_1)\right) F_{n_y}(t_2)\\ &= \sum_{m_x\in\{M_x\}}\left(\sum_{n_x=1}^{N_x}\sum_{n_y=1}^{N_y} W_{n_x}\tilde{f}_{m_x n_x}(t_1)\omega_{n_y m_x}F_{n_y}(t_2)\right).\end{aligned} \tag{5}$$

(The sum $z_x \in \{Z_x\}$ is a sum over a particular group of Z_x substates. In Appendix A, expressions for $\phi_x(t)$ and $\phi_{x,x}(t_1, t_2)$ in path representation are given.) Equation (5) emphasizes the role of the KS's topology in expressing the $\phi_{x,y}(t_1, t_2)$s. N_x and M_x are the numbers of initial and final substates in state x in the KS, respectively. Namely, each event in state x starts at one of the N_x initial substates, labeled, $n_x = 1, \ldots, N_x$, and terminates through one of the M_x final substates, labeled $m_x = 1, \ldots, M_x$, for a reversible *on–off* connection KS (Fig. 2(E)), or $m_x = N_x + 1 - H_x, \ldots, N_x + M_x - H_x$, for an irreversible *on–off* connection KS (Fig. 1(B)), where H_x $(= 0, 1, \ldots, N_x)$ is the number of substates in state x that are both initial and final ones. (In each of the states the labeling of the substates starts from 1.) An event in state x starts in substate n_x with probability W_{n_x}. The first passage time PDF for exiting to substate n_y, conditional on starting from substate n_x $(x \neq y)$, is $f_{n_y n_x}(t)$, and $F_{n_x}(t) = \sum_{n_y} f_{n_y n_x}(t)$. Writing $f_{n_y n_x}(t)$ as $f_{n_y n_x}(t) = \sum_{m_x} \omega_{n_y m_x} \tilde{f}_{m_x n_x}(t)$ emphasizes the role of the *on–off* connectivity, where $\omega_{n_y m_x}$ is the transition probability from substate m_x to substate

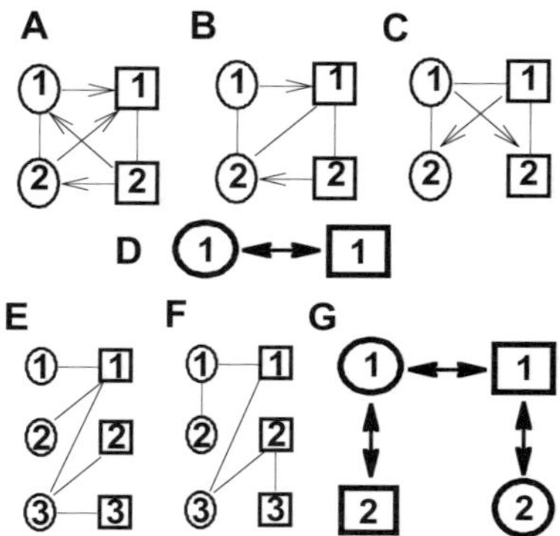

Fig. 2. Indistinguishable KSs. (KSs (A)–(C) have the simplest RD form (D) of one substate in each of the states. KSs (A)–(C) are equivalent when they have the same $\phi_{on}(t)$ and $\phi_{off}(t)$. Equivalent KSs (E)–(F) have $R_{x,y} = 2$, $x, y = on, off$, and tri-exponential $\phi_{on}(t)$ and $\phi_{off}(t)$. The corresponding RD form is shown in (G).)

n_y, and $\tilde{f}_{m_x n_x}(t)\omega_{n_y m_x}$ is the first passage time PDF, conditional on starting from substate n_x, for exiting to substate n_y through substate m_x.

2.3. *Relationships between the master equation and the path representation*

All the factors in Eq. (5) can be expressed in terms of the matrices of Eq. (1). W_{n_x} and $f_{n_y n_x}(t)$ are related to the master equation by

$$W_{n_x} = (\mathbf{V}_y \mathbf{P}_y(ss))_{n_x} / N_x,$$

and

$$f_{n_y n_x}(t) = (\mathbf{V}_x \mathbf{G}_x(t))_{n_y n_x}.$$

$f_{n_y n_x}(t)$ can be further rewritten as

$$f_{n_y n_x}(t) = \sum_{m_x} \omega_{n_y m_x} \tilde{f}_{m_x n_x}(t),$$

and similarly for $(\mathbf{V}_x \mathbf{G}_x(t))_{n_y n_x}$, we have

$$(\mathbf{V}_x \mathbf{G}_x(t))_{n_y n_x} = \sum_k (\mathbf{V}_x)_{n_y k} (\mathbf{G}_x(t))_{k n_x}.$$

Note however that the factors in the right-hand side in the above two sums are not equal but proportional,

$$\tilde{f}_{k n_x}(t) = \alpha_{x,k} (\mathbf{G}_x(t))_{k n_x}; \quad \alpha_{x,k} = -(\mathbf{K}_x)_{kk},$$

and

$$\omega_{n_y k} = (\mathbf{V}_x)_{n_y k} / \alpha_{x,k}.$$

3. RD forms

In this section we present the canonical forms of reduced dimensions and the relationships between the data, the canonical form topology, and the KS. Our canonical forms are based upon the path representation in Sec. 2.2.

3.1. *The rank of $\phi_{x,y}(t_1, t_2)$ and its topological interpretation*

For discrete time, $\phi_{x,y}(t_1, t_2)$ is a matrix. The rank of this matrix, $R_{x,y}$ $(= 1, 2, \ldots)$, which is the number of non-zero eigenvalues (or singular values for a non-square matrix) of its decomposition has an important physical meaning: it determines the topology of the RD form. So, it is important to note that $R_{x,y}$ can be found without the need to find the actual functional form of $\phi_{x,y}(t_1, t_2)$. Importantly, we can relate $R_{x,y}$ $(x \neq y)$ to the topology of the underlying KS. When none of the terms in an external sum on Eq. (5), after the first or the second equality, are proportional, $R_{x,y} = \min(M_x, N_y)$ (Examples in Figs. 2(A)–2(C) and 2(F)). Otherwise, $R_{x,y} < \min(M_x, N_y)$ (Examples in Fig. 2(E) and Appendix B). Then, Eq. (5) is rewritten such that it has the *minimal* number of additives in the external summations,

$$\begin{aligned}
\phi_{x,y}(t_1, t_2) &= \sum_{n_y \in \{\tilde{N}_y\}} \left(\sum_{n_x=1}^{N_x} W_{n_x} f_{n_y n_x}(t_1) \right) F_{n_y}(t_2) \\
&\quad + \sum_{m_x \in \{\tilde{M}_x\}} \left(\sum_{n_x=1}^{N_x} W_{n_x} \tilde{f}_{m_x n_x}(t_1) \right) \left(\sum_{n_y \notin \{\tilde{N}_y\}} \omega_{n_y m_x} F_{n_y}(t_2) \right). \qquad (6)
\end{aligned}$$

This leads to the equality, $R_{x,y} = \tilde{N}_y + \tilde{M}_x$. $\tilde{N}_y$ and $\tilde{M}_x$ can be related to KS's *on–off* connectivity. Consider a case where $M_x < N_y$, and there is a group of final substates in state x, $\{O_x\}$, with connections *only* to a group of initial substates in state y, $\{O_y\}$, and $O_x > O_y$ (see Fig. B4 in Appendix B). Then $\tilde{M}_x = M_x - O_x$ and $\tilde{N}_y = O_y$. (Further discussion and a generalization of this relationship are given in Appendix B.)

3.2. *The RD form*

The $R_{x,y}$ s are obtained from the $\phi_{x,y}(t_1, t_2)$s without the need to find its actual functional forms, thus constituting a fitting-free relationship between an ideal trajectory to the *on–off* connectivity and details of the underlying KS. Utilizing this relationship, the kinetic scheme space is divided into canonical forms, RD forms, using the $R_{x,y}$ s. Excluding KSs with symmetry, $R_{x,y}$ $(x \neq y)$ is the number of substates in state y in the RD form. RD forms can represent underlying KSs with symmetry and irreversible connections because they are built from all four $R_{x,y}$ s. The RD form has the minimal number of substates needed to reproduce the data. Connections in the RD form are only between substates of different states. For each connection in the RD form, there is a WT-PDF that is not necessarily exponential. We denote by $\varphi_{x,ij}(t)$ the WT-PDF in the RD form connecting substates $j_x \to i_y$. $\varphi_{x,ij}(t)$ is the sum of exponentials with the same eigenvalues and maximal terms as in $\phi_x(t)$.

3.3. *Mapping a KS into an RD form*

Mapping a KS into an RD form is based on clustering some of the initial substates in the KS into substates in the RD form, where initial substates in the KS that are not clustered are mapped to themselves. The KS's *on–off* connectivity determines whether an initial substate in the KS is clustered or mapped to itself. For a nonsymmetric KS, initial substates in state y in the KS that contribute to $R_{x,y}$ $(x \neq y)$ are mapped to themselves and those that do not contribute to $R_{x,y}$ are clustered, where initial-y-state substates in a cluster are all connected to the same final-x-state substate that contributes to $R_{x,y}$. (When substate m_x has a single exit-connection to substate n_y, which is its only entry-connection, substate n_y is defined as the one contributing to the rank.) For example, the KS in Fig. 2(E) is mapped into an RD form (Fig. 2(G)) when clustering substates 1_{off}–2_{off} into the RD form's substates 1_{off} where substate 3_{off} is mapped to itself giving rise to substate 2_{off} in the RD form, because only substate 3_{off} contributes to the rank $R_{on,off}$ among the *off* initial substates. A similar scenario is seen in the *on* state in this KS with a change in the labeling; substate 1_{on} is mapped to itself into substate 1_{on}, and substates $2_{on}-3_{on}$ are clustered into RD form substate 2_{on}. Only substate 1_{on} contribute to $R_{off,on}$ among the *on* initial substates.

The clustering procedure determines the coefficients in the exponential expansion of the $\varphi_{x,ij}(t)s$. (Technical details for obtaining these WT-PDFs given a KS are discussed in Appendix B.) Note that the clustering procedure, along with the fact that substates in the KS that are not initial or final ones do not affect the RD form's topology, reduce the KS dimensionality to that of the RD form.

3.4. *Examples and the utility of RD forms*

The simplest topology for an RD form has one substate in each of the states, namely, $R_{x,y} = 1$ $(x, y = on, off)$, and the only possible choice for $\varphi_{x,11}(t)$ is $\phi_x(t)$ (Fig. 2(D)). This means that all the information in the data is contained in $\phi_{on}(t)$ and $\phi_{off}(t)$. Consequently, KSs with $R_{x,y} = 1$ $(x, y = on, off)$ and the same $\phi_{on}(t)$ and $\phi_{off}(t)$ are indistinguishable (assuming no additional information on the mechanism is known). Examples of such KSs are shown in Figs. 2(A)–2(C). The $R_{x,y} = 1$ case was discussed in Refs. 29–31. The generalization of the equivalence of KSs for any case is straightforward using RD forms. KSs with the same $R_{x,y}$ s and the same WT-PDFs for the connections in the RD form cannot be distinguished. Indistinguishable KSs with $R_{x,y} = 2$ $(x, y = on, off)$ and tri-exponential $\phi_{on}(t)$ and $\phi_{off}(t)$ are shown in Figs. 2(E) and 2(F), and their corresponding RD form is shown in Fig. 2(G).

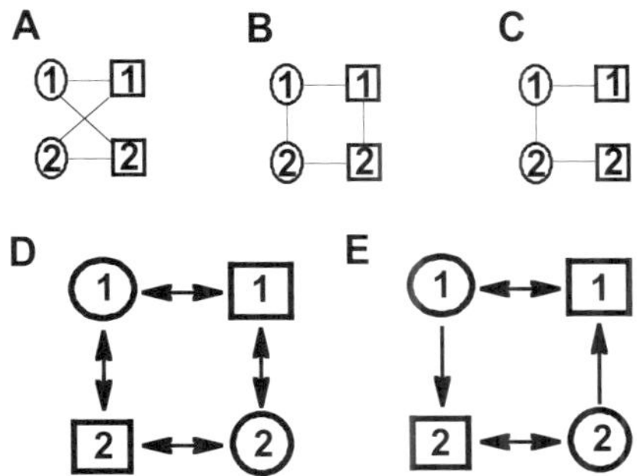

Fig. 3. Distinguishable KSs with $R_{x,y} = 2$, $x, y = on, off$ and bi-exponential $\phi_{on}(t)$ and $\phi_{off}(t)$. (We exclude symmetry in this example; KS 3(C) is distinct from KSs 3(A) and 3(B), because the corresponding RD forms, 3E and 3D, respectively, have different connectivity. KS 3(A) and KS 3(B) are also distinct, because the WT-PDFs for the connections in the RD form of KS 3(A) are exponential, whereas those of KS 3(B) are direction-dependent and bi-exponentials.)

Clearly, two KSs with different $R_{x,y}$ s can be resolved by the analysis of a two-state trajectory. One of the advantage of RD forms is that it can be employed as a powerful tool in resolving KSs with the same $R_{x,y}$ s, and the same number of exponentials in $\phi_{on}(t)$ and $\phi_{off}(t)$, even without the need of performing actual calculations. This is done based only on distinct complexity of the WT-PDFs for the connections in the corresponding RD forms; for example, compare the KSs in Figs. 3(A) and 3(B), or, on different connectivity of RD forms, for example, compare KSs in Figs. 3(A) and 3(B) with the KS in Fig. 3(C). This means that it is impossible to find positive (> 0) transition rates for the KSs in Figs. 3(A)–3(C) that make the $\phi_{x,y}(t_1, t_2)$ s from these KSs the same; so these KSs can be distinguished by analyzing a two-state trajectory (excluding symmetric cases for which the $\phi_{x,y}(t_1, t_2)$ s factorize to the product of $\phi_x(t_1)$ $\phi_y(t_2)$ s).

Note that an RD form can preserve the microscopic reversibility on the *on–off* level even when having irreversible connections. These can be "balanced" by the existence of direction-dependent WT-PDFs for the connections: microscopic reversibility in an RD form means that the $\phi_{x,y}(t_1, t_2)$ s obtained when reading the two-state trajectory in the forward direction are the same as the corresponding $\phi_{x,y}(t_1, t_2)$s obtained when reading the trajectory backward (as suggested in Ref. 40 for aggregated Markov chains). Using matrix notation, microscopic reversibility means, $\phi_{x,y}(t_1, t_2) = [\phi_{y,x}(t_1, t_2)]^{\mathrm{T}}$, where T stands for the transpose of a matrix.

The division of KSs into RD forms is also useful when, on top of the information extracted from the "original" two-state trajectory, complementary details about the observed process are available. (Complementary details are obtained by analyzing different kinds of measurements of the system, e.g., the crystal structure of the biopolymer, or by analyzing two-state trajectories while varying some parameters, e.g., the substrate concentration.[13–16]) Suppose that the connectivity of the underlying KS is unchanged by the manipulation. Then, the additional information can be used to resolve KSs that correspond to the RD form found from the statistical analysis of the "original" two-state trajectory, whereas any KS with a different RD form is irrelevant. Alternatively, when manipulating the system leading to a change in the connectivity of the underlying KS, or even to the addition or removal of substates, the RD forms obtained from the different data sets will almost always be distinct. RD forms can handle both scenarios; in the first case an adequate parameter tuning relates the RD forms obtained from the various sources, whereas in the second case the RD forms cannot be related by a parameter tuning.

4. Constructing the RD form from the data

The RD form is built from finite data following a three-stage algorithm: (1) Obtain the spectrum of the $\phi_x(t)$ s by Padé approximation.[57] The spectrum of the WT-PDFs for the x to y connections in the RD form is the same as that of $\phi_x(t)$, because substates of the same state in the RD form are not connected. Differences lay in the pre-exponential coefficients. (2) Find the ranks of the $\phi_{x,y}(t_1, t_2)$ s, and use it to build the RD form topology. (3) Apply a maximum likelihood procedure for finding the pre-exponential coefficients of the $\varphi_{x,ij}(t)$ s. Our routine for maximizing the likelihood function uses its analytical derivatives.

We use the above three-stage algorithm in the construction of the RD form in Fig. 1(C) from finite data. The KS is shown in Fig. 1(B). A two-state trajectory (Fig. 1(A)) is generated by simulating a random walk in the RD form. This is done by a modified Gillespie Monte Carlo method. Each transition in the simulation happens in two steps. Assume the process starts at substate i_x. The first step chooses the destination of the next location, determined by the weights of making a transition $i_x \rightarrow j_y$: $w_{j_y i_x} = \bar{\varphi}_{x,j_y i_x}(0)/\sum_{j'_y} \bar{\varphi}_{x,j'_y i_x}(0)$. (Note that from the analytical expressions of Appendix B, the sum $\sum_{j'_y} \bar{\varphi}_{x,j'_y i_x}(0)$ is unity, but due to numerical issues it can be smaller than unity.) The second step uses the particular j_y, and draws a random time out of a normalized density $\varphi_{x,j_y i_x}(t)/\bar{\varphi}_{x,j_y i_x}(0)$. The procedure is then repeated at the new location. It is much faster to generate a two-state trajectory using the RD form then to simulate a random walk in the underlying KS.

Figure 4 displays the analytical and the experimental $\phi_{on}(t)$ and $\phi_{off}(t)$. The simulated data contained 10^6 *on–off* events. The $\phi_x(t)$ s are accurately obtained from the data for times such that their amplitudes are two orders of magnitude smaller than its maximal values.

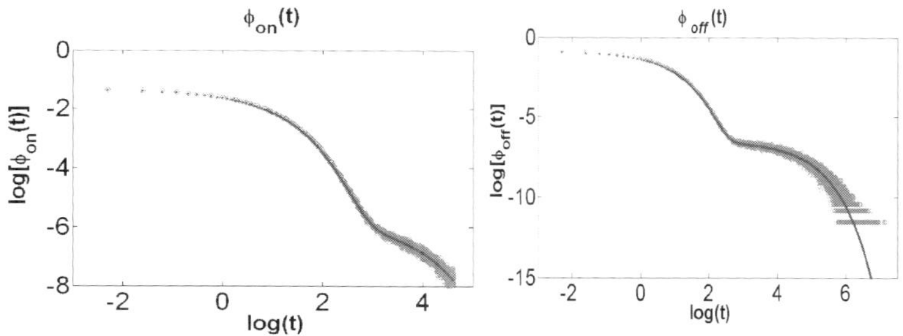

Fig. 4. The WT-PDFs of the *on* (left) and the *off* (right) states on a log–log scale. Shown are both the WT-PDFs obtained from a numerical solution of Eq. (2) (full line), and by constructing these PDFs from a 10^6 *on–off* event trajectory (circled symbols).

The Padé approximation method gives the correct amplitudes, rates, and number of components for these WT-PDFs.

The next stage in the construction of RD form estimates the ranks of the $\phi_{x,y}(t_1, t_2)$ s. The correct values for the $R_{x,y}$s, $R_{x,y} = 2, x, y = on, off$, are easily obtained analytically from the $\phi_{x,y}(t_1, t_2)$ s by applying singular value decomposition (SVD), but the ratio of the large to small singular value is large ($\sim 10^3$). This result means that the contribution from the large singular value contains most of the signals, and corresponds to the limit of an infinitely long trajectory. Thus, one may expect technical difficulties in detecting the exact number of non-zero singular values from an experimental matrix, due to the limited number of events in the trajectory. To deal with this issue, we build a series of cumulative 2D WT-PDFs. The first order cumulative of $\phi_{x,y}(t_1, t_2)$ is defined by

$$c_1\phi_{x,y}(T_1, T_2) = \int_0^{T_1}\int_0^{T_2} dt_1 dt_2 \phi_{x,y}(t_1, t_2),$$

and the generalization to higher order cumulative PDFs naturally follows:

$$c_n\phi_{x,y}(T_1, T_2) = \int_0^{T_1}\int_0^{T_2} dt_1 dt_2 c_{n-1}\phi_{x,y}(t_1, t_2).$$

A cumulative 2D PDF reduces the noise in the *original* PDF, but also preserves the rank of the original PDF. For each 2D PDF we obtain its spectrum of singular values and plot the ratio of successive singular values as a function of the order of the large singular value in the ratio. This plot should show large values for signal ratios, and a constant behavior with a value of about a unity for noise ratios. Figures 5(A) and 5(B) show the singular value ratio method applied on $\phi_{on,off}(t_1, t_2)$ and its first three cumulative PDFs. Both the second and the third cumulative PDFs show large values for the first two ratios and a constant behavior for larger ratios. This is a signature for a rank 2 histogram. (Note that this behavior is not seen in the original PDF and its first cumulative.) For comparison, Fig. 5(C) plots the

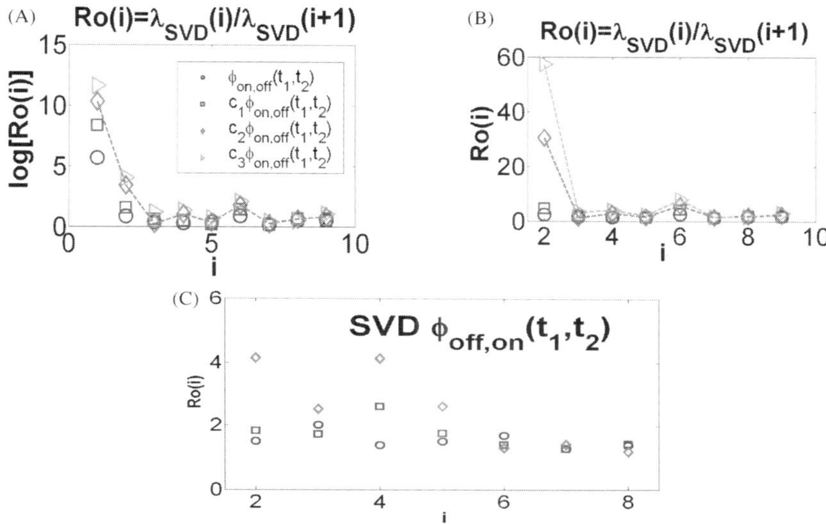

Fig. 5. The successive singular value ratio method for estimating the rank of 2D histograms: (A) The successive singular value ratio method is applied on $\phi_{on,off}(t_1, t_2)$ and its first three cumulative PDFs. The first ratio contains most of the signal in all PDFs. (B) The second and third order cumulative PDFs indicate a rank 2 PDF. (C) The same method is applied on $\phi_{off,on}(t_1, t_2)$, and indicates a rank 1 PDF. Here, the diamond and square symbols correspond to the second and first cumulative PDFs, respectively, and the circled symbols correspond to the original PDF.

same quantities calculated for $\phi_{off,on}(t_1, t_2)$, all of which show the same constant behavior for ratios greater than 1, indicating a rank 1 histogram. A rank behavior is observed also for $\phi_{on,on}(t_1, t_2)$ and $\phi_{off,off}(t_1, t_2)$ (data not shown).

Based on these findings, a low-resolution RD form with one *on* substate and two *off* substates is built. A search for eight parameters, denoted by $\mathbf{E} = \{\alpha_{x,jHi}\}$, in the bi-exponential expansions of the $\varphi_{x,ij}(t)$ s,

$$\varphi_{x,ji}(t) = \sum_{H=1}^{2} \alpha_{x,jHi} e^{-\lambda_{x,H} t}, \tag{7}$$

is performed by a maximum likelihood method with constraints. (Note that some of the $\alpha_{x,jHi}$s are zeros, but this information is not known *a priori*.) The likelihood function is given by

$$L(\hat{\mathrm{E}}|data) = \sum_{x,y}\sum_{i}\log(\phi_{x,y}(t_i, t_{1+i})).$$

The constraints in the algorithm demand that Eq. (7) are positive for every value of t, and also the normalization of the WT-PDFs for the connection

$$\sum_{j}\int_0^\infty \varphi_{x,ji}(t)dt = 1,$$

for every x and i. The analytical derivatives of the likelihood function are used in the maximization procedure. The maximization is performed using the command *fmincon* in the software *Matlab*. (In fact, we minimize, $-L(\hat{\mathrm{E}}|data)$.) The maximization procedure is performed in less than a minute for data made of 10^3 events. In this case, a straightforward optimization gave the correct answer. (Note that the optimization can also find local maximum; so, for a general case, it is required to choose various sets of initial conditions.) For the studied case, the error bars for the elements in $\{\hat{\mathrm{E}}\}$ are always less than a percent of the found values. (The error bars are found by inverting the Hessian matrix of second derivatives of the likelihood function with respect to the unknowns and substituting the solution for the unknowns. The diagonal elements of the obtained matrix give the variance of the fit.[58])

5. Concluding remarks

This paper utilized the information content in a two-state trajectory for an efficient elucidation of a unique mechanism that can generate it. The KS space is partitioned into canonical forms, RD forms, that are (usually) not Markovian. The topology of a canonical form is determined by the ranks $R_{x,y}$ s of the $\phi_{x,y}(t_1, t_2)$ s. An RD form has connections only between substates of different states, which are usually non-exponential WT-PDFs. The relationship between the ranks $R_{x,y}$ s and the KS's *on–off* connectivity is given. This relationship enables mapping a KS into an RD form based only on the *on–off* connectivity of the KS. The relationships between the ranks $R_{x,y}$s and the KS's *on–off* connectivity are based on path representation of the $\phi_{x,y}(t_1, t_2)$s given in Eqs. (5) and (6). One of the advantages of RD forms is that it is constituted as a powerful tool in discriminating among *on–off* KSs.

An example that builds an RD form from a data of 10^6 *on–off* cycles shows the applicability of RD forms in analyzing realistic data. We thus believe that RD forms will be found useful in the analysis of single-molecule measurements that result in two-state trajectories.

Appendix A

In this section, we give expressions for $\phi_x(t)$ and $\phi_{x,x}(t_1, t_2)$ using the path representation. Here, and in Appendix B, $x \neq y$, unless otherwise it is explicitly indicated. The expression for $\phi_x(t)$ is obtained from Eq. (5) by integrating over one time argument, $\phi_x(t) = \int_0^\infty \phi_{x,y}(t, \tau)d\tau = \int_0^\infty \phi_{y,x}(\tau, t)d\tau$, which leads to

$$\phi_x(t) = \sum_{n_y=1}^{N_y} \sum_{n_x=1}^{N_x} W_{n_x} f_{n_y n_x}(t)$$
$$= \sum_{m_x \in \{M_x\}} \sum_{n_x=1}^{N_x} \sum_{n_y=1}^{N_y} W_{n_x} \tilde{f}_{m_x n_x}(t) \omega_{n_y m_x}. \tag{A1}$$

The expression for $\phi_{x,x}(t_1, t_2)$ is obtained from Eq. (5) when introducing an additional summation that represents the random walk in state y that takes place in between the two measured events in state x,

$$\phi_{x,x}(t_1, t_2) = \sum_{n'_x=1}^{N_x} \sum_{n_y=1}^{N_y} \sum_{n_x=1}^{N_x} W_{n_x} f_{n_y n_x}(t_1) p_{n'_x n_y} F_{n'_x}(t_2)$$
$$= \sum_{m_x \in \{M_x\}} \sum_{n'_x=1}^{N_x} \sum_{n_y=1}^{N_y} \sum_{n_x=1}^{N_x} W_{n_x} \tilde{f}_{m_x n_x}(t_1) \omega_{n_y m_x} p_{n'_x n_y} F_{n'_x}(t_2). \tag{A2}$$

Here, $p_{n'_x n_y}$ is the probability that an event that starts at substate n_y exits to substate n'_x, and is given by $p_{n'_x n_y} = \bar{f}_{n'_x n_y}(0)$, where $\bar{g}(s) = \int_0^\infty g(t)e^{-st}dt$ is the Laplace transform of $g(t)$.

Note that higher order successive WT-PDFs, e.g., $\phi_{x,y,z}(t_1, t_2, t_3)$, do not contain additional information on top of the $\phi_{x,y}(t_1, t_2)$ s. When the underlying KS has no symmetry (i.e. the spectrum of $\phi_x(t)$, $x = on, off$, is nondegenerate) and/or irreversible connections, it is sufficient to use $\phi_{x,y}(t_1, t_2)$ for $x \neq y$, where for other cases, $\phi_{x,x}(t_1, t_2)$ s, $x = on, off$, contain complementary information.

Appendix B

In this Appendix, we give expressions for the WT-PDFs for the connections in the RD form, denoted by $\varphi_{x,ij}(t)$ s, for any KS. We do not consider symmetric KSs separately, because symmetry is apparent in the functional form of the $\varphi_{x,ij}(t)$ s. Further discussion regarding the topological interpretation of $\tilde{M}_x$ and $\tilde{N}_y$ in Eq. (6) is also given.

The waiting time PDFs for the connections in the RD form are uniquely determined by the clustering procedure in the mapping of a KS into an RD form. The clustering procedure is based upon the identification of substates in the *on–off* connectivity that contribute to the ranks $R_{x,y}$. The four ranks $R_{x,y}$ s determine the RD form topology, and the mapping determines the incoming flux and outgoing flux for each substate in the RD form. This makes the RD forms legitimate canonical forms that preserve all the information contained in the two-state trajectory.

The technical details to obtain the $\varphi_{x,ij}(t)$s, given a KS, are spelled out below when considering two cases: (1) None of the terms in an external sum in Eq. (5), after the first or second equality, are proportional to each other, and (2) some of the terms in an external sum in Eq. (5), after the first or second equality, are proportional to each other.

B1. *Reversible on–off connection KS*s

Say, $M_x \geq N_y$, or equivalently $N_x \geq M_y$ (Fig. B1(A) with $x = off$). Based on the clustering procedure, there are N_y substates in each of the states in the RD form, and as many as $2N_y^2$ WT-PDFs for the connections in the RD form. Initial substates in state x are clustered, and the expression for $\varphi_{x,n_y i_x}(t)$ reads

$$\varphi_{x,n_y i_x}(t) = \frac{1}{N_{x,m_y}} \sum_{n_x} P_{y,m_y}(ss)(\mathbf{V_y})_{n_x m_y} f_{n_y n_x}(t). \qquad \text{(B1)}$$

In Eq. (B1), we use the normalization N_{x,m_y}, defined through the equations

$$N_x = \vec{\mathbf{1}}_x^T \mathbf{V}_y \vec{\mathbf{P}}_y(ss) = \sum_{m_y, n_x} P_{y,m_y}(ss)(\mathbf{V_y})_{n_x m_y}$$

$$= \sum_{m_y} N_{x,m_y} = \sum_{n_x} N_{x,n_x}.$$

As notation is concerned, we set in Eq. (B1) $j_y \to n_y$, because there are $n_y = 1, \ldots, N_y$ substates in state y in the RD form, and we can

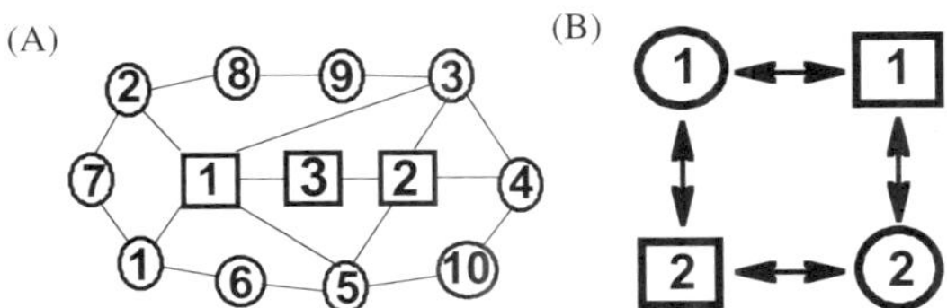

Fig. B1. (A) A reversible connection KS, with $N_{on} = M_{on} = 2$ and $N_{off} = M_{off} = 5$. (B) The RD form of KS (A). The RD form's substate 1_{off} corresponds to the cluster of the KS's *off* substates $1_{off} - 3_{off}$ and 5_{off}, because these are connected to substate 1_{on} in the KS, which contributes to the rank $R_{on,off}$. The RD form's substate 2_{off} corresponds to the cluster of the KS's *off* substates $3_{off} - 5_{off}$, because these are connected to substate 2_{on} in the KS, which contributes to the rank $R_{on,off}$. Note that a particular initial substate can appear in more than a single cluster, which simply means that the overall steady-state flux into the substate is divided into several contributions. The initial *on* substates in the KS both contribute to $R_{off,on}$; so, they are mapped to themselves in the RD form. The WT-PDFs for the connections can be obtained from Eqs. (B1) and (B2).

also employ the meaning of n_y as the initial substates in state y in the underlying KS. Additionally, we associate m_y on the right-hand side (RHS), which has the meaning of final substates in the underlying KS, with i_x on the left-hand side (LHS), i.e. $m_y \to i_x$. Note that for a KS with only reversible *on–off* connections, $m_y = 1, \ldots, M_y$; so, the values of m_y and i_x can be the same.

The expression for $\varphi_{y,i_x n_y}(t)$ is different from that for $\varphi_{x,n_y i_x}(t)$ in both the normalization used and the factors that are summed, which is a result of the mapping of the initial substates in state y to themselves. $\varphi_{y,i_x n_y}(t)$ is given by

$$\varphi_{y,i_x n_y}(t) = \frac{1}{N_{y,n_y}} \sum_{m_x} P_{x,m_x}(ss)(\mathbf{V_x})_{n_y m_x} \tilde{f}_{m_y n_y}(t)\tilde{\omega}_{m_y};$$
$$\tilde{\omega}_{m_y} = \sum_{n_x} \omega_{n_x m_y}. \tag{B2}$$

Note that here, $\varphi_{y,i_x n_y}(t) = \tilde{f}_{m_y n_y}(t)\tilde{\omega}_{m_y} = (\mathbf{G_y}(t))_{m_y n_y} \sum_{n_x} (\mathbf{V_y})_{n_x m_y}$. In Eq. (B2), we associate m_y on the RHS with i_x on the LHS, i.e. $m_y \to i_x$. Again, for a KS with only reversible transitions, the i_x s can have the same values as that of the m_y s.

B2. *Irreversible on–off connection KSs*

Obtaining the $\varphi_{x,ij}(t)$ s for irreversible *on–off* connection KSs is similar to getting these WT-PDFs for reversible *on–off* connection KSs. The reason is that the clustering procedure is based on the *directional* connections between final substates in state x and initial substates in state y. However, some technical details may differ. We consider two cases.

(1) Let $M_x \geq N_y$ and $M_y \geq N_x$ (Figs. B2(A) and B2(B)). Then, the WT-PDFs for the connections are given by

$$\varphi_{x,n_y n_x}(t) = \frac{1}{N_{x,n_x}} \sum_{m_y} P_{y,m_y}(ss)(\mathbf{V_y})_{n_x m_y} f_{n_y n_x}(t) = f_{n_y n_x}(t), \tag{B3}$$

and

$$\varphi_{x,n_x n_y}(t) = \frac{1}{N_{y,n_y}} \sum_{m_x} P_{x,m_x}(ss)(\mathbf{V_x})_{n_y m_x} f_{n_x n_y}(t) = f_{n_x n_y}(t). \tag{B4}$$

Note that for this case any $\varphi_{z,ij}(t)$ is equal to the corresponding $f_{ij}(t)$. This is an outcome of the KS's topology for which in both the *on* to *off* and the *off* to *on* connections, the number of initial substates in a given state is lower than the number of final substates in the other state.

(2) Let $N_x > M_y$ and $N_y > M_x$ (Figs. B3(A) and B3(B)). Then, the WT-PDFs for the connections are given by

$$\varphi_{x,j_y i_x}(t) = \frac{1}{N_{x,m_y}} \sum_{n_x} P_{y,m_y}(ss)(\mathbf{V_y})_{n_x m_y} \tilde{f}_{m_x n_x}(t) \tilde{\omega}_{m_x}, \tag{B5}$$

and

$$\varphi_{y,i_x j_y}(t) = \frac{1}{N_{y,m_x}} \sum_{n_y} P_{x,m_x}(ss)(\mathbf{V_x})_{n_y m_x} \tilde{f}_{m_y n_y}(t) \tilde{\omega}_{m_y}. \tag{B6}$$

In Eqs. (B5) and (B6), we use the mapping $m_y \to i_x$ and $m_x \to j_y$ between the RHS and the LHS indexes. (In particular, $m_y - (N_y - H_y) = i_x$ and $m_x - (N_x - H_x) = j_y$.)

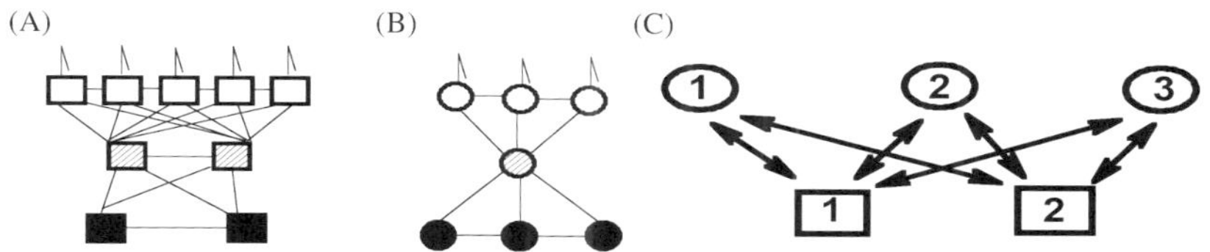

Fig. B2. An example for a KS with irreversible *on–off* connections, and $N_{on} = 2$, $M_{on} = 5$, $N_{off} = 3$, and $M_{off} = 3$. The KS is divided into two parts shown in (A) (*on* state) and (B) (*off* state) for a convenient illustration. The RD form is shown in (C). The WT-PDFs for the connections can be obtained from Eqs. (B3) and (B4).

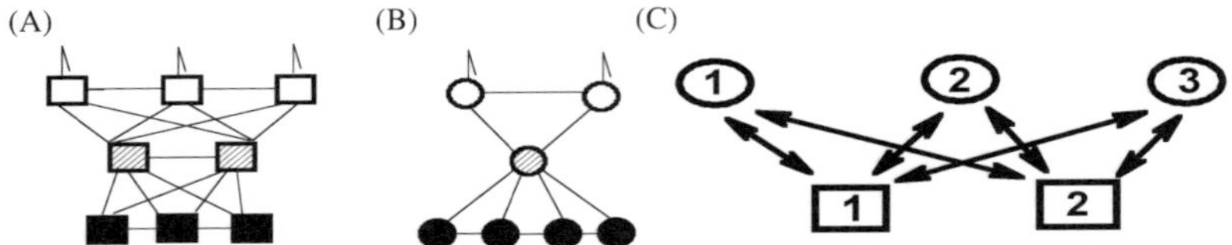

Fig. B3. An irreversible *on–off* connection KS with $N_{on} = 3$, $M_{on} = 3$, $N_{off} = 4$, and $M_{off} = 2$. The panels are divided as in Fig. B2. The WT-PDFs for the connections can be obtained from Eqs. (B5) and (B6).

We turn now to deal with cases in which some of the terms in Eq. (5) are proportional, and therefore Eq. (6) is used to expressing $\phi_{x,y}(t_1, t_2)$. We consider only the KSs with reversible *on–off* connections, but the same analysis is relevant to KSs with irreversible *on–off* connections.

Let $M_x \leq N_y$, or equivalently $N_x \leq M_y$ (see Fig. B4(A) with $x = off$). So it follows that, $R_{x,y} < M_x$, which is a result of a special *on–off* connectivity. In particular, let $\{O_y\}$ and $\{O_x\}$ be the groups of substates in states y and x, respectively, such that the substates in $\{O_x\}$ are connected only to the substates in $\{O_y\}$ and $O_y < O_x$. (In Fig. B4(A), the group $\{O_{off}\}$ contains the substates 1_{off}, 2_{off}, and 3_{off}, and the group $\{O_{on}\}$ contains the substates 1_{on} and 2_{on}).

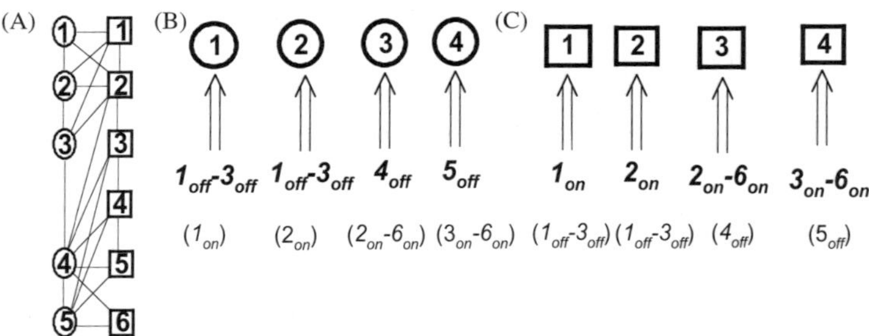

Fig. B4. (A) A reversible connection KS with $R_{x,y} = 4$ ($x \neq y$). The RD form's topology is shown on (B) and (C). The clustering procedure and the parent substates (in the parentheses) are indicated at the base of the double arrows. For example, substate 1_{off} in the RD form corresponds to the cluster of initial-*off*-substates $1_{off} - 3_{off}$ in the KS. These are connected to substate 1_{on} in the KS. The WT-PDFs for the connections in the RD form can be obtained from Eqs. (B7)–(B12).

Thus, both initial and final substates contribute to the rank $R_{z,z'}$ for $z \neq z'$, and the expressions for the $\varphi_{z,ij}(t)$ s are distinct in each of the following three regimes:

(a) For $n_x \notin \{O_x\}$ and $n_y \notin \{O_y\}$,

$$\varphi_{x,j_y i_x}(t) = \frac{1}{N_{x,n_x}} \sum_{m_y} P_{y,m_y}(ss)(\mathbf{V_y})_{n_x m_y} \tilde{f}_{m_x n_x}(t) \sum_{n_y \notin \{O_y\}} \omega_{n_y m_x}, \tag{B7}$$

and

$$\varphi_{y,i_x j_y}(t) = \frac{1}{N_{y \in O_y, m_x}} \sum_{n_y} P_{x,m_x}(ss)(\mathbf{V}_x)_{n_y m_x} f_{n_x n_y}(t), \tag{B8}$$

where $N_{y \in O_y, m_x} = \sum_{n_y \in \{O_y\}} P_{x,m_x}(ss)(\mathbf{V}_x)_{n_y m_x}$, and we associate $n_x \to i_x$ and $m_x \to j_y$.

(b) For $n_x \notin \{O_x\}$ and $n_y \in \{O_y\}$,

$$\varphi_{x,j_y i_x}(t) = \frac{1}{N_{x,n_x}} \sum_{m_y} P_{y,m_y}(ss)(\mathbf{V_y})_{n_x m_y} f_{n_y n_x}(t), \tag{B9}$$

and

$$\varphi_{y,i_x j_y}(t) = \frac{1}{N_{y,n_y}} \sum_{m_x} P_{x,m_x}(ss)(\mathbf{V}_x)_{n_y m_x} f_{n_x n_y}(t), \tag{B10}$$

where we associate $n_y \to j_y$ and $n_x \to i_x$.

(c) For $n_x \in \{O_x\}$ and $n_y \in \{O_y\}$,

$$\varphi_{y,i_x j_y}(t) = \frac{1}{N_{y,n_y}} \sum_{m_x} P_{x,m_x}(ss)(\mathbf{V}_x)_{n_y m_x} \tilde{f}_{m_y n_y}(t) \sum_{n_x \in \{O_x\}} \omega_{n_x m_y}, \tag{B11}$$

and

$$\varphi_{x,j_y i_x}(t) = \frac{1}{N_{x \in O_x, m_y}} \sum_{n_x \in \{O_x\}} P_{y,m_y}(ss)(\mathbf{V}_y)_{n_x m_y} f_{n_y n_x}(t), \tag{B12}$$

where we associate $n_y \to j_y$ and $m_y \to i_x$.

As a final remark, we show how to use O_y and O_x to express $R_{x,y}$. When $M_x < N_y$ and $\{O_x\}$ and $\{O_y\}$ are as defined above,

$$R_{x,y} = M_x - (O_x - O_y). \tag{B13}$$

This result can be generalized to the case of J groups in the underlying KS that are connected in the way defined above for the case of a single pair of groups. The generalized result reads

$$R_{x,y} = M_x - \sum_j (O_{x,j} - O_{y,j}). \tag{B14}$$

These expressions imply that $\tilde{M}_x$ and $\tilde{N}_y$ on Eq. (6) are related to the KS's topology by

$$\tilde{M}_x = M_x - \sum_j O_{x,j}, \tag{B15}$$

and

$$\tilde{N}_y = \sum_j O_{y,j}. \tag{B16}$$

When $M_x > N_y$, and there are groups $\{Z_x\}$ and $\{Z_y\}$ with $Z_x < Z_y$, such that substates in $\{Z_y\}$ are connected only to substates in $\{Z_x\}$, we define $O_x = M_x - Z_x$ and $O_y = N_y - Z_y$, and Eq. (B13) holds. For J such groups, we define $O_{x,j} = M_x/J - Z_{x,j}$ and $O_{y,j} = N_y/J - Z_{y,j}$, and Eqs. (B14)–(B16) hold.

For a KS with symmetry, $\tilde{M}_x$ and $\tilde{N}_y$ are chosen in a different way than the one that relies on the *on–off* connectivity; for such a case, the choice that makes the number of additives in the external sums of Eq. (6) minima simply groups the identical PDFs. The topology of the RD form is determined by the largest $R_{x,y}$.

References

1. W. E. Moerner and M. Orrit, *Science* **283** (1999) 1670–1676.
2. S. Weiss, *Science* **283** (1999) 1676–1683.
3. E. Neher and B. Sakmanm, *Nature* **260** (1976) 799–802.

4. J. J. Kasianowicz, E. Brandin, D. Branton and D. W. Deamer, *Proceedings of the Natural Acadamy of Sciences USA* **93** (1996) 13770–13773.
5. L. Kullman, P. A. Gurnev, M. Winterhalter and S. M. Bezrukov, *Physical Review Letters* **96** (2006) 038101–038104.
6. B. Schuler, E. A. Lipman and W. A. Eaton, *Nature* **419** (2002) 743–747.
7. H. Yang, G. Luo, P. Karnchanaphanurach, T. Louie, I. Rech, S. Cova, L. Xun and X. S. Xie, *Science* **302** (2003) 262–266.
8. W. Min, G. Lou, B. J. Cherayil, S. C. Kou and X. S. Xie, *Physical Review Letters* **94** (2005) 198302.
9. E. Rhoades, E. Gussakovsky and G. Haran, *Proceedings of the Natural Academy of Sciences USA* **100** (2003) 3197–3202.
10. X. Zhuang, H. Kim, M. J. B. Pereira, H. P. Babcock, N. G. Walter and S. Chu, *Science* **296** (2002) 1473–1476.
11. H. Lu, L. Xun and X. S. Xie, *Science* **282** (1998) 1877–1882.
12. L. Edman, Z. Földes-Papp, S. Wennmalm and R. Rigler, *Chemical Physics* **247** (1999) 11–22.
13. K. Velonia, O. Flomenbom, D. Loos, S. Masuo, M. Cotlet, Y. Engelborghs, J. Hofkens, A. E. Rowan, J. Klafter, R. J. M. Nolte *et al.*, *Angewandte Chemie International Edition* **44** (2005) 560–564.
14. O. Flomenbom, K. Velonia, D. Loos *et al.*, *Proceedings of the Natural Academy of Sciences USA* **102** (2005) 2368–2372.
15. O. Flomenbom, J. Hofkens, K. Velonia *et al.*, *Chemical Physics Letters* **432** (2006) 371.
16. B. P. English, W. Min, A. M. van Oijen, K. T. Lee, G. Luo, H. I. Sun, B. J. Cherayil, S. C. Kou and X. S. Xie., *Nature Chemical Biology* **2** (2006) 87–94.
17. R. J. Davenport, G. J. L. Wuite, R. Landick and C. Bustamante, *Science* **287** (2000) 2497.
18. S. Nie, D. T. Chiu and R. N. Zare, *Science* **266** (1994) 1018–1021.
19. R. Shusterman, S. Alon, T. Gavrinyov and O. Krichevsky, *Physical Review Letters* **92** (2004) 048303-1-4.
20. G. Zumofen, J. Hohlbein and C. G. Hübner, *Physical Review Letters* **93** (2004) 260601-1-4.
21. A. E. Cohen and W. E. Moerner, *Proceedings of the Natural Academy of Sciences USA* **103** (2006) 4362–4365.
22. R. M. Dickson, A. B. Cubitt, R. Y. Tsien and W. E. Moerner, *Nature* **388** (1997) 355–358.
23. I. Chung and M. G. Bawendi *Physical Review B* **70** (2004) 165304-1-5.
24. E. Barkai, Y. Jung and R. Silbey, *Annual Review of Physical Chemistry* **55** (2004) 457–507.
25. J. Tang and R. A. Marcus, *Journal of Chemical Physics* **123** (2005) 204511-1-6.

26. R. Horn and K. Lange, *Biophysical Journal* **43** (1983) 207–223.
27. F. Qin, A. Auerbach and F. Sachs, *Biophysical Journal* **79** (2000) 1915–1927.
28. F. G. Ball and M. S. P. Sansom, *Proceedings of the Royal Society of London B* **236** (1989) 385.
29. O. Flomenbom, J. Klafter and A. Szabo, *Biophysical Journal* **88** (2005) 3780–3783.
30. O. Flomenbom and J. Klafter, *Acta Physica Polonica B* **36** (2005) 1527–1535.
31. O. Flomenbom and J. Klafter, *Journal of Chemical Physics* **123** (2005) 064903-1-10.
32. J. B. Witkoskie and J. Cao, *Journal of Chemical Physics* **121** (2004) 6361–6372.
33. W. J. Bruno, J. Yang and J. Pearson, *Proceedings of the National Academy of Sciences USA* **102** (2005) 6326–6331.
34. O. Flomenbom and R. J. Silbey, *Proceedings of the National Academy of Sciences USA* **103** (2006) 10907.
35. R. J. Bauer, B. F. Bowman and J. L. Kenyon, *Biophysical Journal* **52** (1987) 961–978.
36. P. Kienker, *Proceedings of the Royal Society of London B* **236** (1989) 269–309.
37. D. R. Fredkin and J. A. Rice, *Journal of Applied Probability* **23** (1986) 208–214.
38. D. Colquhoun and A. G. Hawkes, *Philosophical Transations of the Royal Society of London B Biological Sciences* **300** (1982) 1–59.
39. J. Cao, *Chemical Physics Letters* **327** (2000) 38–44.
40. L. Song and K. L. Magdeby, *Biophysical Journal* **67** (1994) 91–104.
41. M. O. Vlad, F. Moran, F. W. Schneider and J. Ross, *Proceedings of the National Academy of Sciences USA* **99** (2002) 12548–12555.
42. S. Yang and J. Cao, *Journal of Chemical Physics* **117** (2002) 10996–11009.
43. N. Agmon, *Journal of Physical Chemistry B* **104** 7830–7834 (2000).
44. I. V. Gopich and A. Szabo, *Journal of Chemical Physics* **124** (2006) 154712-1.
45. H. Qian and E. L. Elson, *Biophysical Chemistry* **101** (2002) 565–576.
46. S. C. Kou, B. J. Cherayil, W. Min, B. P. English and X. S. Xie, *Journal of Physical Chemistry B* **109** (2005) 19068–19081.
47. Y. Kafri, D. K. Lubensky and D. R. Nelson, *Biophysical Journal* **86** (2004) 3373.
48. J. Y. Sung and R. J. Silbey, *Chemical Physics Letters* **415** (2005) 10–14.
49. R. Granek and J. Klafter, *Physical Review Letters* **95** (2005) 098106-1-4.
50. J. W. Shaevitz, S. M. Block and M. J. Schnitzer, *Biophysical Journal* **89** (2005) 2277–2285.
51. V. Barsegov and D. Thirumalai, *Physical Review Letters* **95** (2005) 168302-1-4.

52. F. Šanda and S. Mukamel, *Journal Chemical Physics* **108** (2006) 124103-1-15.
53. P. Allegrini, G. Aquino, P. Grigolini, L. Palatella and A. Rosa, *Physical Review E* **68** (2003) 056123–1-11.
54. A. B. Kolomeisky and M. E. Fisher, *Journal of Chemical Physics* **113** (2000) 10867–10877.
55. I. Goychuk and P. Hänggi, *Physical Review E* **70** (2004) 051915-1-9.
56. O. Flomenbom and J. Klafter, *Physical Review Letters* **95** (2005) 098105-1-4.
57. E. Yeramian and P. Claverie, *Nature* **326** (1987) 169–174.
58. D. R. Cox and D. V. Hinkley, *Theoretical Statistics* (Chapman & Hall/CRC, USA, 1974).

CHAPTER 11

Weak Ergodicity Breaking in Single-Particle Dynamics

E. Barkai

Department of Physics
Bar Ilan University, Ramat-Gan 52900, Israel

1. Introduction

In nature, one encounters many phenomena in which some quantity varies in time in a complicated way. Usually there is no hope of determining this variation in detail, except for certain averaged features. The average over time is an awkward procedure; one therefore replaces the irregularly varying function of time by an ensemble of functions. All averages are redefined as averages over a suitable ensemble rather than over some time interval of the single realization of the time-varying quantity. One may actually observe a large number of particles and average the result; this means that one has a physical realization of the ensemble. One may also observe one and the same particle on a long-time interval and then the time average is the observed quantity, and not the ensemble average.

The condition for the ergodicity of the system, namely that the time averages and ensemble averages coincide, is that the behavior of the measured signal during one interval is uncorrelated with the measured signal during the next interval.[1] The time average of any physical quantity is defined as

$$\bar{I} = \frac{1}{t} \int_0^t I(t')\mathrm{d}t'. \tag{1}$$

In order for the ergodic hypothesis to hold, it should be possible to represent it also as

$$\bar{I} = \frac{1}{N} \sum_{i=1}^{N} I_i, \tag{2}$$

where I_i is a short-time average

$$I_i = \frac{1}{\tau} \int_{(i-1)\tau}^{i\tau} I(t')\mathrm{d}t' \tag{3}$$

and $\tau = t/N$. As usual we take the limit of $t \to \infty$ and $N \to \infty$ such that τ remains finite. If I_i are independent random variables, the sum in Eq. (2) can be replaced with an average over a suitable ensemble. This means that there exists a τ which is longer than a characteristic timescale of the problem, the latter time is the time when roughly speaking the trajectory becomes uncorrelated with its past history. On the other hand, τ should be much smaller than the total measurement time t. The basic assumption is that for the underlying dynamics, some finite characteristic timescale actually exists, and then it is in principle possible to find an intermediate timescale τ. If this is not the case ergodicity is broken and new methods are needed. In particular if the characteristic timescale diverges, we expect a nonergodic behavior.

What exactly is the meaning of a diverging characteristic timescale in single-particle experiments? In Sec. 2 we consider the example of blinking nanocrystals (NC) under continuous wave illumination. At random times the NC jumps between a bright state in which it emits many photons and a dark state in which it is turned off. This instability prevents many possible applications in nanotechnology and hence it is widely investigated though not yet completely understood. It is found that the probability density function (PDF) of the on and the off times are described by power laws, for example, the off times[2] $\psi(t_{\mathrm{off}}) \propto t^{-(1+\alpha)}$ and in many cases $0 < \alpha < 1$. It is easy to see that the average off (and on) times diverge

$$\langle t_{\mathrm{off}} \rangle = \int_0^\infty t\psi(t)\mathrm{d}t = \infty, \tag{4}$$

because of the slow decay of the PDF. In this case we identify the characteristic timescales with the averaged on and the averaged off times.

The concept of power-law sojourn or trapping time PDF $\psi(t)$ is very old,[3] and was introduced in the context of charge transport in amorphous semi-conductors, and later found applications in diverse fields of Physics and Chemistry. In single-particle experiments power-law waiting times describe the sticking time statistics of particles diffusing in actin network, a system important for our understanding of physical properties of the cell environment.[4] Several excellent popular papers,[5,6] reviews,[7,8] and books,[9,10] explain the physical origin of the power-law waiting time PDF $\psi(t)$ and the physical consequences of it, in different physical systems. The new aspect of the problem is that these days single-molecule and single-particle experiments, observe power-law sojourn times on the single-particle level. Since in single-particle experiments we perform time averages, and not ensemble averages, and since the characteristic timescale is infinite, we are dealing with the problem of ergodicity breaking. In particular, the theoretical question of interest is: What theory replaces the standard ergodic Boltzmann–Gibbs theory for systems described by this type of anomalous kinetics?

Time averages of physical observables in nonergodic single-particle experiments are nonreproducible quantities. For example, if one records the time average of the intensity of a blinking quantum dot, the average is random even if the dot showed thousands of transitions between the state on and state off. This situation is very unusual in Physics, since we are used to physical observables which change randomly but usually assume that their long-time average is reproducible. Experimentalists should obviously ask themselves if the time average they are recording is reproducible or not and theorists should worry whether the ensemble average actually represents the measured time average. Even though the time average is random, the situation is not totally hopeless and, at least for simple cases, one may construct a theory which predicts the distribution of time-averaged quantities, showing precisely the deviations of the time average from the ensemble average.

Ergodicity is the pillar on which Boltzmann's statistical mechanical theory is built. For a system with Hamiltonian H, and within the canonical ensemble,

$$P_B(\sigma) = \frac{\exp\left[-\frac{H(\sigma)}{k_b T}\right]}{Z} \tag{5}$$

is the probability in ensemble sense of finding the system in state σ and Z is the partition function. Another important quantity is the occupation fraction, which is the fraction of time the system occupies a state σ during a measurement,

$$\bar{p}(\sigma) = T_\sigma / t, \tag{6}$$

where T_σ is the total occupation time in state σ and t is the measurement time. The time-averaged observable is

$$\bar{I} = \sum_\sigma \bar{p}(\sigma) I(\sigma) \tag{7}$$

while the ensemble average of the same quantity is

$$\langle I \rangle = \sum_\sigma P_B(\sigma) I(\sigma). \tag{8}$$

For an ergodic canonical system in thermal equilibrium, $\bar{p}(\sigma) = P_B(\sigma)$, and the two averages coincide. For nonergodic systems the occupation time of each state is needed in order to predict the statistical behavior of the system. Hence, in this chapter we investigate the problem of statistics of occupation times for dynamics governed by power-law statistics, showing the deviations from usual ergodic behavior.

Following Bouchaud,[11] we distinguish between two different types of ergodicity breaking. The most common situation is the case where the system is decomposed into several regions in phase space, and the system starting in one region cannot reach another. Then exploration of phase space is limited according to the particular starting point of the system. The more interesting case, according to the authors opinion, is weak ergodicity breaking. In this case the phase space is fully connected; the system does visit all its phase space; however, the dynamics is nonstationary, and the occupation fraction of a state is not equal to the fraction of systems in the ensemble occupying the same state. This type of weak ergodicity breaking is related to power-law waiting times, Lévy statistics, anomalous diffusion, fractional calculus, and statistical aging, which are well established mathematically and found in many physical systems.

We now discuss weak ergodicity breaking for three cases: blinking NCs which is a driven system far from thermal equilibrium (Sec. 2), the continuous time random walk model (Sec. 3), and systems with quenched disorder, in particular, the quenched trap model (Sec. 4). Recently we reviewed other sources of fluctuations in single-molecule spectroscopy,[12] which is also the topic of other chapters in this book. The study is intended to survey the group's work on weak ergodicity breaking for a general audience, skipping the main technical details and the proofs, which can be found in the literature.

2. Blinking nanocrystals

Experiments[2, 13–15] show fluorescence blinking for a single semiconductor NC such as CdSe illuminated with a continuous-wave laser field. As shown in Fig. 1 the NC is found in two states, either on

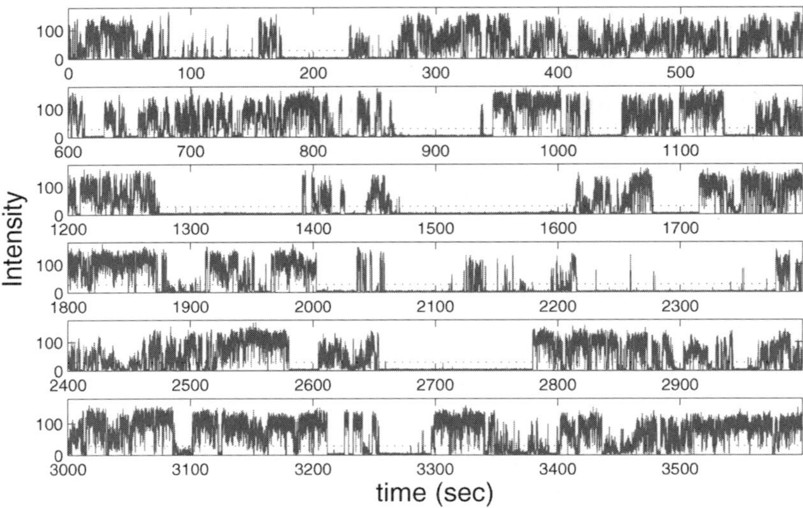

Fig. 1. Intensity fluctuations in a CdSe–ZnS NC under continuous laser illumination at room temperature.[19,20] The dotted horizontal line was selected as a threshold to divide off and on states. Notice the large variations of the on and off times, which is a manifestation of the power-law statistics.

or off. The on (off) times are believed to correspond to a neutral (charged) nanocrystal, respectively.[16] Thus the switching sequence on–off–on–off... corresponds to the sequence of chemical reactions; neutral NC $\rightarrow$ charged NC $\rightarrow$ neutral NC $\rightarrow$ charged NC, etc. Briefly, the idea is that once a charge is located on the dot, the nonradiative Auger process of electron–hole recombination is faster than the process of spontaneous emission; hence, a charge on the dot quenches the radiation and the dot is in an off state.

In contrast to expectation, distributions of on times and off times exhibit power-law statistics. The PDFs of on and off sojourn times exhibit

$$\psi_{\text{on}}(t) \sim A_{\text{on}} t^{-(1+\alpha_{\text{on}})} \quad \text{and} \quad \psi_{\text{off}}(t) \sim A_{\text{off}} t^{-(1+\alpha_{\text{off}})}. \tag{9}$$

In many cases it is found that $0 < \alpha_{\text{on}} < 1$ and $0 < \alpha_{\text{off}} < 1$. For example, in Ref. 39 215 NCs were measured and the exponents

$\alpha_{\rm on} = 0.58 \pm 0.17$ and $\alpha_{\rm off} = 0.48 \pm 0.15$ were found (note that within the error of measurement $\alpha_{\rm on} = \alpha_{\rm off} = \alpha = 1/2$). As mentioned in the introduction, the average on and off time is infinite and the process has no finite characteristic timescale. The exponents α do not depend on the temperature of the system, and the radius of the nanocrystals; they appear universal in the sense that for a given sample they are the same for all the quantum dots. Some variation of $\alpha_{\rm off}$ and $\alpha_{\rm on}$ is found, which depends on the dielectric constant of the environment.[17] However, for nearly all cases $\alpha < 1$. In these experiments there exists a long time window where power-law sojourn times are observed, a time window where hundreds or thousands of transitions between on and off and vice versa are observed.

For simplicity, let us assume that the process is a two-state process with $\alpha_{\rm on} = \alpha_{\rm off} = \alpha$. We normalize the intensity of the bright state to unity so that $I(t) = 0$ or $I(t) = 1$. Data analysis[18] shows that the sojourn times in the on and off states are independent random variables. It is easy to see that the time average intensity of the NC is

$$\bar{I} = \frac{\int_0^t I(t)\mathrm{d}t'}{t} = \frac{T^+}{t}, \tag{10}$$

where T^+ is the total time the NC is in state on. For the case $0 < \alpha_{\rm on} = \alpha_{\rm off} = \alpha < 1$ the time average intensity remains random even in the limit of $t \to \infty$. The PDF of the time average intensity is the Lamperti PDF[21,22]

$$f(\bar{I}) = \delta_{\alpha,\mathcal{R}}(\bar{I}), \tag{11}$$

where

$$\delta_{\alpha,\mathcal{R}}(x) = \frac{\sin \pi\alpha}{\pi} \frac{\mathcal{R} x^{\alpha-1}(1-x)^{\alpha-1}}{\mathcal{R}^2(1-x)^{2\alpha} + x^{2\alpha} + 2\mathcal{R}(1-x)^{\alpha} x^{\alpha} \cos \alpha\pi}, \tag{12}$$

where $\mathcal{R} = A_{\rm on}/A_{\rm off}$. The ensemble average intensity is $\langle I \rangle = A_{\rm on}/(A_{\rm off} + A_{\rm on})$. If $\alpha_{\rm on} = \alpha_{\rm off} = \alpha > 1$ we have usual ergodic behavior $f(\bar{I}) = \delta(\bar{I} - \langle I \rangle)$, namely the time average intensity is equal to the ensemble average. For a detailed derivation of Eq. (12), see Appendix in Ref. 23.

The $\delta_{\alpha,\mathcal{R}}(x)$ distribution has the following behaviors. If $\alpha \to 1$ we have the usual ergodic behavior $f(\bar{I}) = \delta(\bar{I} - \langle I \rangle)$, namely if $\alpha = 1$ the $\delta_{\alpha,\mathcal{R}}(x)$ is a delta function. This is the expected behavior; when the average on and off time is finite, we can make time averages for long enough times so that ergodicity is eventually obtained. If $\alpha \to 0$, then either the NC is in state on or off during the whole measurement, corresponding to the situations $\bar{I} = 1$ or $\bar{I} = 0$, respectively. In this case PDF $f(\bar{I})$ is composed of two delta functions on $\bar{I} = 1$ and $\bar{I} = 0$. For $\alpha \ll 1$ the PDF $f(\bar{I})$ has a U shape since trajectories are either in on state or in off state for very long periods, while when $\alpha \to 1$ from below, the shape of $f(\bar{I})$ is of a distorted W. $\mathcal{R}$ is the asymmetry parameter. When $\mathcal{R} = 1$, the PDF $f(\bar{I})$ is symmetric. In Figs. 2 and 3, in the upper left panel (with $r = 0$), we show $\delta_{\alpha,\mathcal{R}}(x)$ for $\alpha = 0.3$ (U shape) and $\alpha = 0.8$ (W shape), respectively.

The special case of $\mathcal{R} = 1$ and $\alpha = 1/2$ is the arcsine law. Consider a Brownian motion where $\dot{x}(t) = \eta(t)$ and $\eta(t)$ is a Gaussian white noise. Lévy investigated the residence time T^+ in the domain $x > 0$ in half space $x > 0$, when the motion is unbounded and the total observation time is t. Naive expectation is that $T^+/t = 1/2$, with small fluctuations when $t \to \infty$, namely the particle occupies $x > 0$ for half of the observation time, and half of the time it resides in $x < 0$. Instead the PDF of $0 < T^+/t < 1$ is given by the famous arcsine law

$$f(T^+/t) = \delta_{1/2,1}\left(\frac{T^+}{t}\right) = \frac{1}{\pi\sqrt{\left(\frac{T^+}{t}\right)\left(1 - \frac{T^+}{t}\right)}}, \tag{13}$$

which has a U shape. This means that for a typical realization of the Brownian trajectory, the particle spends most of the time in one half of the space (say $x > 0$) and not in the other $x < 0$. This behavior is related to the observation that the PDF of first return time, for simple Brownian motion, from $x_0 \neq 0$ to $x = 0$ follows a long tailed

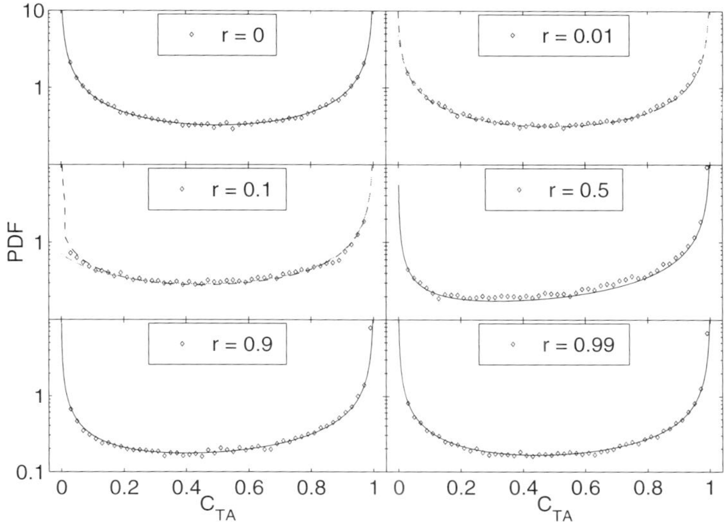

Fig. 2. The PDF of the time average intensity correlation function $C_{\text{TA}}(t', t)$ for $\alpha = 0.3$ and for varying values of $r = t'/t$. The lines are analytical results,[22] and the diamonds are numerical simulations. Not shown is a delta function contribution on $C_{\text{TA}} = 0$. This contribution is zero when $r = 0$; so $\langle C_{\text{TA}} \rangle = 1/2$ when $r = 0$ while when $r \to 1$ Eq. (19) is valid and $\langle C_{\text{TA}} \rangle = 1/4$. For an ergodic process the PDF of the time average correlation function would be a delta function centered on the ensemble average, which is clearly not the case here.

PDF $\psi(t) \propto t^{-(1+\alpha)}$ and $\alpha = 1/2$. Hence, the average return time is infinite, and there is no finite characteristic timescale to the problem.

Several models for the blinking dots have appeared in the literature: Fluctuating barrier models,[15] trap models,[24,25] deep surface state model,[26] diffusion-controlled reaction models,[19,27,28] and an electron transfer model.[29] It is not our aim to survey these models here; further, the physical picture for blinking dots is not yet complete. We only briefly mention that first passage times for a charge carrier ejected from the dot to the medium (corresponding to an off-state), where the ejected charge is following a random walk (diffusion) in the surrounding medium of the dot, follows a power-law decay[19,27] even if the diffusion is three-dimensional. An important ingredient of the theory is that the electron–hole interaction on the nanometer scale is

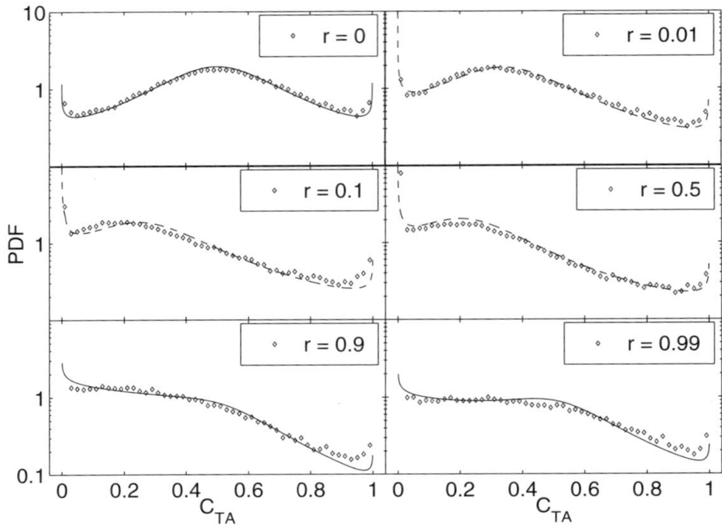

Fig. 3. The PDF of the time average intensity correlation function $C_{\mathrm{TA}}(t', t)$ for $\alpha = 0.8$ and for varying values of $r = t'/t$. A peak appears close to the value predicted by ergodic theory, for example a peak on $C_{\mathrm{TA}} = 1/2$ when $r = 0$, however the distribution of the time-averaged correlation function is still non-trivial. As we increase r the PDF tends to concentrate more on smaller values of C_{TA} reflecting the trend of the ensemble average $\langle C_{\mathrm{TA}} \rangle = 1/2$ when $r = 0$ while $\langle C_{\mathrm{TA}} \rangle = 1/4$ when $r \to 1$.

strong enough so that the ejected electron or hole does not manage to escape to infinity (Ref. 19 and references therein), since the Coulomb potential is a bias of the electron toward the hole. Similar mechanism was used to explain the power-law like decay of photoluminescence for geminate electron–hole recombination.[30] Combined contributions from electron–hole interaction,[19] tunneling, and diffusion may lead to deviations from the universal exponent $\alpha = 1/2$ predicted based on simple diffusion models. Similar ideas hold also for the on-state.[19,24] Thus, simple generic models, for example, diffusion models, can be used to explain the power-law behavior and hence weak ergodicity breaking in single-particle experiments. We therefore expect that weak ergodicity breaking will be found in other single-molecule systems. Indeed, power-law statistics and intermittency is not limited to single quantum dots. It is also found in single organic molecules.[31–33] For example, recent experiments using Raman spectroscopy exhibit temporal power-law behavior for organic molecules,[34] while ensemble measurements predict a Lévy type of temporal behavior for the dynamics of green fluorescence protein.[35]

Remark 1. A method to detect the strange kinetics, already on the level of an ensemble of NCs was suggested by Jung *et al.*[36] It was shown that the photon statistics of an ensemble of NCs obeys Lévy statistics, not the usual Gaussian statistics, in such a way that the fluctuations of photon counts are very large. In particular, the Mandel Q parameter, describing the fluctuations of the photon counts, increases with time even in the limit of long time (the variance of photon counts exhibits ballistic behavior $\langle n^2 \rangle - \langle n \rangle^2 \sim t^2$ instead of the usual t^1 Gaussian behavior). Recent measurements confirm this prediction.[20] In principle, the methods in Refs. 20 and 36 can be used even for an ensemble of $N \to \infty$ dots, at least for an ensemble of identical dots. A different method to detect the strange kinetics already for a collection of dots was considered by the Cao-Bawendi groups.[37] Profound issues on the differences between time averages of single-particle experiments and time averages of $N > 1$ particle experiments are under current research. In particular, the question remains: is there a transition at some critical N from a nonergodic to ergodic behavior?

Remark 2. The assumption of a two-state blinking process is an approximation; in reality, the on-state is composed of several on-states.[20,38] Even so the qualitative picture of nonergodicity based on a two-state process, while a simplification, is rather good.

Remark 3. In experiments there is in many cases an exponential cutoff time on the *on*-time PDF. In the limit of weak laser field and low temperature this cutoff time is very long. If the waiting time PDF of both on- and off-times, has a power-law behavior followed by an exponential cutoff, one expects to see weakly nonergodic behavior, followed by a transition to ergodic behavior on longer measurement time. In the experiments discussed in Sec. 2.2 these cutoffs do not influence the ergodicity breaking for measurements of 1 h, where a very large number of transitions between on- and off-states are observed.

2.1. *Distribution of the intensity correlation function*

From a single intensity trajectory $I(\tau)$, recorded in the time interval $(0, t)$, we construct a time-averaged intensity correlation function

$$C_{\text{TA}}(t, t') = \frac{\int_0^{t-t'} I(\tau)I(\tau + t')\mathrm{d}\tau}{t - t'}. \tag{14}$$

When $t \to \infty$ and for ergodic and stationary stochastic process $C_{\text{TA}}(t, t') = \langle C(t') \rangle$ is a function of t' only, and $\langle \cdots \rangle$ is an ensemble average. For blinking NCs the time-averaged correlation function is an irreproducible random function.[42] The analysis of its distribution, for a two-state on- and off-process, is nontrivial when the sojourn times in state on and off have power-law distributions. Excellent approximations for this distribution were obtained, using a nonergodic mean field theory.[22,39]

Here we will only briefly survey the theory for simple limiting cases. We use a two-state renewal model $I(\tau) = 1$ (on) or $I(\tau) = 0$ (off), PDFs of waiting times in these states are described by $\psi_{\text{on}}(t) = \psi_{\text{off}}(t) = \psi(t)$, for simplicity. If $t' = 0$ we use $I^2(\tau) = I(\tau)$ (since $I(\tau) = 1$ or $I(\tau) = 0$) and then

$$C_{\text{TA}}(t, t' = 0) = \frac{T^+}{t}. \tag{15}$$

Therefore, using Eq. (10) the time-averaged correlation function is equal to the time average intensity when $t' = 0$. It follows that when $t' = 0$ the PDF of C_{TA} is

$$P_{t'=0}(C_{\text{TA}}) = \delta_{\alpha,1}(C_{\text{TA}}). \tag{16}$$

The index 1 is due to the symmetry of the process under consideration $\psi_{\text{on}}(t) = \psi_{\text{off}}(t) \sim t^{-(1+\alpha)}$.

Now consider the case $t'/t \to 1$, and t large. Since t' is also large we use a decoupling approximation, which means that events at the beginning of the time trace are independent of those at its end. Mathematically we use the approximation

$$C_{\mathrm{TA}}(t,t') \simeq \bar{I}_{(0,t-t')}\bar{I}_{(t',t)}, \tag{17}$$

where the random time average

$$\bar{I}_{(a,b)} \equiv \frac{\int_a^b I(t)\mathrm{d}t}{a-b}. \tag{18}$$

So $\bar{I}_{(0,t-t')}$ $(\bar{I}_{(t',t)})$ in Eq. (17) is the time-averaged intensity at the beginning (end) of the time trace since $t' \to t$. It is reasonable to assume that these two random variables are independent when $t \to \infty$. If $\psi(t)$ has a finite first moment, for example, the exponential PDF, we may replace the time averages in (17) with the ensemble average. In that case use the assumed symmetry of on and off waiting times $\langle I \rangle = 1/2$, and hence $C_{\mathrm{TA}} = 1/4$ when $t' \to \infty$ which is the usual ergodic behavior. For a nonergodic process $0 < \alpha < 1$, the PDF of $\bar{I}_{(0,t-t')}$ in Eq. (17) is the $\delta_{\alpha,1}$ function, as shown in the previous section. Due to the nonstationarity of the process $\bar{I}(t',t) = 1$ or $\bar{I}(t',t) = 0$ with probability one half. In the limit under investigation $I(\tau)$ is equal one or equal zero in the interval (t',t), since very large on and off times tend to appear at the end of the measurement process, which is a statistical effect. Hence, according to these arguments

$$P_{t'\to t}(C_{\mathrm{TA}}) \simeq \frac{1}{2}\delta_{\alpha,1}(C_{\mathrm{TA}}) + \frac{1}{2}\delta(C_{\mathrm{TA}}). \tag{19}$$

Note that the ensemble average is $\langle C_{\mathrm{TA}} \rangle = 1/4$ in this limit.

The interesting cases where t' is neither small nor large are treated in Ref. 22. Here we will present the PDF of C_{TA}, using numerical simulations, which agree very well with the theory. In Fig. 2 we show the PDF of C_{TA} for $\alpha = 0.3$ and varying $r = t'/t$. As mentioned, when $r = 0$ the PDF of C_{TA} is a $\delta_{\alpha,1}(C_{\mathrm{TA}})$ function, and since

$\alpha = 0.3$ we observe a U shape. This means that many trajectories are either completely on while others are off, for very long durations. The ensemble average $\langle C_{\mathrm{TA}} \rangle = 1/2$ is the least likely value obtained in a measurement. For $\alpha = 0.8$ and $r = 0$ we see in Fig. 3, a W-shaped distribution. Since we are now approaching the ergodic phase $\alpha = 1$ a peak appears close to the ensemble average value $\langle C_{\mathrm{TA}} \rangle = 1/2$; however, still large fluctuations are observed. When we increase r the PDF tends to concentrate more to the left. This is easier to see for $\alpha = 0.8$. The latter behavior reflects the trend of the ensemble average which for $r \to 1$ is on $\langle C_{\mathrm{TA}} \rangle = 1/4$. When $r \to 1$ Eq. (19) is valid, though in the figures we do not present the delta function on $C_{\mathrm{TA}} = 0$.

Remark 4. The ensemble average intensity correlation function exhibits aging and nonstationarity,[27,40,41] namely, it depends both on the measurement time and on t', while in ordinary processes the ensemble average correlation function is a function of the time difference t' only. More generally, aging observed for ensemble averages and weak ergodicity breaking are related.

2.2. *Experimental evidence for weak ergodicity breaking*

Experimental evidence for ergodicity breaking in blinking NCs was demonstrated first by Dahan and coworkers.[40,42] In Fig. 4 we show the distribution of the occupation fraction in state up for 100 CdSe–ZnS NCs, at room temperature, following the works in Refs. 19 and 20 where experimental details are provided. Since data analysis of only 100 dots was made, the statistics is not excellent, and the data is noisy. Hence, the experimental data is compared with simple stochastic numerical simulations of the on and off processes, showing that the noise in the figure is most likely due to the finite number of dots in the sample. It is clear that the occupation time in state on, of the blinking dots, remains random even in the limit of long measurement time. The measurement time is long in the sense that hundreds of transitions between on and off during the observation time are found.

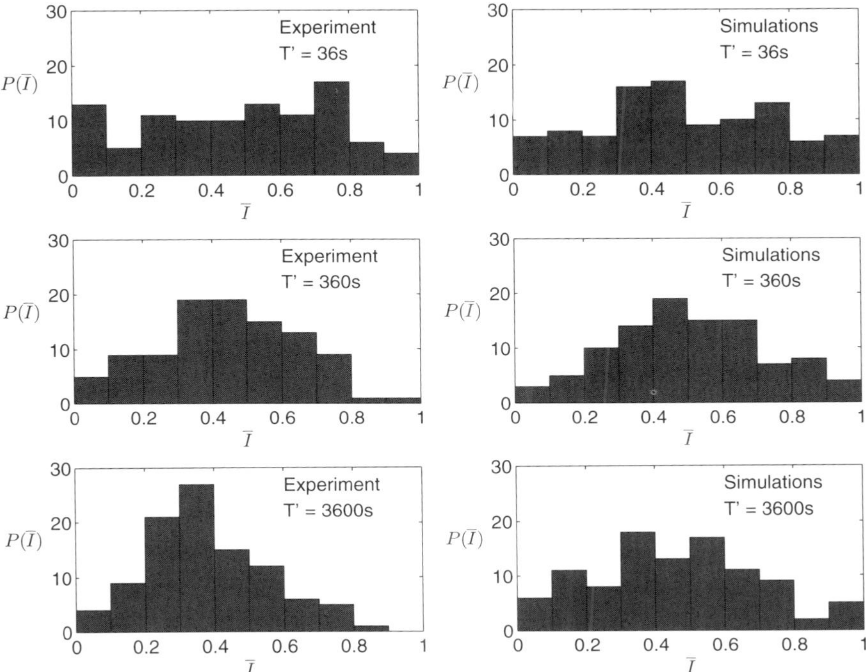

Fig. 4. The probability density function of the occupation fraction in state on for 100 blinking NCs, experiment versus simple simulation. The occupation fraction is the time average intensity for a simple on and off process. The figure demonstrates that these time averages are irreproducible random variables, indicating ergodicity breaking.

Note that when we increase the measurement time the picture does not change much, reflecting the scale-free nature of the on and off blinking processes. Measurement of intensity correlation functions for the same dot, or an ensemble of identical dots was presented previously showing very clearly the deviations from the ensemble average, and the invalidity of the ergodic hypothesis.[19,42]

Remark 5. Note that Fig. 4 also illustrates an interesting symmetry between on and off times, since the distribution of the on time occupation is uniform (i.e. $A_{\rm on} \simeq A_{\rm off}$). This surprising result is discussed in Ref. 20.

3. Continuous time random walk

The continuous time random walk (CTRW) described normal and anomalous diffusion in many systems.[7,8] Its anomalous version was introduced in the seventies to describe sub-diffusion $\langle x^2 \rangle \sim t^\alpha$ with $0 < \alpha < 1$ of charge carriers in amorphous semiconductors.[3] More recently CTRW type of dynamics was investigated in single-particle experiments, for example, the group of Weitz[4] measure the diffusion of a small bead in an actin network. In this experiment, the particle is trapped in the polymer network, and it is then released from the trap, diffuses and then gets trapped once more. It is shown that the PDF of trapping times follow a power law $\psi(t) \propto t^{-(1+\alpha)}$, which in turn is related to the subdiffusive motion of an ensemble of particles $\langle x^2 \rangle \sim t^\alpha$ if $0 < \alpha < 1$. In the experiments,[4] α depends on the ratio of the size of the bead R and the typical size of the mesh l of the polymer network. When $R \ll l$ diffusion is normal, since the bead does not interact with the polymer network, while in the opposite limit $R \gg l$, the particle is localized $\alpha \to 0$; in between, an anomalous diffusion characterized by power-law waiting times with $0 < \alpha < 1$ is found.

Following Ref. 46 we consider a one-dimensional CTRW on a lattice; each step is independent of the previous steps, and the step length is equal to the lattice spacing. The sojourn times on lattice sites are independent, identically distributed random variables, with a PDF

$$\psi(t) \sim A t^{-(1+\alpha)}/|\Gamma(-\alpha)|, \quad \text{for } t \to \infty \tag{20}$$

and $0 < \alpha < 1$. The lattice points are $x = 0, \ldots, L$. The boundaries are reflecting. The probability of jumping left from x is $q_l(x)$, and that of jumping right is $1 - q_l(x)$. This probability may change from site to site and reflects local bias, for example external fields. The goal is to find the occupation fraction T_x/t, where T_x is the total time the particle spends on site x.

We consider the indicator function $\theta(t)$ which is equal to unity if the particle is on x; otherwise it is zero. Clearly, the occupation time is $T_x = \int_0^t \theta(t)\mathrm{d}t$. The indicator function is jumping between two states, similar to the blinking NC. The sojourn times in state $\theta(t) = 1$ are distributed according to $\psi(t)$. The PDF of times in state $\theta(t) = 0$ is

$$\psi_0(t) = q_l(x)\psi_{Lx}^{\text{fpt}}(t) + [1 - q_l(x)]\psi_{\text{Rx}}^{\text{fpt}}(t). \tag{21}$$

Here $\psi_{Lx}^{\text{fpt}}(t)$ is the PDF of the first passage time from site $x - 1$ to x (the index L is for left random walk). Similarly for ψ_{Rx}^{fpt} for random walks starting on $x + 1$.

To obtain statistics of occupation times we have to first solve the corresponding first passage time of the CTRW problem. For general transition probabilities $q_l(x)$ this might seem a difficult task at first. The problem was treated in Ref. 41 for a CTRW on a lattice, and the following behavior was found. Let $P_{\text{eq}}(x)$ be the probability in ensemble sense that a particle occupies lattice point x in equilibrium. For a finite system, this equilibrium is always found for long times provided that $q_l(x) \neq 1, 0$ except for the reflecting boundaries. In this case the PDF of the occupation fraction $\bar{p}_x = T_x/t$ is given by[43]

$$f(\bar{p}) = \delta_{\alpha,\mathcal{R}}(\bar{p}) \tag{22}$$

with

$$\mathcal{R} = \frac{P_{\text{eq}}(x)}{1 - P_{\text{eq}}(x)}. \tag{23}$$

If we assume that the system is in thermal equilibrium, $P_{\text{eq}}(x)$ is equal to Boltzmann's probability $P_{\text{B}}(x)$. Detailed balance conditions were applied on the dynamics which means that the transition probabilities $q_l(x)$ are related to the temperature of the system, and the energy of the particle on site x, $V(x)$, where $V(x)$ is an external potential field acting on the system.[23,43] In this case

$$\mathcal{R} = \frac{P_{\rm B}(x)}{1 - P_{\rm B}(x)}, \tag{24}$$

where as usual $P_{\rm B}(x) = \exp[-V(x)/k_b T]/Z$. Equation (24) gives a relation between the nonergodic dynamics, statistics of occupation times, and the partition function of the problem. Here it is emphasized that the condition of detailed balance used to derive this relation means that for an ensemble of noninteracting CTRW particles, thermal equilibrium is reached. In this sense there is no conflict between CTRW theory and Boltzmann statistics, as long as we limit our observations to ensembles.[23]

Detailed derivation of Eq. (22) and its generalization to more general occupation times, for example, occupation times in more than one lattice point, or occupation times in disconnected regions in space, are given in Ref. 23. The core idea behind this equation is that due to the power-law trapping times of the CTRW model, the PDF of the first passage times also decays as a power law $\psi^{\rm fpt} \propto t^{-(1+\alpha)}$, unlike the exponential decay found for normal diffusion of a bounded Brownian particle. Hence the indicator function exhibits power-law intermittency, with no characteristic timescale since $\alpha < 1$. As for the asymmetry parameter, from Eq. (22), we have

$$\langle \bar{p}_x \rangle = \frac{\mathcal{R}}{1 - \mathcal{R}}. \tag{25}$$

However, the averaged occupation time must be equal to the ensemble population $\langle \bar{p}_x \rangle = P_{\rm eq}(x)$. Hence, the relation Eq. (24) must hold.

The physical paths of the CTRW particle lead to weak ergodicity breaking. The power-law waiting times of the particle means that while the particle makes many jumps and visits all lattice points in the limit of long time, the particle will typically locate on one lattice point for a duration which is of the order of the measurement time. In different trajectories this very long sojourn time can be in any of the lattice points in the system. This behavior is very different from strong ergodicity breaking, where a particle is limited in its motion to a finite region in space (for example, if we have some $q_l(x)$ equal zero or unity in the system).

One may quantify this behavior using the concept of the visitation fraction. Let n_x be the number of visits in cell x, and n the total number of jumps made during the measurement time t. For a single trajectory, n_x and n are random numbers. However, for weak ergodicity breaking[44]

$$n_x/n \sim P_{\text{eq}}(x) \tag{26}$$

in the limit of long times, while for strong ergodicity breaking, this relation does not hold. In usual ergodic systems with a finite average time between jumps $\langle\tau\rangle$ we have $(n_x\langle\tau\rangle)/(n\langle\tau\rangle) \sim T_x/t \sim P_{\text{eq}}(x)$; so both the occupation fraction and the visitation fraction are given by the fraction of systems from the ensemble in the state x.

Remark 6. The CTRW is a stochastic model, hence it is of interest to obtain Eq. (22) starting with deterministic dynamics, which was done in Ref. 45 using nonlinear maps.

Remark 7. Instead of the CTRW model, one can use a fractional Fokker–Planck equation approach to investigate statistics of occupation times and weak ergodicity breaking.[46] Such an approach is based on the continuum approximation and differential equations, and in this sense simpler than random walk theory on a lattice.

4. The quenched trap model

So far we considered models governed by renewal dynamics, sometimes called annealed dynamics, where once a transition is made the waiting time does not depend on the previous history of the dynamics. We now consider the quenched trap model. This is a well investigated model which leads to anomalous diffusion[7] and aging[47] behaviors.

Since the disorder is quenched (namely, it does not change in time) the renewal property of the CTRW walk is not valid.

We consider a particle undergoing a one-dimensional random walk on a quenched energy landscape on a lattice. The lattice is of finite length L and to each lattice point $x = 0, a, 2a, \ldots, L$ a random variable $E_x > 0$ is assigned, which is minus the energy of site x; so E_x is the depth of the trap on site x. The energies $\{E_x\}$ are independent, identically distributed (IID) random variables with a common PDF $\rho(E) = (1/T_g)\exp(-E/T_g)$. Such a density of states leads to anomalous behaviors when the temperature $T < T_g$, and was investigated by many (see Refs. 7, 47, 48), for example, in recent single-molecule pulling experiments.[49] Due to an interaction with a heat bath the particle may escape the trap x, with an escape time given by Arrhenius law $\tau_x = \exp(E_x/T)$. The particle can jump only to one of its nearest neighbors. Since E_x does not change in time the diffusion process is correlated with the specific realization of disorder, and to this day there does not exist an exact solution of the problem of diffusion in the trap model. It is easy to show that the PDF of the waiting times averaged over disorder is

$$\psi(\tau) = \frac{T}{T_g}\tau^{-(1+T/T_g)}, \qquad \tau > 1. \tag{27}$$

So, when $T < T_g$ the average waiting time diverges. It is tempting though wrong to assume that we can now replace the dynamics of the quenched model with that of a CTRW model with a power-law waiting time (27). Such a mean field approximation is known to fail, since it neglects the correlations between the underlying random walk and the specific realization of disorder the particle is interacting with. In our context we will show (without giving the proof which is in Ref. 48) that the distribution of occupation times in the trap model, behaves differently from that found for the annealed CTRW.

In addition to the random field, a nonrandom potential field is acting on the system so that the total energy of the particle on site $x\widetilde{E}_x$ is

$$\widetilde{E}_x = U_x^{\text{det}} - E_x. \tag{28}$$

For finite L and long measurement times, a thermal equilibrium is reached. The random variable of interest is the occupation fraction $\bar{p} = T_{[0,L_1]}/t$, where $T_{[0,L_1]}$ is the time the particle spends in the domain $[0, L_1]$ where $L_1 \leq L$. According to Boltzmann–Gibbs statistics, within the canonical formalism of statistical mechanics, we have in the limit of long measurement time $t \to \infty$ and L finite

$$\bar{p} = \frac{t^{\text{Occ}}}{t} \sim \frac{\sum_{x=1}^{L_1} e^{-(U_x^{\text{det}} - E_x)/T}}{\sum_{i=1}^{L} e^{-(U_x^{\text{det}} - E_x)/T}}. \tag{29}$$

The occupation fraction $\bar{p}$ is a random variable since it is a function of the random energies $\{E_x\}$. Note that we assume that for a specific realization of disorder, namely for a single system, the dynamics is ergodic and described by the standard canonical ensemble. A more detailed discussion on the dynamics which leads to this rather general type of equilibrium is found in Ref. 48.

The PDF of the occupation fraction, in a continuum limit, was obtained very recently.[48] For the glassy phase $T < T_g$, it was found that

$$f(\bar{p}) \sim \delta_{T/T_g, \mathcal{R}}(\bar{p}) \tag{30}$$

with

$$\mathcal{R} = \frac{P_B(T_g)}{1 - P_B(T_g)}, \tag{31}$$

where the Boltzmann factor is defined with the temperature T_g, and not T:

$$P_B(T_g) = \frac{\int_{x_1}^{L} \exp(-U^{\text{det}}(x)/T_g)\mathrm{d}x}{Z(T_g)}, \tag{32}$$

where $Z(T_g)$ is the normalizing partition function of the deterministic part of the potential. In the normal phase $T > T_g$ we recover usual

Boltzmann statistics where the PDF of the occupation fraction is

$$f(\bar{p}) \sim \delta[\bar{p} - P_B(T)]. \tag{33}$$

We see that unlike the CTRW model the relevant temperature is T_g not T when $T < T_g$. Roughly speaking, in this type of model, we have two temperatures, the temperature of the disorder T_g and the thermal temperature T. When $T < T_g$ the relevant temperature is T_g, which enters into the Boltzmann factor $P_B(T_g)$ (Eq. (32)). In the limit $T \to 0$, the behavior described by Eq. (30) is expected. At $T = 0$, the particle is located at point x with the minimum energy of the system. This lattice point can be either in the observation zone, or out of it, and then $f(\bar{p})$ is composed of two delta functions on $\bar{p} = 0$ or $\bar{p} = 1$. When $T \to \infty$ we also expect that the disorder is not important and the Boltzmann statistics is valid. Our results show the exact transition between these two extreme limits, followed by the $\delta_{\alpha,\mathcal{R}}(\bar{p})$ function with a sharp transition at T_g.[48]

The quenched trap model describes a single particle in a random environment. A similar situation is common in single-molecule spectroscopy, where a molecule is interacting with a specific random environment, hence its Hamiltonian is random. In this case the occupation time is random due to the randomness of the Hamiltonian, and not due to ergodicity breaking. Still, at least for the trap model, the $\delta_{\alpha,\mathcal{R}}(\bar{p})$ function describes the distribution of the occupation times, when $T < T_g$, while Boltzmann statistic is valid for weak disorder $T > T_g$. One should wonder how general are the result Eqs. (30), (31)? As briefly discussed in Ref. 46, the $\delta_{\alpha,\mathcal{R}}(\bar{p})$ function describes occupation times in other models of quenched disorder, for example, random comb models which are random walks on a loopless fractal structure, where the origin of the anomalous kinetics is random geometry. Thus, while the exponent $\alpha = T/T_g$ found for the trap model is not universal, our main result is valid for other models of quenched disorder. The common feature of these models is a random Hamiltonian leading to anomalous diffusion with power-law sojourn times, which in turn leads to the $\delta_{\alpha,\mathcal{R}}(\bar{p})$ function describing the occupation fraction in equilibrium. Note that in Ref. 50 an investigation of occupation times in the Sinai model was considered, which also leads to nontrivial distributions, which are different from that found for the trap model.

5. Discussion

Scale-free dynamics, with a power-law distribution of trapping times, leads to weak ergodicity breaking. This behavior is found in blinking NCs, and in the CTRW model. Previous works used the concept of power-law waiting times to describe the evolution of ensembles of particles. Now with new technology of video-microscopy and single-molecule spectroscopy, this type of dynamics is found to be common in laboratory. In such systems time averages are not described by ensemble averages, and usual Boltzmann statistics does not hold. Still, in the limit of long measurement times, a general statistical law for the occupation times gives the deviations from ergodicity. We also showed that the distribution of occupation times in models of quenched disorder behaves in a way similar to the CTRW model; however, now we obtain an effective temperature T_g while for the CTRW it was the usual temperature T. Mathematically, all our results are related to Lévy's central limit theorems. Standard statistical mechanics is based on the ergodic assumption, and the Gaussian central limit theorem,[51] while our approach is based on the more general central limit theorem of Lévy.

In some cases weak ergodicity breaking is not in conflict with usual statistical mechanics. In statistical mechanics one considers a single system described by a Hamiltonian, in thermal equilibrium, and in the thermodynamic limit, where the measurement time is made long before the size of the system is made large. Of course, the assumptions behind usual ergodic statistical mechanics are rather strong. Let us see in what ways it fails in the cases investigated so far.

1. For nanosystems and single molecules this thermodynamic limit is not always relevant, for example, for the blinking NCs. As we explained briefly in the text, one possible mechanics of the power-law statistics is the diffusion process of a charge carrier ejected from the NC to the environment. If this diffusion is limited by a finite system, this will introduce cutoffs on the first passage time, and then the dynamics will turn ergodic. However, these cutoffs are clearly not relevant, since the time it takes for the charge carrier to reach the boundary of the system is always larger than the measurement time. Further, the NCs are driven by a laser field, and are not in thermal equilibrium.
2. The nontrivial occupation times we found for the quenched trap are the effects of the nonself averaging property of the model. This corresponds to a situation where the Hamiltonian is random, and the distribution of occupation times is built from a distribution of Hamiltonians. Hence no conflict with usual statistical mechanics arises since the latter considers a single Hamiltonian.
3. In nonlinear dynamical systems where weak ergodicity breaking is found,[45] the dynamics is not governed by a Hamiltonian.

Thus, the surprise in the authors opinion is not so much in the breaking of ergodicity, but rather in that for these very different models one can construct a rather general statistical theory of occupation times, which reduces to usual ergodic behavior when $\alpha \to 1$. Since the particle explores its phase space, unlike strong ergodicity breaking where the particle is limited to a few states which depend on the initial condition, we can construct a general statistical theory. In particular, relation between the partition function and the distribution of the occupation time was found in Eqs. (24) and (31), even though the statistics is nonBoltzmann.

As we showed the power-law distribution of trapping times leads to weak ergodicity breaking. In some systems, the power laws are truncated after a long time. These cutoffs may induce a transition between weak ergodicity breaking and usual ergodic behavior. For blinking

NCs, for low laser intensity, and low temperatures, the cutoffs are very large. As mentioned, measurements of blinking dots for 1 h, with hundreds of transitions between off and on times, exhibited weak ergodicity breaking. It is possible that for longer measurement times a transition to ergodicity is found. The important point is that weak ergodicity breaking is a useful concept, for many decades in time, and for the measurement time relevant in laboratory. And, at least in some models weak ergodicity breaking is also an asymptotic theory. The investigation of the transition from weak ergodicity breaking to usual ergodicity due to cutoffs, for blinking NCs, CTRWs, and the quenched trap model, could be a topic of future research.

Acknowledgment

This work was supported by the Israel Science Foundation. The chapter is a brief review of work with G. Bel, S. Burov, K. Kuno, G. Margolin, and V. Protasenko.

References

1. N. G. van Kampen, *Stochastic Processes in Physics and Chemistry* (North Holland, Amsterdam, 1992).
2. M. Kuno, D. P. Fromm, H. F. Hamann, A. Gallagher and D. J. Nesbitt, *Journal of Chemical Physics* **112** (2000) 3117.
3. H. Scher and E. W. Montroll, *Physical Review B* **12** (1975) 2455.
4. I. Y. Wong *et al.*, *Physical Review Letters* **92** (2004) 178101.
5. J. Klafter, M. F. Shlesinger and G. Zumofen, *Physics Today* **49**(2) (1996) 33.
6. I. M. Sokolov, J. Klafter and A. Blumen, *Physics Today* **55** (2002) 48.
7. J. P. Bouchaud and A. Georges, *Physics Reports* **195** (1990) 127.
8. R. Metzler and J. Klafter, *Physics Reports* **339** (2000) 1.
9. G. H. Weiss, *Aspect and Applications of the Random Walk* (North Holland, Amsterdam, 1994).
10. B. D. Hughes, *Random Walks and Random Environments* (Clarendon Press, Oxford, 1996).
11. J. P. Bouchaud, *Journal of De Physique I* **2** (1992) 1705.
12. E. Barkai, Y. Jung and R. Silbey, *Annual Review of Physical Chemistry* **55** (2004) 457.

13. M. Nirmal, B. O. Dabbousi, M. G. Bawendi, J. J. Macklin, J. K. Trautmanm, T. D. Harris and L. E. Brus, *Nature* **383** (1996) 802.
14. M. Kuno, D. P. Fromm, H. F. Hamann, A. Gallagher and D. J. Nesbitt, *Journal of Chemical Physics* **115** (2001) 1028.
15. K. T. Shimizu, R. G. Neuhauser, C. A. Leatherdale, S. A. Empedocles, W. K. Woo and M. G. Bawendi, *Physical Review B* **63** (2001) 205316.
16. A. L. Efros and M. Rosen, *Physical Review Letters* **78** (1997) 1110.
17. A. Issac, C. von Borczyskowski and F. Cichos, *Physical Review B* **71** (2005) 161302(R).
18. S. Bianco, P. Grigolini and P. Paradisi, *Journal of Chemical Physics* **123** (2005) 17404.
19. G. Margolin, V. Protasenko, M. Kuno and E. Barkai, *Advances in Chemical Physics* **133** (2006) 327, cond-mat/0506512.
20. G. Margolin, V. Protasenko, M. Kuno and E. Barkai, *Journal of Physical Chemistry B* **110** (2006) 19053.
21. J. Lamperti, *Transactions of American Mathematical Society* **88** (1958) 380.
22. G. Margolin and E. Barkai, *Physical Review Letters* **94** (2005) 080601.
23. G. Bel and E. Barkai, *Physical Review E* **73** (2006) 016125.
24. R. Verberk, A. M. van Oijen and M. Orrit, *Physical Review B* **66** (2002) 233202.
25. M. Kuno, D. P. Fromm, S. T. Johnson, A. Gallagher and D. J. Nesbitt, *Physical Review B* **67** (2003) 125304.
26. P. A. Frantsuzov and R. A. Marcus, *Physical Review B* **72** (2005) 155321.
27. G. Margolin and E. Barkai, *Journal of Chemical Physics* **121** (2004) 1566.
28. G. Margolin and E. Barkai, *Physical Review E* **72** (2005) 025101(R).
29. J. Tang and R. A. Marcus, *Physical Review Letters* **95** (2005) 107401.
30. K. M. Hong, J. Noolandi and R. A. Street, *Physical Review B* **23** (1981) 2967.
31. M. Haase, C. G. Hubner, E. Reuther *et al.*, *Journal of Physical Chemistry B* **108** (2004) 10455.
32. J. Schuster, F. Cichos and C. von Borczykowski, *Applied Physics Letters* **87** (2005) 051915.
33. J. P. Hoogenboom, E. M. H. P. van Dijk, J. Hernando, N. F. van Hulst and M. F. Garcia-Parajo, *Physical Review Letters* **95** (2005) 097401.
34. A. R. Bizzarri and S. Cannistraro, *Physical Review Letters* **94** (2005) 068303.
35. P. Didier, L. Guidoni and F. Bardou, *Physical Review Letters* **95** (2005) 090602.
36. Y. Jung, E. Barkai and R. J. Silbey, *Chemical Physics* **284** (2002) 181.
37. I. Chung, J. B. Witkoskie, J. Cao and M. Bawendi, *Phys. Rev. E* **73** (2006) 011106.
38. K. Zhang, H. Chang, A. Fu, A. P. Alivisatos and H. Yang, *Nano Letters* **6** (2006) 843.
39. G. Margolin and E. Barkai, *Journal of Statistical Physics* **122** (2006) 137.

40. X. Brokmann, J. P. Hermier, G. Messin, P. Desbiolles, J. P. Bouchaud and M. Dahan, *Physical Review Letters* **90** (2003) 120601.
41. J. B. Witkoskie and J. Cao, *Journal of Chemical Physics* **125** (2006) 244511.
42. G. Messin, J. P. Hermier, E. Giacobino, P. Desbiolles and M. Dahan, *Optics Letters* **26** (2001) 1891.
43. G. Bel and E. Barkai, *Physical Review Letters* **94** (2005) 240602.
44. G. Bel and E. Barkai, *Journal of Physics Condensed Matter* **17** (2005) S4287–S4304.
45. G. Bel and E. Barkai, *Europhysics Letters* **74** (2006) 15.
46. E. Barkai, *Journal of Statistical Physics* **123** (2006) 883.
47. E. M. Bertin and J. P. Bouchaud, *Physical Reviews E* **67** (2003) 026128.
48. S. Burov and E. Barkai, *Physical Review Letters* **98** (2007) 250601.
49. J. Brujic *et al.*, *Nature Physics* **2** (2006) 282.
50. S. N. Majumdar and A. Comtet, *Physical Review Letters* **89** (2002) 060601.
51. A. Y. Khintchine, *The Mathematical Foundations of Statistical Mechanics* (Dover, New York, 1948).

About the Editors

Eli Barkai received his BSc, MSc, and PhD (with Victor Fleurov) in Physics from Tel-Aviv University. While at Tel Aviv, he developed with Yossi Klafter a fractional kinetic framework describing anomalous transport in dynamical systems. In 1998, he joined the Chemistry department in Massachusetts Institute of Technology for his postdoctoral research, where he developed the theory of single-molecule spectroscopy with Bob Silbey. Dr Barkai joined the Physics department in Notre Dame University, Indiana, in the capacity of an Assistant Professor, in 2002. In 2004, he returned to Israel to join the Physics department at Bar-Ilan University. His main research interests today are weak ergodicity breaking and photon counting statistics for single-molecule spectroscopy.

Frank L. H. Brown received his BS in Chemistry and BA in Applied Mathematics from the University of California, Berkeley, and his PhD in Physical Chemistry from the Massachusetts Institute of Technology in 1998. While at the Massachusetts Institute of Technology, he worked with Robert Silbey on theoretical problems related to single-molecule spectroscopy in low-temperature systems. He then moved to the University of California, San Diego, to complete postdoctoral training in the labs of Kent Wilson and J Andrew McCammon, where his research focus shifted toward more biologically inspired problems. In 2001, Dr Brown joined the faculty at the University of California, Santa Barbara. Dr Brown's current research interests include a variety of problems in physical chemistry and biophysics, mostly related to single-molecule statistics and lipid bilayer membranes.

Michel Orrit works in the field of the interaction of light with organic condensed matter. From 1979 onwards, he worked on surface excitons in molecular crystals with Ph Kottis in Bordeaux. During a postdoctoral stay in Göttingen in 1985 with H Kuhn and D Möbius, he worked on Langmuir–Blodgett films doped with dyes. Back in Bordeaux, he used spectral hole burning to study low-temperature dynamics and molecular orientation in ultrathin molecular films. With J Bernard, he observed the fluorescence of immobilized single molecules for the first time in 1990. Since then, single-molecule fluorescence has developed quickly in several groups throughout the world, in particular towards room temperature from 1993. Since then, Dr Orrit's group, first in Bordeaux, then in Leiden after 2001, has applied single-molecule spectroscopy to molecular photophysics, solid state dynamics, nonlinear optics, and to other single nano-objects, semiconductor nanocrystals and metal nanoparticles.

Haw Yang received his BS in Chemistry from National Taiwan University, and his PhD in Physical Chemistry from the University of California, Berkeley in 1999. While at the University of California, Berkeley, he worked with Charles Harris using femtosecond IR spectroscopy to probe the mechanism of chemical bond activation by organometallic compounds in solution. He then moved to Harvard University to complete postdoctoral training in Sunney Xie's lab, where he was exposed to single-molecule spectroscopy. In 2002, Dr Yang joined the faculty at the University of California, Berkeley. His current research interests include protein structure-function dynamics, semiconductor nanocrystals, and the physical and chemical states inside a cell.

Index

activation free energy, 167–169, 175
AFM, *see* atomic force microscopy
aging, 126, 369, 378, 383
  non-stationarity, 369, 377, 378
anomalous relaxation, 95, 96, 380, 383
antibunching, 99, 105
atomic force microscopy, 139, 141, 153

Bayesian, 31
  theorem, 33
  inference, 173
  statistics, 246, 247, 278, 279
  information criterion, 9, 10
  Monte Carlo algorithm, 279
  priori probability, 279
  statistical likelihood, 278
Bell's formula, 161, 166, 168, 170, 175
Bell–Evans model, 163, 164
binding constant, 140
blinking, 62, 63, 90, 369
bunching, 95, 99

central limit theorem, 95
Chapman–Kolmogorov equation, 114
change-point method, 13–15
chi-squared distribution, 37, 49
coherent trapping, 85
conformational, 245–249, 258, 259, 263–265, 267, 268, 272–274, 276
  channel, 248, 249, 258, 263, 265, 276
  fluctuations, 245–247, 249, 258, 259, 263–265, 268, 272–274, 276
  inter-conversion rate, 248, 259
constant-force experiment, 155, 158, 161, 166, 174, 175
continuous time random walks, 124–126, 131, 133, 134, 380–383
control parameters, 288, 299, 300, 302, 303, 305, 309, 310
convolution theorem, 129
coordinate stochastic, 113, 134
correlation function, 93, 94, 96, 98–104, 106–120, 123, 126, 127, 131–134, 250, 252, 254, 258, 263, 268, 271, 273, 276, 289, 290, 292, 293, 295–301, 304, 307, 310, 311, 376–378
  equivalence with factorial moments, 93, 94, 96, 98, 101–103, 123, 134
  intensity correlation, 263
  multipoint, 94, 96, 98, 100, 108–110, 113, 117, 119, 120, 123–125, 127, 134
  on-time correlation, 254, 258, 263, 268
Crooks relation, 149
CTRW, *see* random walk (continuous time)

data analysis, 246, 271, 278, 339, 365, 371
  complete time series analysis, 279
  event histogram, 279
  indicator analysis, 279
likelihood function, 33
density matrix, 80, 83, 85, 90
detailed balance, 257, 259, 260, 262, 280, 290, 298–302, 381
  average rate, 259, 260, 263

chemical balance, 259, 260
non-equilibrium steady state, 281
phenomenological chemical kinetics, 259
signature of violations, 260, 280
detuning frequency, 83, 88
diffusion through the laser spot, 225–227
diffusion-controlled reaction, 249, 256, 270, 277, 278, 373
Wilemski–Fixman expression, 269, 277, 278
analogy, 229, 233, 236
dissociation, 139–141, 155
Distribution of the number of photons in a bin, 190
Mandel's formula, 222
Poisson distribution, 202
two conformations, 208–210
two-state fluorophores, 194
see also photon statistics
DNA hairpin, 159
driven two-level system
dephasing, 267
optical Bloch equation, 274
population relaxation, 245
spontaneous decay, 245, 246, 265, 266, 268, 273
dynamics Gaussian, 117, 120, 133
dynamics Markovian, 114, 121
dynamics nonMarkovian, 94, 124, 369–375, 380–383

electronic transfer, 111, 113, 115, 117, 123, 133
ergodic theorem
ensemble average, 251, 270, 277, 278
individual transition events, 251
ergodicity, 126, 365–369, 372, 374, 376, 378, 379, 382, 383, 386–389, *see also* weak ergodicity breaking
event statistics
on-time correlation function, 254, 258, 263, 268
two-event number density, 258, 268, 269
excitation quenching, 111
excited state lifetime, 99, 104

factorial moments, 93, 94, 96, 98, 101–103, 123, 134
Feynman-Kac theorem, 141
Fisher information, 6, 11, 15, 19
fluorescence, 94, 99, 101, 102, 107, 108, 110, 111, 115, 250, 252, 263, 270, 274–278, 280, 369, 375
emission intensity, 263, 273, 277
frequency resolved, 107, 110
intermittency, 12, 245, 374, 382
lifetime, 264
fluorescence resonance energy transfer, 111, 113
Fokker–Planck equation, 116, 117, 121
force spectroscopy, 139, 140, 175, 176
force-ramp experiments, 157, 158, 161, 162, 172, 175
four wave mixing, 99, 107
free energy, 140, 141, 143–146, 148–155, 163–169, 172, 174, 175
FRET, *see* fluorescence resonance energy transfer

generating function, 61–64, 73–75, 79, 191–193, 217
diffusing fluorophores, 229, 236
steady-state approximation, 207, 219
stochastic rate, 246, 247, 255–258, 262–264
generating functional, 117–119
glassy dynamics, 82
Green's function, 94, 97, 113–116, 121, 124, 127, 187, 220, 232

Hamiltonian, 141–143
harmonic-cusp model, 160, 168, 170, 174
hidden Markovian process, 246, 248
hidden states, 248–250, 253, 254, 264
monitored transitions, 255
resolved state, 248, 249, 258–260
histogram, 37, 39, 47, 53, 147

hypothesis testing, 33, 40
  sequential test, 22
hysteresis, 140

ideal gas, 64–66, 68, 69, 71
indicator
  Mandel's Q parameter, 272, 278, 381
  Poisson indicator, 251, 255, 269, 271, 272, 274, 275
  renewal indicator, 268, 269, 271, 272, 275
informational content of the data, 33
initial conditions, 246, 247, 250, 252, 253, 256, 262, 268, 271, 272, 275, 276
  initial event averaging, 253, 272
  initial time averaging, 252, 272
  stationary flux, 252, 253, 259, 265, 274, 276
  steady-state population, 252
instrument noise, 150, 151
intensity correlation, *see* correlation
intensity correlation function, 195, 196, 217, 376–378
  diffusing fluorophores, 232, 237
  separation of time scales, 211–214, 220–221
  two-state fluorophores, 197
interphoton time distribution, 184, 197–199, 214–215, 217, 221
  diffusing fluorophores, 235
  relation to the intensity correlation function, 199, 201
  separation of time scales, 221

Jarzynski's identity, 141–143, 146, 148–151, 175

kinetic schemes, 337–342, 344–349, 352–358, 360
kinetics, 62, 63, 68–70, 72, 73, 76, 82, 140, 141, 155, 156, 161, 162, 175, 176
Kramers theory, 164–170
  high-barrier approximation, 165, 168, 173
Kramers–Kronig relation, 119
Kubo–Anderson theory, 141

Lévy statistics, 369, 375
life time, 156, 157, 159, 160, 162, 173–175
likelihood of photon trajectory, 199
linear-cubic model, 168, 170, 174
linker, 150, 152–154, 158–162, 174
  anharmonic, 159, 161, 162, 171
  molecular, 139, 159, 160, 162
Liouville space notation, 96

Mandel parameter, 71, 75, 85, 89, 196, 197
marginalization rule, 33
master curve, 159, 162, 172–174
master equation, 70, 73, 74, 77–79, 90, 91, 339, 340, 343
  quantum master equation, 95
matrix of monitored transitions, 189
  relation to renewal theory, 200
  steady-state approximation, 208
maximum entropy, 9, 11, 19
maximum likelihood, 156, 160, 172, 174
mean number of photons, 195, 218
  diffusing fluorophores, 231
  two-state fluorophore, 197
measurement, 96, 99, 109, 117–119, 121
  classical, 96, 97, 99, 103, 105, 106, 108, 110, 118, 119, 134
memory effects, 245–247, 263, 266, 267, 279, 280
  memory time, 264
  non-renewal process, 276, 278, 384
  renewal process, 268, 271, 273–275, 277, 278, 380–383, 369
modulated reaction model, 246–249, 254, 256, 263, 273
  hidden Markovian process, 246, 248
  resolved (observed) transitions, 248
  unresolved transitions (hidden state), 248

nano-crystals, 366, 369–371
  blinking, 366, 367, 369, 373, 375, 376, 378, 379, 381, 387–389
  CdSe, 369, 370, 378
  quantum dots, 367, 371, 375
nanopore, 159, 160

nonequilibrium, 140–143, 148, 150, 151, 175
nonself averaging, 388
normal modes, 293
normally distributed noise, 34
number of states analysis, 41

on/off times, 289, 304, 307, 311, 370
optical Bloch equations, 82, 88, 90
oscillations, 287, 288, 290, 292, 293, 296–298, 301, 302, 304, 307, 310, 311
overdamped oscillator, 120
  spectral density, 118, 120, 121

paradox of two envelopes, 44
parameter estimation, 33, 50
partition function, 143, 144
path integral, 141, 150
path representation, 340–344, 353
phase matching, 107, 108
photon counting, 63, 68, 73–76, 78, 79, 81, 83, 87, 90, 94, 95, 99, 100, 102, 105, 107, 134
  statistics, 71–74, 79, 81, 84, 88, 245, 248, 269, 270, 274, 275
  counting moment, 270, 271, 277
  number density, 258, 268–271, 274, 275
  survival probability, 256, 263, 264, 270, 277
  waiting time distribution function, 247, 250, 253, 260, 263, 269, 280
photon arrival time, 68, 93, 110, 111, 113, 115, 117, 123, 132, 277
  burst, 21, 225
Poisson process, 62–64, 71, 72, 271, 272
  Poisson indicator, *see* indicator
Poisson statistics, 63, 68, 69, 76, 201, 203
posterior likelihood, 33
probability density distribution (PDF), 34, 109–111, 113, 114, 116, 122–124, 127–132
process Markovian, 114, 115, 122, 123, 127
pulling, 139–144, 146, 149–157, 159–164, 171–173, 175

quantum beats, 85, 87
quantum dynamics, 63, 72, 79, 81, 85
  Bloch equation, 274
  coherence, 273, 274
  dephasing, 267
  population, 245, 248, 250–252, 260, 274–276
quantum master equation, *see* master equation
quasi stationary states, 38
quasi-adiabatic approximation, 157, 162
quasi-harmonic approximation, 144, 146, 153–155
quenched trap model, 369, 383, 386, 389

Rabi frequency, 82, 84, 86, 87
  oscillations, 84, 85
random function, 288
random walk, 62, 373, 381, 383, 384, 386
  continuous time random walk, 369, 380
rare event, 32, 52
reaction events, 289, 307–311
Redfield dynamics, 81
reduced dimensions canonical forms, 339, 340, 344
reliability, 36, 54
renewal theory, 124, 125, 127, 200, 201
renormalization group, 302, 304
rotating wave approximation (RWA), 82, 83, 86

separability condition, 290, 291, 311
Shannon entropy, 33
single molecule echo, 254, 265
  difference function, 266, 267, 279
  echo time, 265–267
  focal time, 265, 266
single-molecule time series, 245–247, 250, 251, 254, 255, 260, 268, 278, 280
  on–off blinking traces, 245, 246, 258, 260, 269, 280
  photon emission (fluorescence) traces, 245–247, 269, 272, 275, 280

spectroscopy
  single-molecule, 61–64, 66, 67, 72, 73, 81, 90
spontaneous emission, 61, 62, 68, 75, 80, 82, 83, 87
stiff-spring approximation, 145, 146, 153, 154
stochastic dynamics, 79–81
stochastic Gaussian rate model, 264
  cumulant expansion, 258, 263
  memory function, 254, 258, 263–265, 268, 269, 277
  memory time, 264, 278
stochastic Liouville equation (SLE), 97, 105, 106, 116, 299, 301, 302, 304, 310
stochastic rate model, 246, 247, 255, 256, 262–264
  interaction picture, 255, 256, 263
  stochastic averaging, 263, 270
  time ordering, 256
  time-dependent rate, 256, 263
subdiffusion, 126, 380–386
superoperator, 97, 100–102, 118
  resetting, 97, 99, 105, 125
survival function, 307

temperature effects, 105
thermodynamics, 140, 141, 176
three-level system, 85, 86
time series, 63, 64, 67, 72
TLS, *see* two-level system
trajectory picture, 96
transfer matrix formalism, 246, 247, 260, 269, 273, 274, 280
  interconversion rate matrix, 248
  reaction rate matrix, 248, 257
transition dipole, 85, 87–89
transition state, 163, 164, 166–168, 172, 174, 175
Triplet blinking, 182, 183, 214, 237
tweezers
  magnetic tweezers, 139
  optical tweezers, 139
two-level system, 75, 77, 78, 82, 84, 85
two-state trajectories, 338, 339, 347–349, 352–354

Uhlenbeck-Ornstein process, 120
umbrella sampling, 148, 150
unfolding, 139–141, 152–155, 160
unified theory of molecular rupture, 170

V-system, 85, 86, 89
variance, 196, 203, 209, 232

waiting time, 64, 68, 69, 72
  distribution function, 125, 126, 133
  waiting time probability distribution functions, 338–341, 345–350, 352, 354–358
weak ergodicity breaking, 365, 369, 374, 378, 382, 383, 387–389
Wigner distribution function, 118
WLC, *see* worm-like chain model
worm-like chain model, 151, 158, 159, 161, 162